中　外　物　理　学　精　品　书　系

本书出版得到“国家出版基金”资助

中 外 物 理 学 精 品 书 系

引 进 系 列 · 8

Finite Size Effects in Correlated Electron Models:

Exact Results

关联电子模型中的有限尺度效应

——精确结果

(影印版)

〔乌克兰〕兹维亚金(A. A. Zvyagin)著

著作权合同登记号　图字:01-2012-2825
图书在版编目(CIP)数据

关联电子模型中的有限尺度效应:精确结果:英文/(乌克兰)兹维亚金(Zvyagin, A. A.)著.—影印本.—北京:北京大学出版社,2012.12
(中外物理学精品书系·引进系列)
书名原文:Finite Size Effects in Correlated Electron Models: Exact Results
ISBN 978-7-301-21554-8

Ⅰ.①关…　Ⅱ.①兹…　Ⅲ.①凝聚态-物理学-研究-英文　Ⅳ.①O469

中国版本图书馆 CIP 数据核字(2012)第 274429 号

书　　名: **Finite Size Effects in Correlated Electron Models: Exact Results(关联电子模型中的有限尺度效应——精确结果)**(影印版)
著作责任者:〔乌克兰〕兹维亚金(A. A. Zvyagin)　著
责 任 编 辑:刘　啸
标 准 书 号:ISBN 978-7-301-21554-8/O·0896
出 版 发 行:北京大学出版社
地　　址:北京市海淀区成府路 205 号　100871
网　　址:http://www.pup.cn
新 浪 微 博:@北京大学出版社
电 子 信 箱:zpup@pup.cn
电　　话:邮购部 62752015　发行部 62750672　编辑部 62752038　出版部 62754962
印 刷 者:北京中科印刷有限公司
经 销 者:新华书店
730 毫米×980 毫米　16 开本　24 印张　445 千字
2012 年 12 月第 1 版　2012 年 12 月第 1 次印刷
定　　价:64.00 元

《中外物理学精品书系》
编 委 会

序　　言

物理学是研究物质、能量以及它们之间相互作用的科学。她不仅是化学、生命、材料、信息、能源和环境等相关学科的基础，同时还是许多新兴学科和交叉学科的前沿。在科技发展日新月异和国际竞争日趋激烈的今天，物理学不仅囿于基础科学和技术应用研究的范畴，而且在社会发展与人类进步的历史进程中发挥着越来越关键的作用。

我们欣喜地看到，改革开放三十多年来，随着中国政治、经济、教育、文化等领域各项事业的持续稳定发展，我国物理学取得了跨越式的进步，做出了很多为世界瞩目的研究成果。今日的中国物理正在经历一个历史上少有的黄金时代。

在我国物理学科快速发展的背景下，近年来物理学相关书籍也呈现百花齐放的良好态势，在知识传承、学术交流、人才培养等方面发挥着无可替代的作用。从另一方面看，尽管国内各出版社相继推出了一些质量很高的物理教材和图书，但系统总结物理学各门类知识和发展，深入浅出地介绍其与现代科学技术之间的渊源，并针对不同层次的读者提供有价值的教材和研究参考，仍是我国科学传播与出版界面临的一个极富挑战性的课题。

为有力推动我国物理学研究、加快相关学科的建设与发展，特别是展现近年来中国物理学者的研究水平和成果，北京大学出版社在国家出版基金的支持下推出了《中外物理学精品书系》，试图对以上难题进行大胆的尝试和探索。该书系编委会集结了数十位来自内地和香港顶尖高校及科研院所的知名专家学者。他们都是目前该领域十分活跃的专家，确保了整套丛书的权威性和前瞻性。

这套书系内容丰富，涵盖面广，可读性强，其中既有对我国传统物理学发展的梳理和总结，也有对正在蓬勃发展的物理学前沿的全面展示；既引进和介绍了世界物理学研究的发展动态，也面向国际主流领域传播中国物理的优秀专著。可以说，《中外物理学精品书系》力图完整呈现近现代世界和中国物理科学发展的全貌，是一部目前国内为数不多的兼具学术价值和阅读乐趣的经典物理丛书。

《中外物理学精品书系》另一个突出特点是，在把西方物理的精华要义“请进来”的同时，也将我国近现代物理的优秀成果“送出去”。物理学科在世界范围内的重要性不言而喻，引进和翻译世界物理的经典著作和前沿动态，可以满足当前国内物理教学和科研工作的迫切需求。另一方面，改革开放几十年来，我国的物理学研究取得了长足发展，一大批具有较高学术价值的著作相继问世。这套丛书首次将一些中国物理学者的优秀论著以英文版的形式直接推向国际相关研究的主流领域，使世界对中国物理学的过去和现状有更多的深入了解，不仅充分展示出中国物理学研究和积累的“硬实力”，也向世界主动传播我国科技文化领域不断创新的“软实力”，对全面提升中国科学、教育和文化领域的国际形象起到重要的促进作用。

值得一提的是，《中外物理学精品书系》还对中国近现代物理学科的经典著作进行了全面收录。20 世纪以来，中国物理界诞生了很多经典作品，但当时大都分散出版，如今很多代表性的作品已经淹没在浩瀚的图书海洋中，读者们对这些论著也都是“只闻其声，未见其真”。该书系的编者们在这方面下了很大工夫，对中国物理学科不同时期、不同分支的经典著作进行了系统的整理和收录。这项工作具有非常重要的学术意义和社会价值，不仅可以很好地保护和传承我国物理学的经典文献，充分发挥其应有的传世育人的作用，更能使广大物理学人和青年学子切身体会我国物理学研究的发展脉络和优良传统，真正领悟到老一辈科学家严谨求实、追求卓越、博大精深的治学之美。

温家宝总理在 2006 年中国科学技术大会上指出，“加强基础研究是提升国家创新能力、积累智力资本的重要途径，是我国跻身世界科技强国的必要条件”。中国的发展在于创新，而基础研究正是一切创新的根本和源泉。我相信，这套《中外物理学精品书系》的出版，不仅可以使所有热爱和研究物理学的人们从中获取思维的启迪、智力的挑战和阅读的乐趣，也将进一步推动其他相关基础科学更好更快地发展，为我国今后的科技创新和社会进步做出应有的贡献。

《中外物理学精品书系》编委会　主任

中国科学院院士，北京大学教授

王恩哥

2010 年 5 月于燕园

FINITE SIZE EFFECTS IN CORRELATED ELECTRON MODELS

EXACT RESULTS

Andrei A. Zvyagin
National Academy of Sciences, Ukraine

ICP
Imperial College Press

To My Family

Preface

Low-dimensional correlated electron systems are a fascinating topic in condensed matter physics which started in the 1930s, developed in the 1960s and received great attention from physicists during the last decade. Research into low-dimensional magnetism occupied an important part of this field, because it originated from electron correlations due to Coulomb interaction. In the last twenty years the problems of low-dimensional correlated electron systems has turned from a narrow, special topic of mathematical physics into one of the central problems of condensed matter physics. This became true due to the great progress in the miniaturization of technology, and many great findings in the composition of new materials like metal-oxides, carbon nanotubes, organic compounds, optical lattices for ultra-cold quantum gases, *etc.*, in most of which the scale of research is micro- and nano-physics. New mesoscopic and nano-devices are based on quantum dots, wires, rings in which the low-dimensionality and quantum nature are basic features. This, in turn, resulted in the creation and development of powerful theoretical and mathematical approaches, like Bethe's ansatz, bosonization and conformal field theory. These approaches are not only important to the pure quantum many-particle theory, but also happen to be extremely useful in a number of areas related to this theory. The reason for such an application is that certain important phenomena, like high-T_c superconductivity, physics of magnetic impurities, *etc.*, cannot seem to be explained in the framework of weak couplings, *i.e.*, perturbative theory, or mean-field-like approach. One-dimensional exact solutions provide a complete and unambiguous picture of correlated electron systems and play a role in the basis to further applications of perturbative methods.

The aim of the book is to present an introduction to recent achievements in theoretical studies of exactly integrable low-dimensional models of strongly correlated electrons and spin models. The central topic of this book is finite size effects in lattice exactly solvable spin and electron models. However, this book is not a review. A great number of papers were published since the 1920s till now on exact solutions in one-dimensional quantum models and it would be completely hopeless to discuss and even mention them all. (I would like to use this opportunity to sincerely apologise to those authors whose important contributions are not mentioned here.) Unfortunately, it is impossible to write about the many very important aspects of the Bethe's ansatz, the main subject of this book, like its purely mathematical developments, exact solutions of field theory models, continuous models of interacting electrons, models of systems with more than one internal degree of freedom (*e.g.*, orbital moments of electrons), systems with lower symmetries, models with long-range interactions, multi-chain quantum models, magnetic and hybridization impurities in three-dimensional metals, *etc.*, though they are related to the topic of the book. Some effects, like the behaviour of elementary excitations in quantum correlated electron and spin chains are also not presented here. I can only refer the interested reader to some excellent monographs, review articles, collections of reprints, like [Baxter (1982); Gaudin (1983); Andrei, Furuya and Lowenstein (1983); Tsvelick and Wiegmann (1983); Schlottmann (1989); Izyumov and Skryabin (1990); Jimbo (1990); Korepin, Bogoliubov and Izergin (1993); Schlottmann and Sacramento (1993); Korepin and Eßler (1994); Ha (1996); Schlottmann (1997); Takahashi (1999)].

The structure of this book is as follows. After a short introduction to statistical mechanics and thermodynamics, I remind the reader of some important facts about thermodynamics of quantum spins and free electrons in crystals in Chapter 1. Very important Mermin–Wagner and Hohenberg theorems are also presented in this chapter, to explain to the reader the importance of exact studies in low-dimensional quantum systems. In Chapter 2 several exact results of one-dimensional theory of quantum spins are presented. Those theories are relatively simple, but their knowledge permits us to understand the deeper nature of homogeneous quantum spin chains. Chapter 3 is devoted to the description of the main aim of this book — the Bethe's ansatz in its most known form, the co-ordinate Bethe ansatz. The development of this method for models of correlated electrons, the nested Bethe ansatz, is presented in Chapter 4. Chapter 5 explains features of the elegant algebraic Bethe ansatz, or the quantum inverse scattering method to the reader. Hence, the reader experienced in Bethe ansatz is, probably, aware of those studies. The main results of the book are presented in Chapters 6, 7 and 8. Chapter 6 describes the difference in thermodynamic behaviours of bulk particles with those, situated at edges (boundaries) of open chains. Similar effects of isolated impurities are presented in Chapter 7. The reader can see how great the difference in behaviours of host electrons (spins), and "surface" or impure ones is. Very often effects for homogeneous hosts and "distinguished" sites are qualitatively different, and they have to be taken into account when one interprets data of experiments in low-dimensional electron systems. I present results for various hosts and impurities, from the simplest ones, to the more complicated. Chapter 8 gives the description for thermodynamic behaviour of finite concentration of impurities in correlated electron and quantum spin hosts. To the best of my knowledge, Bethe ansatz solvable models are the *only* example, where it is possible to obtain exact thermodynamic characteristics for correlated electron and spin systems with *ensembles* of impurities, *e.g.*, to investigate the interplay between correlation effects and disorder exactly. The important method of modern Bethe ansatz thermodynamics, the quantum transfer matrix approach, is also presented in that chapter. In Chapter 9 other finite size effects are described. For example, recent experiments drew attention to studies of quantum topological effects, like persistent currents in quantum rings with or without embedded quantum dots. Another aspect of similar finite size effects is the possibility to extract from them the information about the asymptotic behaviour of correlation functions, using the conformal field theory approach. This is why, in Chapter 9 a short introduction to the scaling theory and conformal field theory is given. Finally, in Chapter 10 I discuss which methods can be used beyond exact ones. Here some short descriptions of scaling theory of quantum phase transitions and bosonization are given. However, all these theories cannot be presented in detail, and I refer the reader to several excellent books and review articles like [Ma (1976); Sólyom (1979); Cardy (1996); Di Franchesco, Mathieu and Sénéshal (1997); Sondhi, Girvin, Carini and Shahar (1997); Gogolin, Nersesyan and Tsvelik (1998); van Delft and Schoeller (1998); Nagaosa (1998); Sachdev (1999); Kadanoff (2000)]. I hope that those readers, who are familiar with exact solutions, will find some new interesting facts about finite size effects in one-dimensional quantum spin and electron systems, while the book can serve as an introduction for beginners to introduce them to the beautiful world of exact solutions.

My understanding of quantum low-dimensional magnetic systems and correlated metallic systems has grown over many years. It is a great pleasure for me to thank the many friends and colleagues who contributed to it. First of all I would like to thank my father, who introduced me to the physics of low-dimensional systems, and V. M. Tsukernik, whose deep knowledge of quantum physics of magnetism supported me so much. I especially want to thank my long-term co-authors Pedro Schlottmann, Andreas Klümper, Henrik Johannesson, and Holger Frahm for stimulating interaction, helpful suggestions and essential support. I have benefited from and appreciate interesting discussions with I. Affleck, B. L. Altshuler, N. Andrei, R. Z. Bariev, B. Douçot, U. Eckern, F. H. L. Eßler, P. Fulde, T. Giamarchi, L. I. Glazman, F. D. M. Haldane, A. G. Izergin, Y. S. Kivshar, V. E. Korepin, I. V. Krive, V. E. Kravtsov, H.-J. Mikeska, P. Nozières, A. A. Ovchinnikov, L. A. Pastur, N. M. Plakida, A. S. Rozhavsky, E. Runge, F. Steglich, P. Thalmeier, A. M. Tsvelik, P. B. Wiegmann, Yu Lu, and J. Zittartz. Finally, I am very grateful for all the valuable support as well as suggestions from my wife, Natasha.

A. A. Zvyagin

Contents

Preface vii

1. Introduction 1

1.1 Why is the Topic of the Book Worthwhile Studying? 1
1.2 Thermodynamics 5
1.3 Statistical Mechanics: Simple Models 7
1.4 Mermin–Hohenberg Theorem 17

2. Quantum Spin-$\frac{1}{2}$ Chain with the Nearest-Neighbour Couplings 25

2.1 One-Dimensional Spin Hamiltonian 26
2.2 Ising Chain 27
2.3 Isotropic XY Ring 30
2.4 Ising Chain in a Transverse Magnetic Field 35
2.5 Dimerized XY Chain 38

3. Co-ordinate Bethe Ansatz for a Heisenberg–Ising Ring 49

3.1 Bethe Ansatz 49
3.2 Simple Solutions of the Bethe Ansatz Equations: Strings . 55
3.3 Thermodynamic Bethe Ansatz 57
3.4 The Ground State Behaviour 62
3.5 Magnetic Field Behaviour in the Ground State: Wiener–Hopf Method 66

4. Correlated Electron Chains: Co-ordinate Bethe Ansatz 73

4.1 Hubbard Chain 74
4.2 t-J Chain 97

5. Algebraic Bethe Ansatz 115

5.1 The Algebraic Bethe Ansatz for a Spin-$\frac{1}{2}$ Chain 115
5.2 SU(2)-Symmetric Spin-S Chain 124
5.3 The Algebraic Bethe Ansatz for Correlated Electron Models 137

6. Correlated Quantum Chains with Open Boundary Conditions 149

6.1 Open Boundaries. XY and Ising Chains 149
6.2 Open Boundaries: Co-ordinate Bethe Ansatz for the Heisenberg–Ising Chain 153
6.3 Open Boundaries. The Algebraic Bethe Ansatz 164
6.4 Open Hubbard Chain 171
6.5 Open Supersymmetric t-J Chain 181

7. Correlated Quantum Chains with Isolated Impurities 189

7.1 Impurities in XY Chains 189
7.2 Impurities in Spin Chains: Bethe Ansatz 200
7.3 Impurity in Correlated Electron Chains 219

8. Correlated Quantum Chains with a Finite Concentration of Impurities 243

8.1 Impurities' Bands 243
8.2 Disodered Ensembles of Impurities in Correlated Chains. "Quantum Transfer Matrix" Approach 262

9. Finite Size Corrections in Quantum Correlated Chains 285

9.1 Finite Size Corrections for Quantum Spin Chains 285
9.2 Finite Size Corrections for Correlated Electron Chains . . . 294
9.3 Elements of Conformal Field Theory 304
9.4 Asymptotics of Correlation Functions 316
9.5 Persistent Currents in Correlated Electron Rings 323

10. Beyond the Integrability: Approximate Methods 335

10.1 Scaling Analysis 335
10.2 Bosonization 339

Bibliography 355

Index 365

Chapter 1

Introduction

In this chapter we shall remind the reader of some basic ideas of thermodynamics and statistical physics of interacting electron and spin systems. We shall show how thermal fluctuations destroy long-range order in low-dimensional quantum interacting systems at any nonzero temperature if only short-range interactions are present.

1.1 Why is the Topic of the Book Worthwhile Studying?

Low-dimensional electron systems (insulating magnets and conductors) have been an active topic of scientific research long before their experimental realization in organic conductors, polymers, Peierls insulators and nanoscale and mesoscopic systems, *e.g.*, quantum wires and edge states of the fractional quantum Hall effect devices. There are several principal differences between one space dimension and higher dimensions, most of which can be traced back to the reduced phase space in one dimension. Key properties distinguishing one-dimensional systems used to be connected with thermal fluctuations destroying long-range order at any nonzero temperature if only short-range interactions are present and quantum fluctuations tending to suppress a broken continuous symmetry, the *spin-charge separation* (the charge and spin content of wave functions of electrons move with different speeds), the breakdown of the *Fermi liquid* description, *i.e.*, absence of Fermi liquid quasiparticle pole in the Green's function (it becomes a *marginal Fermi liquid* or *Tomonaga–Luttinger liquid* with collective excitations due to global *conformal symmetry*), and the localization of electrons with even a small amount of disorder.

During recent years the interest in the strongly correlated electron and quantum spin systems has grown considerably. Usually low-lying electron-hole excitations of three-dimensional metals are successfully described within the phenomenological Landau's Fermi liquid theory. A Fermi liquid is the Fermi sphere and a gas of weakly interacting between each other *quasiparticles* defined *via* poles in one-particle Green's functions. Quasiparticles continuously evolve from free fermions when the interaction is adiabatically switched on. This is why, they have the same sets of quantum numbers and statistics as noninteracting electrons. In one space dimension the residue of the Fermi liquid quasiparticle pole vanishes and it is replaced by incoherent collective excitations, which follow from the global conformal symmetry. These excitations involve non-universal power-law singularities, which, in turn, determine the asymptotic behaviour of low-energy correlation functions. Although the Fermi surface is still properly defined, the discontinuity (jump) of the momentum distribution at the Fermi surface disappears, due to the above mentioned singularities. Systems displaying such breakdown of the Fermi liquid picture and exotic low-energy spectral properties are known as Tomonaga–Luttinger liquids.

On the other hand, in a number of recent experiments on the low-temperature behaviour of the rare-earth compounds and alloys, which are essentially three-dimensional, the *non Fermi liquid* character of the behaviour of the specific heat, magnetic susceptibility and (magneto) resistivity has also been observed during last couple of decades. It was pointed out recently that these features of the *non Fermi liquid* characteristics can be explained using the concept of the disordered behaviour of magnetic impurities in such systems. Superconductivity and antiferromagnetism in low dimensions has regained interest with the discovery of high-T_c superconductors and new heavy fermion superconductors. Very anisotropic magnetic and transport properties of the former arise primarily from the CuO planes there. Many of normal state properties of the two-dimensional high-T_c superconductors are very different from normal metals and cannot be reconciled with a standard Fermi liquid theory. A marginal Fermi liquid picture, similar to the one of one-dimensional electron systems, has been proposed to explain some of these features. Models of stripe-like effectively one-dimensional structures were proposed to explain some essential properties of high-T_c cuprates and heavy fermion Kondo lattices. The one-dimensional Kondo lattice model, realization of which is often considered as realistic model for heavy fermion materials, is still poorly understood. One can say that the finite concentration of magnetic impurities and random distribution of their characteristics (Kondo temperatures) will give rise to frustration, spin gap, non Fermi liquid critical behaviour and possible additional magnetic phase transitions (similar to metamagnetic ones). Possible implications of one-dimensional strongly correlated electron systems other than high-T_c superconductors could be new metal-oxides with ladder structure, and the edge states of the fractional quantum Hall effect, heavy fermion systems, *etc.* Ladder spin or correlated electron structures are non-trivial examples of quantum systems with the properties of both one- and two-dimensional models.

A substantial level of understanding of one-dimensional quantum correlated electron and/or spin systems has been reached over the past years. The exact solution with the help of the *Bethe's ansatz* of numerous models together with field-theoretical studies have provided deep insight into the ground state of systems, the complete classification of states, thermodynamic properties, and an asymptotic behaviour of correlation functions. Within the Bethe ansatz method the eigenfunctions and eigenvalues of the stationary Schrödinger equation are parametrized by a set of parameters known as *rapidities.* A system with internal degrees of freedom (such as a spin) requires a sequence of additional, nested Bethe ansatz for the wave function. In fact, each internal degree of freedom gives rise to one set of rapidities. Independently of the symmetry of the wave function and spin, eigenstates are occupied according to the Fermi-Dirac statistics, because hard-core particles are considered. Usually, in the ground state (and at low temperatures) each internal degree of freedom contributes with one Fermi (Dirac) sea. Fermi velocities of these Fermi seas are, generally speaking, different from each other, giving rise to the effect of charge and spin separation.

Over the past decade finite size effects in one-dimensional systems have been of considerable interest. The finite size of a system (*e.g.*, an electron conductor or a magnetic chain) manifests itself in several ways. First, impurities are important in low-dimensional quantum systems, since their contribution to extensive quantities (*e.g.*, the energy, charge and magnetic susceptibilities, resistance and specific heat) can become relatively large and observable. Boundaries and edges of open chains behave as some special sort of impurities with many similarities and differences in their behaviours. Second, the finite length of a mesoscopic (nanoscale) quantum chain (wire) or ring gives rise to quantum topological effects, *i.e.*, to *persistent currents* with oscillation periods given by interferences of the Aharonov–Bohm and Aharonov–Casher type. Finally, finite size corrections to the energy of a one-dimensional system determine critical exponents of the asymptotic dependence at large distances and long times of correlation functions via conformal field theory.

Integrable impurity models that represent realistic situations are a spin (magnetic) impurity in quantum spin chains and the Anderson hybridization impurity, in correlated electron chains. Similar behaviour was observed in Kondo, mixed-valent and heavy-fermion systems. Persistent currents have been observed experimentally in small quantum metal or semiconductor rings in the geometry of the Aharonov–Bohm effect. Persistent charge and spin currents arise from the magnetic flux and a radial electric field through the ring, respectively. Quantization of fluxes leads to periodic oscillations of currents. It is often important to know the influence of magnetic impurities on persistent currents — this situation was recently realized experimentally in nano-size devices, in which so called *quantum dots* were embedded into quantum rings, where electron–electron correlations play an essential role. The study of impurities or boundary potentials in strongly correlated electron systems is a relatively new subject of investigations of the past decade. In one-dimensional quantum models, due to the singularities in the density of states and collective excitations (Tomonaga–Luttinger liquid properties) both impurities and interactions in the host have to be examined with exact methods, because a perturbative approximate approach or a mean field-like theory could provide even qualitatively incorrect results there. The interplay of interactions between electrons and magnetic impurities can dramatically change low-energy properties in one-dimensional quantum systems. Recent experiments on nanoscale quantum rings with quantum dots embedded (applied point contact voltage) connect the problem of impurities with that of conductance (closely related to recent developments of theory of open low-dimensional electron systems), mesoscopic persistent current and *Coulomb blockade* oscillations. The problem of recent experiments on conductance oscillations for chiral edge state currents of the fractional quantum Hall effect in two-dimensional electron gas and singularities of angle-resolved photoemission spectra of one-dimensional electron systems are also deeply connected with low-temperature characteristics of low-dimensional highly correlated electron systems.

In the present book we want to introduce the reader to the interesting world of Bethe ansatz solvable (sometimes one uses the word *integrable*) models of correlated electron systems in one space dimension. Our goal is to give the reader descriptions of the main methods of the Bethe ansatz and to show how the finite size effects, which we discussed above, can be described in the framework of the Bethe ansatz theories. Unfortunately, we cannot present all results here, which were published during recent years, and follow only some, important (in our mind), main steps of the development of such theories. This is why, we want to apologize to those authors, whose (very important) papers are not cited in this book, only because of the lack of space and time. We also would like to apologize to the reader, who looks for the proper description of approximate methods, used to describe one-dimensional correlated electron systems. We shall mention those methods only briefly, because those results have been reviewed in several brilliant books and review articles. Finally, the book is devoted only to theoretical description of one-dimensional correlated electron systems, as a result applications of the results to very interesting experiments (as well as the description of those experiments themselves) are beyond the scope of our book.

1.2 Thermodynamics

Thermodynamic potentials as a function of its *natural (independent) variables* completely determine the total thermodynamics of any system. Other important quantities can be expressed as derivatives of thermodynamic potentials. There are several such potentials. The most known is the *internal energy*, E, with the natural variables $\mathcal{S}$, the entropy of our system, and V, the volume, with

$$dE = Td\mathcal{S} - pdV \ , \tag{1.1}$$

where T is the temperature and p is the pressure. The other thermodynamic potentials are obtained from the internal energy by Legendre transformations. If one has a convex function $f(x)$ (*i.e.*, such that its second derivative with respect to x is positive, $f''(x) > 0$), the Legendre transform $\bar{f}(x)$ is $\bar{f}(x) = xy - f(x)$, where $x(y)$ is defined as a root for given y of the equation $y = \partial f/\partial x$ (the convexity guarantees that this equation can be solved for any y that lies between the maximum and minimum gradients of f). Obviously Legendre transformations are invertable. It also turns out that $\bar{\bar{f}}(x) = f(x)$. Then the *Helmholtz free energy*, F, which is the function of the natural variables T and V, can be obtained as

$$dF = d(E - T\mathcal{S}) = -\mathcal{S}dT - pdV \ . \tag{1.2}$$

The *Gibbs free energy*, G, is the function of the natural variables T and p. [The fourth thermodynamic potential, H, which should not be confused with the value of the magnetic field, see below, is the function of the natural variables $\mathcal{S}$ and p.] The connection between the Gibbs free energy and the

Helmholtz free energy is

$$G = F + pV \ . \tag{1.3}$$

The temperature is the natural variable of the Helmholtz free energy, and also other macroscopic parameters, which determine the energy levels of the system, are natural variables too. For example, a simple magnetic system has the Helmholtz free energy $F(T, H)$, where H is an external magnetic field. In such a case the Gibbs free energy is the Legendre transform of F with respect to its second natural variable (*e.g.*, $G(T, M) = F + HM$, where M is the magnetization of the system).

Usually one distinguishes *extensive* (*i.e.*, proportional to the mass of the system) and *intensive* variables. The Helmholtz and Gibbs free energies are both extensive variables. There are two different ways the energy of the system can be changed: one can work on the system, or one can supply heat to it. The change of, *e.g.*, the internal energy, is $dE = TdS - pdV$, where the first term is the supplied heat and the second term manifests the work done on the system, both in a reversible change.

For any system, the natural variables for the Helmholtz or Gibbs free energies are the temperature and an intensive variable, while the natural variables of the other are T and an extensive variable (*cf.* $H \leftrightarrow M$ for a magnetic system). The per-particle value of the free energy whose natural variables are intensive is usually called the *chemical potential*, μ. Often one also introduces free energies and other extensive variables per unit mass (e, f and g or s and m, respectively).

For the magnetic system one has

$$\begin{aligned} s = -\left(\frac{\partial g}{\partial T}\right)_m = -\left(\frac{\partial f}{\partial T}\right)_H \ , \\ e = f - T\left(\frac{\partial f}{\partial T}\right)_H = \left(\frac{\partial \beta f}{\partial \beta}\right)_H \ , \end{aligned} \tag{1.4}$$

and

$$\begin{aligned} m = -\left(\frac{\partial f}{\partial H}\right)_T \ , \\ H = -\left(\frac{\partial g}{\partial m}\right)_T \ , \end{aligned} \tag{1.5}$$

where the subscript denotes the fixed variable and $\beta = (k_B T)^{-1}$ (k_B is the Boltzmann's constant, equal to 1.38×10^{-23} $\mathrm{JK^{-1}}$; in what follows we shall put it equal to 1, except for specially mentioned places, *i.e.*, we shall measure the temperature in energy units). The magnetic susceptibility χ_T and the specific heat, c_T, at constant temperature are then

$$\chi_T = -\left(\frac{\partial^2 f}{\partial H^2}\right)_T \tag{1.6}$$

and

$$c_H = -T\left(\frac{\partial^2 f}{\partial T^2}\right)_H \,, \tag{1.7}$$

respectively. One can also define the specific heat at constant volume, c_V. By writing $dE = c_V dT - p_T dV$ (where p_T is the isothermal pressure), we get $c_V = T(\partial \mathcal{S}/\partial T)_V$, *i.e.*, the entropy is $\mathcal{S}(T,V) = \int^T dT' c_V(V,T')/T'$ (up to some arbitrary function of the volume).

1.3 Statistical Mechanics: Simple Models

When discussing the properties of quantum exactly solvable models we shall mostly be concerned with *thermal equilibrium*. Here, together with the thermodynamics, we need the help of statistical mechanics, because it predicts the behaviour of a system with a great number of degrees of freedom, given the laws ruling its microscopic behaviour. For systems in thermal equilibrium a simple relationship between microscopic properties and macroscopic behaviour takes place. Let us label the different microscopic configurations of the system by the index α. These configurations used to be called *microstates*. Depending on a system, microstates may form either a continuous set (common for classical systems), or discrete one (usual for a quantum system, *e.g.*, in a finite volume: an example of a discrete set of microstates are the eigenstates of a quantum Hamiltonian; we shall mostly consider that situation in what follows). Even a system whose Hamiltonian possesses discrete eigenstates can also be in a continuum of states. The reason for such a property is due to a state being specified by giving the magnitude for the considered system to be in each one of a complete set of basic states. It turns out that these magnitudes are continuous variables. Nonetheless, in the framework of statistical mechanics one usually operates with averages, and averaging the expectation value of some quantum operator over all possible states produces the same result as taking the average of only the members of a complete set of basis state.

Denote the internal energy of some microstate α with E_α and let the system be in thermal equilibrium with a heated bath (sometimes called a *thermostat*) at temperature T. The probability p_α of the system being found in the microstate α is

$$p_\alpha = Z^{-1} \exp(-\beta E_\alpha) . \tag{1.8}$$

All time-independent properties of the considered system in thermal equilibrium follow from this fundamental formula. The normalizing factor Z is usually called the *partition function*: any system must always be in some state, naturally, so that the total of p_α must be one (*i.e.*, $\sum_\alpha p_\alpha = 1$). It yields for the partition function

$$Z = \sum_\alpha \exp(-\beta E_\alpha) . \tag{1.9}$$

This formula is known as the *Gibbs probability distribution*. The thermal average $\langle X \rangle$ of any operator X of the system can be found if one knows microstates and energies of them, and knows the value X_α in each microstate. Then we obtain

$$\langle X \rangle = \sum_\alpha X_\alpha p_\alpha = Z^{-1} \sum_\alpha X_\alpha \exp(-\beta E_\alpha) . \tag{1.10}$$

Actually the Gibbs distribution provides a connection between microscopic laws in a system and its behaviour in a thermal equilibrium.

The partition function Z is a function of temperature and of the parameters which determine the energies of microstates (usually called *constraints*). For instance, an external magnetic field H is such a constraint for a magnetic system. A mechanical work is done on a system by varying the constraints. In fact the reader can see that all the properties of the system can be obtained from the functional dependence of the partition function on T and the constraints.

According to the above, the mean (internal) energy of a system in thermal equilibrium is

$$E(\equiv \langle E \rangle) = Z^{-1} \sum_\alpha E_\alpha \exp(-\beta E_\alpha) . \tag{1.11}$$

It is the sum of all microstates of energies of all microstates weighted by the probabilities that the system is in those microstates. Naturally, it follows that

$$E = -\left(\frac{\partial \ln Z}{\partial \beta}\right)_V \tag{1.12}$$

(*i.e.*, no mechanical work is done on the system). The specific heat of the system in a change at which constraints do not vary, *e.g.*, at constant volume, is

$$c_V = \left(\frac{\partial E}{\partial T}\right)_V = \beta^2 \left(\frac{\partial^2 \ln Z}{\partial \beta^2}\right)_V . \tag{1.13}$$

Hence, changes in the energy of the system must be due to a supplied heat. Therefore, $c_V dT = Td\mathcal{S}$, and one gets $c_V = -\beta(\partial \mathcal{S}/\partial \beta)_V$. Finally, the entropy of the system can be written as

$$\mathcal{S} = -\beta \left(\frac{\partial \ln Z}{\partial \beta}\right)_V + \ln Z \tag{1.14}$$

up to an arbitrary function of the volume (which is a function of the constraints on the considered system; the third law of the thermodynamics requires the entropy of the system to be a constant at $T \to 0$ which determines that function to be zero). It is easy to find the Helmholtz free energy, for which the natural variables are the temperature and the constraints,

$$F = -T \ln Z \tag{1.15}$$

which can also be re-written as $Z = \exp(-\beta F)$.

It is possible to introduce many constraints. For a magnetic system the magnetization M_i (conjugate to the constraint H_i) is $M_i = T(\partial \ln Z/\partial H_i)_{H_{j \neq i},\beta}$ with other constraints fixed. Such equations are known as the *equations of state* for the considered system. This implies that a system with N constraints has $2^N - 1$ possible Gibbs free energies. The latter differs in the choice of constraints.

As a result, the knowledge of the partition function (which can be found from the knowledge of all the energies of microstates of a system), means the knowledge of all thermodynamic potentials and other thermodynamic characteristics. The Hamiltonian then is the function $\mathcal{H}$ of states α and constraints. The fundamental task of quantum mechanics is to find the states of the Hamiltonian and related energies E_α. Then statistical mechanics finds the partition function of a system in thermal equilibrium. Finally, with the help of thermodynamics, one can find all necessary thermodynamic potentials and other thermodynamic characteristics of the considered quantum system.

Let us consider as an example L quantum spins $\frac{1}{2}$ situated in an external magnetic field. The Hamiltonian of this system is called the *Zeeman* Hamiltonian. Suppose the magnetic field H is directed along the axis z, then the Zeeman Hamiltonian is

$$\mathcal{H}_Z = -H \sum_{j=1}^{L} S_j^z \ , \tag{1.16}$$

where S_j^z is the operator of z-projection of the j-th spin, and one has L spins. Notice, that in what follows we shall use the effective magneton, *i.e.*, the coefficient between the mechanical (spin) moment and magnetic moment equal to 1, except for specially mentioned cases. Usually these values are related *via* $g\mu_B$, where g is the effective g-factor of the magnetic ion and $\mu_B = \hbar e/2m$ is the Bohr's magneton, where $-e$ denotes the charge of an electron and m denotes its mass. The value of the Bohr's magneton is equal to 9.27×10^{-24} JT^{-1}.

The partition function of these noninteracting spins can be written in a simple form

$$Z = \left(2\cosh\frac{H}{2T}\right)^L \ . \tag{1.17}$$

The Helmholtz free energy per spin is equal to

$$f = -T\ln[2\cosh(H/2T)] \ . \tag{1.18}$$

It is the smooth function of temperature, and, therefore, there is no phase transition for noninteracting quantum spins. One can calculate the specific heat per spin:

$$c_H = \frac{H^2}{4T^2\cosh^2(H/2T)} \ , \tag{1.19}$$

which shows that $c_H \to 0$ for $T \to 0$ and $T \to \infty$ and has a maximum as a function of temperature. The magnetization per spin m^z is calculated to be

$$m^z = \frac{1}{2}\tanh(H/2T) \ , \tag{1.20}$$

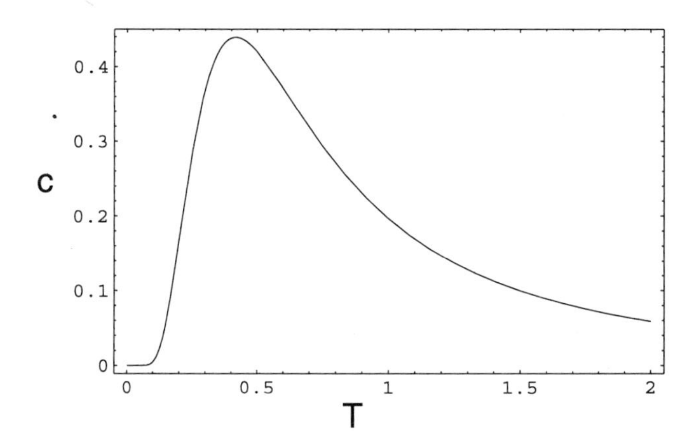

Fig. 1.1 The dependence of the specific heat on temperature for a spin-$\frac{1}{2}$ in a magnetic field $H = 1$, arbitrary units.

i.e., $m^z \to 0$ for $H \to 0$ no matter what the value of T is and $m^z \to \pm\frac{1}{2}$ for $H \to \pm\infty$. The magnetic susceptibility is

$$\chi = [4T\cosh^2(H/2T)]^{-1} , \tag{1.21}$$

which tends to zero for high and small temperatures. In the ground state (*i.e.*, at $T = 0$), the magnetization per site is zero at zero value of the magnetic field and takes its nominal value (*i.e.*, $\pm\frac{1}{2}$ for positive and negative values of H, respectively) at any, even infinitesimally small value of $|H|$. The illustration of the temperature behaviour of the Zeeman system is presented in Figs. 1.1 and 1.2.

As the second simple example, let us consider the thermodynamics of N noninteracting electrons. The Hamiltonian of these electrons is equal to

$$\mathcal{H}_{free} = -\frac{\hbar^2}{2m}\left(\frac{\partial^2}{\partial x^2} + \frac{\partial^2}{\partial y^2} + \frac{\partial^2}{\partial z^2}\right) , \tag{1.22}$$

where m is the mass of a free electron. The wave function of this equation $\psi = V^{-1/2}\exp(i\mathbf{kr})$, where $V = L^3$ is the volume of the system and $\mathbf{k}$ is the wave vector of an electron. The eigenfunction of the stationary Schrödinger equation is

$$E = \frac{\hbar^2}{2m}k^2 . \tag{1.23}$$

The value of the wave vector is related to the wave length λ as $|\mathbf{k}| = k = 2\pi/\lambda$. Connecting the momentum p and the wave length with the de Broglie

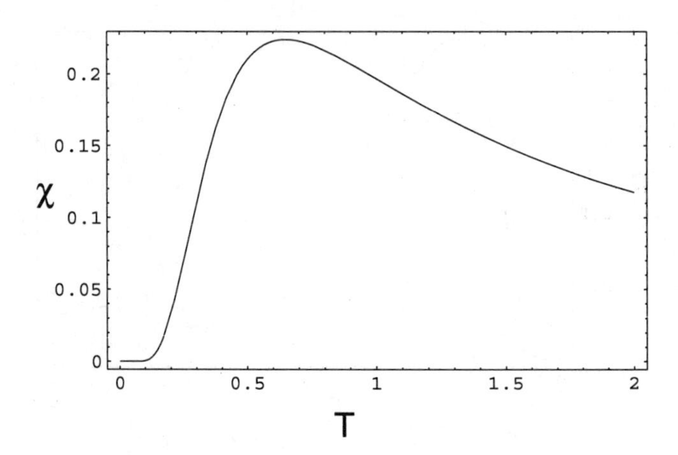

Fig. 1.2 The dependence of the magnetic susceptibility on temperature for a spin-$\frac{1}{2}$ in a magnetic field $H = 1$.

relation $\lambda = 2\pi\hbar/p$, one obtains

$$E = \frac{1}{2m}p^2 = \frac{1}{2}mv^2 , \tag{1.24}$$

which is nothing other than the classical energy for a free particle of the mass m and velocity v. The value of k is determined from the *boundary conditions. Periodic* boundary conditions require $\psi(x+L, y, z) = \psi(x, y, z)$ with similar relations for y and z. It defines the wave vectors as

$$k_{x,y,z} = \frac{2\pi}{L} n_{x,y,z} , \tag{1.25}$$

where $n_{x,y,z} = 0, \pm 1, \pm 2, \ldots$. On the other hand, for *open* boundary conditions $\psi(0) = \psi(L) = 0$; in each direction the eigenfunctions are standing waves $\psi = (\sqrt{2/L})^3 \sin(n_x \pi x/L) \sin(n_y \pi y/L) \sin(n_z \pi z/L)$, where $n_{x,y,z} = 1, 2, \ldots$. It implies that boundary conditions introduce a quantization into such a simple problem: not every value of k (related to the momentum of a particle) pertains to eigenfunctions of the Hamiltonian.

Actually we see that the distribution of numbers $n = \{n_{x,y,z}\}$ determines states of free electrons. However, the reader remembers that due to the Pauli's principle each electron can only be in one state. A very important level is the one which divides the filled and vacant levels of free electrons in the ground state (known as the *Fermi level* n_F), as the function of the number of electrons N. The number of states, which have n less than a certain value n_F is $2 \times (4\pi/3)n_F^3$ in a three-dimensional space, where the first multiplier takes into account the degeneracy due to two possible values of spins of electrons (for $H = 0$). That implies $(8\pi/3)n_F^3 = NL^3$. At $T = 0$, the energy, related to n_F, is equal to

$$E(n_F) \equiv E_F = \frac{\hbar^2}{2m}\left(\frac{2\pi}{L}\right)^2 n_F^2 = \frac{\hbar^2}{2m}(3\pi^2 N)^{2/3} . \tag{1.26}$$

It is known as the *Fermi energy* (sometimes it is used to speak about the Fermi temperature $T_F = E_F/k_B$). The Fermi velocity of free electrons can be defined as

$$v^F = \sqrt{\frac{2E_F}{m}} = \frac{\hbar}{m}(3\pi^2 N)^{1/3} . \tag{1.27}$$

It is often important to find the number of states per unit energy range per unit volume as a function of the energy. Let us denote the *density of states* $g(E)$. One gets in the ground state

$$\int g(E)dE = N = \frac{1}{3\pi^2}(2mE/\hbar^2)^{3/2} \tag{1.28}$$

and, hence,

$$g(E) = \frac{1}{2\pi^2}(2m/\hbar^2)^{3/2}\sqrt{E} . \tag{1.29}$$

By the way, for the two-dimensional free electron system one has $n_F^2 = NL^2/2\pi$, $E_F = (\hbar^2/m)\pi N$, $v^F = (\hbar/m)\sqrt{2\pi N}$ and $g(E) = m/\pi\hbar^2$, while for the one-dimensional one we get $n_F^2 = N^2L^2/4$, $E_F = (\hbar^2/2m)\pi^2N^2$, $v^F = (\hbar/m)\pi N$, and $g(E) = (1/\pi\hbar)\sqrt{m/2}E^{-1/2}$, *i.e.*, for small E the density of states of low-dimensional electron systems is enhanced in comparison with the usual three-dimensional situation.

In the thermal equilibrium the partition function of free electrons with energies E is $Z = \sum(1+\exp[(\mu - E)/T])$, where the sum is over all possible states of electrons. Then, the Helmholtz free energy per state is

$$f_{state} = -T\ln(1 + \exp[(\mu - E)/T]) . \tag{1.30}$$

The mean number of particles per state in the thermal equilibrium is equal to

$$n_F = -\frac{\partial f_{state}}{\partial \mu} = \frac{1}{\exp[(E - \mu)/T] + 1} . \tag{1.31}$$

It is called the *Fermi-Dirac distribution function.* This function defines the distribution of an ideal gas obeying Fermi statistics, *i.e.*, for which the Pauli principle works. In the thermal equilibrium the number of electrons dN with the energies between E and $E + dE$ is given by

$$dN = n_F(E)g(E)dE = \frac{1}{2\pi^2}\left(\frac{2m}{\hbar^2}\right)^{3/2}\frac{\sqrt{E}dE}{e^{(E-\mu)/T}+1} . \tag{1.32}$$

One can introduce the chemical potential μ as the Lagrangian multiplier, and it is determined from the condition that the total number of electrons must be constant $\int dN = N$ (and the total energy should also be constant). The chemical potential plays the role of the Fermi energy of electrons at nonzero temperatures. In a high temperature limit $n_F(E)$ reduces to the classical Boltzmann distribution law. In the ground state it is a step function, *i.e.*, $n_F(E < E_F) = 1$ for $E < E_F \equiv \mu(T = 0)$, and $n_F(E > E_F) = 0$ otherwise. Hence, at $T = 0$, one gets $N = (1/3\pi^2)(2mE_F/\hbar^2)^{3/2}$, in agreement with the above. At any temperature one has $n_F(E = \mu) = \frac{1}{2}$. The Fermi-Dirac distribution is called degenerate when $T \ll \mu$ and non-degenerate when $T \gg \mu$, *i.e.*, in the classical Boltzmann limit.

We can write the internal energy per unit volume of the free electron gas as

$$e = \int_0^\infty En_F(E)g(E)dE . \tag{1.33}$$

For standard metallic systems the Fermi energy E_F is often of the order of several electron-Volts. Hence, it is important to know the behaviour of the thermodynamic characteristics at low temperatures. Consider the integral $I = \int_0^\infty n_F(E)[dF(E)/dE]dE$, where $F(E = 0) = 0$ is any function. Integrating by parts one gets $I = -\int_0^\infty n'_F(E)F(E)dE$. Expanding $F(E)$ by Taylor's theorem $F(E) = F(\mu) + (E-\mu)F'(\mu) + (1/2)(E-\mu)^2F''(\mu) + \cdots$ one obtains $I = L_0F(\mu) + L_1F'(\mu) + L_2F''(\mu) + \cdots$, where $L_n = -\int_0^\infty (E-\mu)^n n'_F(E)dE$. It is possible to replace the lower limits on the integrals by $-\infty$ at low temperatures. Then the reader can see that $L_0 = 1$, $L_1 = 0$ (as well as for any odd n, because $n'_F(\mu)$ is an even function of $E-\mu$), and

$$L_2 = \frac{T^2}{2}\int_{-\infty}^\infty \frac{x^2e^xdx}{(1+e^x)^2} = \frac{\pi^2T^2}{6} . \tag{1.34}$$

This expansion is known as the *Sommerfeld expansion.* To determine the number of electrons N one can use the function $F(E) = \int_0^E g(E)dE$, so that

$$N = \int_0^\infty n_F(E) g(E) dE = \int_0^\mu g(E) dE + \frac{\pi^2 T^2}{6} g'(\mu) \ . \tag{1.35}$$

Taking into account that $\int_0^{E_F} g(E) dE = N$, one gets

$$\mu \approx E_F \left(1 - \frac{\pi^2 T^2}{12 E_F^2} \right) \ , \tag{1.36}$$

which describes the low temperature corrections to the free energy. The function $F(E) = \int_0^E E g(E) dE$ can yield the internal energy of free electrons per site at low temperatures

$$e \approx e_0 + (\mu - E_F) E_F g(E_F) + \frac{3\pi^2 T^2}{12} g(E_F) + \cdots =$$
$$e_0 + \frac{\pi^2}{6} g(E_F) T^2 + \cdots \ , \tag{1.37}$$

where $e_0 = 3NE_F/5$ is the ground state energy. The low temperature specific heat per unit volume of the free electron gas can be written as

$$c = \frac{\pi^2}{3} g(E_F) T + \cdots = \gamma T \ , \tag{1.38}$$

where the low temperature *Sommerfeld coefficient* of the specific heat per unit volume is equal to

$$\gamma = \frac{\pi^2 g(E_F)}{3} = \frac{\pi^2 N}{2E_F} \ . \tag{1.39}$$

It is possible to introduce the Zeeman effect of an external magnetic field H on spins of electrons as the renormalization

$$g(E) \to \frac{1}{2} (g[E + (H/2)] - g[E - (H/2)]) \ , \tag{1.40}$$

which implies that the magnetization of the free electron gas per site is

$$m^z = \int \frac{1}{2} (g[E + (H/2)] - g[E - (H/2)]) n_F(E) dE \ . \tag{1.41}$$

At small values of H one obtains $m^z \sim H \int g'(E) n_F(E) dE \sim H g(\mu)$. Then we can use the function $F(E) = \int_0^E g'(E) dE$, which implies that at low temperatures the magnetic susceptibility of the free electron gas is proportional $\chi \sim g(E_F) = 3N/2E_F$, *i.e.*, it is finite and proportional to the density of states at zero temperature, or inversely proportional to the Fermi energy. This is the famous Pauli paramagnetism of the free electron gas (naturally, the orbital moment of charged electrons produces the Landau's diamagnetic term, the derivation of which we shall not reproduce here).

It is also worthwhile mentioning that the energy of the tight-binding lattice model of electrons with the Hamiltonian

$$\mathcal{H}_{tb} = -t\sum_{\delta}\sum_{\sigma}(a^{\dagger}_{\mathbf{r},\sigma}a_{\mathbf{r}+\delta,\sigma} + \text{H.c.}) \ , \tag{1.42}$$

where $a^{\dagger}_{\mathbf{r},\sigma}$ ($a_{\mathbf{r},\sigma}$) creates (destroys) an electron with the spin projection σ at the lattice site $\mathbf{r}$, t is the hopping integral, and the sum is over the nearest neighbour sites, is reduced to the energy of the free electron gas in the limit of small inter-site distances. Hence, in this limit one can use the above results for the lattice tight-binding model of noninteracting electrons, too. On the other hand, lattice effects can be included into the renormalization of the values of $g(E)$ *etc.* For example, the density of states of the d-dimensional lattice can be obtained due to the equality

$$g(E)dE = \frac{V_0}{dZ}\int d^d\mathbf{k} \tag{1.43}$$

where Z is the number of electrons per unit cell, $\mathbf{k}$ is the wave vector defined up to the addition of an arbitrary linear combination of basic vectors $\mathbf{h}$ of the reciprocal lattice of the crystal,

$$E(\mathbf{k}) = E\left(\mathbf{k} + \sum_{\alpha=1}^{d} n_\alpha h_\alpha\right) \ , \tag{1.44}$$

(n_α are integers), and the integral is extended to the region within one cell of the reciprocal space, in which $E \le E(\mathbf{k}) \le E + dE$. Then an easy calculation yields

$$g(E) = \frac{V_0}{dZ}\sum_{br}\int_{S(E)} \frac{dS}{\sqrt{\sum_{\alpha=1}^{d}\left(\frac{\partial E(\mathbf{k})}{\partial k_\alpha}\right)^2}} \ , \tag{1.45}$$

where the summation $\sum_{br}$ extends over all branches of $E(\mathbf{k})$, $S(E)$ for each branch is defined by $E(\mathbf{k}) = E$, and dS is the length of an infinitesimal portion of $S(E)$. Then the analytic singularities of the density of states can originate from the so-called critical points of $E(\mathbf{k})$, *i.e.*, the points (let us call them E_c) where the derivatives $\partial E/\partial k_\alpha$ vanish. The analysis shows that in one space dimension such critical points are related to the extreme energies (edges) of each branch, where the singularities of the density of states are $g(E) \sim |E - E_c|^{-1/2}$. For the two-dimensional lattice there is a logarithmic feature $g(E) \sim \ln|E - E_c|$ at saddle points in the two-dimensional space, while there are no singularities in the three-dimensional case. These features of the behaviour of the density of states are known as van Hove features due to L. van Hove.

1.4 Mermin–Wagner–Hohenberg Theorem

In our book we want to review some exact results for interacting electron and quantum spin systems. Most of these results are obtained for low-dimensional quantum systems. Why is it important to study low-dimensional quantum systems exactly? It is clear that at low temperatures most thermal fluctuations of quantum systems are frozen, and they do not play an important role for such a situation. However, as the reader knows from quantum mechanics, there can exist zero temperature quantum fluctuations which mostly determine the ground state (for $T = 0$) and low-temperature behaviour of electron systems. These quantum fluctuations are enhanced in low-dimensional systems, due to features of the low-dimensional density of states. If one considers a many-body system, in which an ordering takes place, then it is relatively simple to study the low-temperature thermodynamics of that system, because one can describe the many-body system using a simple wave function of a single particle. The nonzero order parameter means that it has the same value at any place of the system, and, hence, many-body effects are not so important (although those many-body effects produce an ordering). On the other hand, for a system in which there is no ordering, but an interaction between electrons exists, one cannot use a single-particle description *a priori*.

This is why it is very important to present here the theorems due to N. D. Mermin and H. Wagner and to P. C. Hohenberg, who stated that isotropic Heisenberg systems with nearest-neighbour exchange interaction, Bose liquid and interacting electron system do not manifest spontaneous ordering for any nonzero temperature.

For example, for a low-dimensional magnetic system it means that there is no spontaneous, at $H \to 0$, magnetization in it.

The proof of this statement is based on the famous inequality due to N. N. Bogolyubov for operators A and C and some (yet undefined) Hamiltonian:

$$\frac{1}{2T}\langle\{A, A^\dagger\}\rangle\langle[[C, \mathcal{H}], C^\dagger]\rangle \geq |\langle[C, A]\rangle|^2 \ . \tag{1.46}$$

Here we use the standard notations of statistical mechanics from the previous section. Also, $[.,.]$ ($\{.,.\}$) denote (anti)commutator, and $A^\dagger$ is the operator, Hermitian conjugated to A. Let us define

$$(A,B) = \sum_{\alpha,\alpha'}' \langle\alpha|A|\alpha'\rangle^* \langle\alpha|B|\alpha'\rangle \frac{p_\alpha - p_{\alpha'}}{E_{\alpha'} - E_\alpha}, \tag{1.47}$$

where $p_\alpha = Z^{-1}\exp(-\beta E_\alpha)$ and the summation is over all pairs of states, except those with $E_\alpha = E_{\alpha'}$. The reader can observe that

$$0 < \frac{p_\alpha - p_{\alpha'}}{E_{\alpha'} - E_\alpha} < \frac{p_\alpha + p_{\alpha'}}{2T}, \tag{1.48}$$

which implies $(A,A) \le \langle\{A,A^\dagger\}\rangle/2T$. Then, by using the Schwartz inequality $(A,A)(B,B) \ge |(A,B)|^2$ for $B = [C^\dagger,\mathcal{H}]$ and $(A,B) = \langle[C^\dagger,A^\dagger]\rangle$ with the equality $(B,B) = \langle[C^\dagger,[\mathcal{H},C]]\rangle$, one proves Eq. (1.46).

To prove the Mermin–Wagner theorem let us apply the Bogolyubov inequality to the quantum Hamiltonian of an interacting spin system

$$\mathcal{H} = -\sum_{\mathbf{r},\mathbf{r}'} J(\mathbf{r}-\mathbf{r}')\vec{S}_{\mathbf{r}}\vec{S}_{\mathbf{r}'} - H\sum_{\mathbf{r}} S^z_{\mathbf{r}} \exp(-i\mathbf{K}\mathbf{r}) , \tag{1.49}$$

where $\mathbf{r}$ and $\mathbf{r}'$ run over lattice sites of some multi-dimensional lattice, periodic boundary conditions are used for L spins, $S^{x,y,z}_{\mathbf{r}}$ are operators for the projections of spin S in the $\mathbf{r}$-th site of the lattice, and the sum $\sum_{\mathbf{r}} r^2 J(\mathbf{r})$ converges (this is, naturally, the case for the nearest neighbour coupling). As for $\mathbf{K}$, it is zero for the ferromagnetic situation, and $\exp(i\mathbf{K}\mathbf{r}) = \pm 1$ for the same (different) sublattices in the antiferromagnetic situation. Then we take the standard Fourier transforms of all three projections of $\vec{S}_{\mathbf{r}}$ and $J(\mathbf{r})$, and define $C = S^+_{\mathbf{k}}$ and $A = S^-_{-\mathbf{k}-\mathbf{K}}$. The application of the Bogolyubov inequality yields

$$\frac{1}{2}\langle\{S^+_{\mathbf{k}+\mathbf{K}}, S^-_{-\mathbf{k}-\mathbf{K}}\}\rangle \ge 4L^2T(m^z)^2\Big(L^{-1}\sum_{\mathbf{k}'}[J(\mathbf{k}') - J(\mathbf{k}'-\mathbf{k})]$$
$$\times \langle 4S^z_{-\mathbf{k}'}S^z_{\mathbf{k}'} + \{S^+_{\mathbf{k}'}, S^-_{-\mathbf{k}'}\}\rangle + 2LHm^z\Big)^{-1} , \tag{1.50}$$

where $m^z = L^{-1}\sum_{\mathbf{r}}\langle S^z_{\mathbf{r}} \exp(i\mathbf{K}\mathbf{r})\rangle$. The denominator of the last formula is positive. One can also show that

$$\begin{aligned}
L^{-1}\sum_{\mathbf{k}'}[J(\mathbf{k}') - J(\mathbf{k}'-\mathbf{k})]\langle 4S^z_{-\mathbf{k}'}S^z_{\mathbf{k}'} + \{S^+_{\mathbf{k}'}, S^-_{-\mathbf{k}'}\}\rangle + 2LHm^z \\
< L^{-1}|\sum_{\mathbf{r}} J(\mathbf{r})[1-\exp(i\mathbf{k}\mathbf{r})]\sum_{\mathbf{k}'}\langle 4S^z_{-\mathbf{k}'}S^z_{\mathbf{k}'} + \{S^+_{\mathbf{k}'}, S^-_{-\mathbf{k}'}\}\rangle| \\
+ 2L|Hm^z| < 4L\sum_{\mathbf{r}} J(\mathbf{r})[1-\cos(\mathbf{k}\mathbf{r})]S(S+1) + 2L|Hm^z| \\
< 2L\left(k^2 S(S+1)\sum_{\mathbf{r}} r^2|J(\mathbf{r})| + |Hm^z|\right) .
\end{aligned} \tag{1.51}$$

Then we replace the denominator by its upper bound from the last inequality and sum both sides of Eq. (1.50) over $\mathbf{k}$. It yields

$$S(S+1) > \frac{2T(m^z)^2}{L}\sum_{\mathbf{k}}\frac{1}{S(S+1)k^2\sum_{\mathbf{r}} r^2|J(\mathbf{r})| + |Hm^z|} . \tag{1.52}$$

Then, in the so-called *thermodynamic limit*, $L \to \infty$, we can proceed with the calculation of the sums (integrals in the thermodynamic limit), integrating only over the first Brillouin zone, which produces

$$(m^z)^2 < \frac{2\pi S(S+1)I}{k_0^2 T\ln(1+I/|Hm^z|)} \tag{1.53}$$

for the two-dimensional case, and

$$(m^z)^3 < |H|I\left(\frac{S(S+1)}{2T\tan^{-1}\sqrt{I/|Hm^z|}}\right)^2 \tag{1.54}$$

for the one-dimensional case, where $I = S(S+1)k_0^2\sum_{\mathbf{r}} r^2|J(\mathbf{r})|$ and k_0 is the characteristic vector in the k-space, related to the distance from the origin to the nearest Bragg plane. Taking the limit of small values of the magnetic field H, we finally reach the following inequalities

$$|m^z| < \frac{C}{\sqrt{T|\ln|H||}} \tag{1.55}$$

for two space dimensions and

$$|m^z| < \frac{C'|H|^{1/3}}{T^{2/3}} \tag{1.56}$$

for one space dimension, where C and C' are some constants.

These inequalities actually state that for any nonzero temperature there is no spontaneous magnetization (or spontaneous magnetizations of any magnetic sublattices) for one- and two-dimensional Heisenberg magnets for any value of site spins (including classical spins $S \to \infty$). However, this theorem does not rule out the possibility of other phase transitions, and can be applied for anisotropic spin systems only for a special direction of the magnetic field with respect to the anisotropy axis. Also, it does not give any conclusion about the behaviour of spin systems in one and two space dimensions in the ground state (at $T = 0$). Hence, the possibility of *quantum phase transitions*, which take place in the ground state, exists. In the following chapters we shall consider many examples of such quantum phase transitions.

One can generalize the construction to the case of bosons in a superfluid liquid or superconducting pairs of correlated electrons. This statement is known as the *Hohenberg theorem*.

First, we remind the reader that

$$\begin{aligned}\tau^{AB}_{\mathbf{k},\mathbf{k}'}(t-t') &= L^{-d}\langle[A_{\mathbf{k}}(t),[B_{\mathbf{k}'}(t')]]\rangle \\ &= \int_{-\infty}^{\infty} \frac{d\omega}{2\pi}\tau^{AB}_{\mathbf{k},\mathbf{k}'}(\omega)\exp[-i\omega(t-t')] \end{aligned} \tag{1.57}$$

is the *spectral weight function*, where d is the space dimension. If one defines the *response function* as

$$\chi^{AB}_{\mathbf{k},\mathbf{k}'}(z) = \int_{-\infty}^{\infty} \frac{d\omega}{2\pi}\frac{\tau^{AB}_{\mathbf{k},\mathbf{k}'}(\omega)}{\omega - z} \tag{1.58}$$

and the *static response function* $\hat{\chi}^{AB}_{\mathbf{k},\mathbf{k}'} = \chi^{AB}_{\mathbf{k},\mathbf{k}'}(z=0)$, then the equal-time correlation function can be written as a function of the spectral weight function

$$\begin{aligned}C^{AB}_{\mathbf{k},\mathbf{k}'} &= L^{-d}\langle\{A_{\mathbf{k}}(t) - \langle A_{\mathbf{k}}(t)\rangle, B_{\mathbf{k}}(t) - \langle B_{\mathbf{k}}(t)\rangle\}\rangle \\ &= \int_{-\infty}^{\infty} \frac{d\omega}{2\pi}\tau^{AB}_{\mathbf{k},\mathbf{k}'}(\omega)\coth(\omega/2T)\ , \end{aligned} \tag{1.59}$$

which is the *fluctuation-dissipation theorem*.

Taking into account that $(\hat{\chi}^{AB}_{\mathbf{k},\mathbf{k}'})^* = \hat{\chi}^{A^\dagger B^\dagger}_{\mathbf{k}',\mathbf{k}}$ and $\hat{\chi}^{AA^\dagger}_{\mathbf{k},\mathbf{k}} \equiv \hat{\chi}^{AA^\dagger}_{\mathbf{k}} \geq 0$, one can consider static response functions as scalar products. It is easy to show that equal time correlation functions satisfy the Schwartz inequality

$$|\hat{\chi}^{AB}_{\mathbf{k}',\mathbf{k}}|^2 \leq \hat{\chi}^{AA^\dagger}_{\mathbf{k}}\hat{\chi}^{B^\dagger B}_{\mathbf{k}'}\ , \tag{1.60}$$

which, due to $|\coth(\omega/2T)| \geq 2T/|\omega|$ means that

$$2T\hat{\chi}_{\mathbf{k}}^{AA^\dagger} \leq C_{\mathbf{k}}^{AA^\dagger} . \tag{1.61}$$

It is the other manifestation of the Bogolyubov inequality.

For the emergence of the ordered superfluid in a Bose liquid one needs to have a nonzero order parameter, the amplitude of zero mode, $L^{-d/2}\langle a_{\mathbf{k}}\rangle = L^{-d/2}\langle a_{\mathbf{k}}^\dagger\rangle = L^{-d/2}\langle a_0\rangle\delta(\mathbf{k}) = \sqrt{n_0}\delta(\mathbf{k})$, where n_0 is the order parameter, and $a_{\mathbf{k}}^\dagger$ ($a_{\mathbf{k}}$) is the creation (destruction) Bose operator. To use the Bogolyubov inequality we determine the operators $A_{\mathbf{k}'}(t) = i(\partial\rho_{-\mathbf{k}}(t)/\partial t)$ and $B_{\mathbf{k}}(t) = L^{-d/2}a_{\mathbf{k}}(t)$, where $\rho_{\mathbf{k}}$ is the Fourier transform of the density operator $a^\dagger(\mathbf{r},t)a(\mathbf{r},t)$ which has the property $\langle\rho_0\rangle = N$ (N is the number of Bose particles). It is easy to show that

$$\begin{aligned} &\tau_{\mathbf{k}',\mathbf{k}}^{AB}(\omega) = \omega\tau_{-\mathbf{k},\mathbf{k}}^{\rho B}(\omega) \ , \ \tau_{-\mathbf{k}}^{AA^\dagger}(\omega) = \omega^2\tau_{-\mathbf{k}}^{\rho\rho^\dagger}(\omega) \ , \\ &\hat{\chi}_{-\mathbf{k},\mathbf{k}}^{AB} = \int_{-\infty}^{\infty}\frac{d\omega}{2\pi}\tau_{-\mathbf{k},\mathbf{k}}^{\rho B}(\omega) = -\sqrt{n_0} \ . \end{aligned} \tag{1.62}$$

From the *continuity equation*

$$\frac{\partial\rho}{\partial t} + \nabla\cdot\mathbf{j} = 0 \tag{1.63}$$

and the f-sum rule it follows that

$$\frac{k^2 n}{m} = \int_{-\infty}^{\infty}\frac{d\omega}{2\pi}\omega\tau_{-\mathbf{k}}^{\rho\rho^\dagger}(\omega) = P\int_{-\infty}^{\infty}\frac{d\omega}{2\pi}\frac{\tau_{-\mathbf{k}}^{AA^\dagger}(\omega)}{\omega} = \hat{\chi}_{-\mathbf{k}}^{AA^\dagger} \ , \tag{1.64}$$

where P denotes the principal part of the integral. Since $\mathbf{k} \neq 0$ and $C_{\mathbf{k}}^{BB^\dagger} = 2\langle a_{\mathbf{k}}^\dagger a_{\mathbf{k}}\rangle + 1$, we obtain

$$n_{\mathbf{k}} \geq -\frac{1}{2} + \frac{Tmn_0}{k^2 n} . \tag{1.65}$$

The value $L^{-d}\sum_{\mathbf{k}} n_{\mathbf{k}} = n - n_0$ must be finite. But, as follows from the above analysis, it is incompatible with Eq. (1.65) in one- and two-dimensional cases, except for the situation, where $n_0 = 0$, *i.e.*, there is no superfluid ordering in one- and two-dimensional Bose liquid for any nonzero temperature.

For a superconducting Fermi system one can introduce the order parameter

$$\Delta = L^{-d}\sum_{\mathbf{q}} f_{\mathbf{q}}\langle a_{\uparrow\mathbf{q}}a_{\downarrow-\mathbf{q}}\rangle \ , \tag{1.66}$$

where $a_{\sigma\mathbf{q}}$ destroys an electron with the momentum $\mathbf{q}$ and spin σ, and $f_{\mathbf{q}}$ is an arbitrary function with the properties $f_0 = 1$ and $\sum_{\mathbf{q}} f_{\mathbf{q}} < \infty$.

The Bogolyubov inequality is applied in this case to the operators $A_{\mathbf{k}'}(t) = i(\partial\rho_{-\mathbf{k}}(t)/\partial t)$ and $B_{\mathbf{k}}(t) = L^{-d}\sum_{\mathbf{q}} f_{\mathbf{q}}\langle a_{\downarrow\mathbf{k}-\mathbf{q}}a_{\uparrow\mathbf{q}}\rangle$, where one considers the density operator for electrons. We see that

$$L^{-d}\langle[B_{\mathbf{k}}, \rho_{-\mathbf{k}}]\rangle = \Delta + \eta(\mathbf{k}) \ , \tag{1.67}$$

where $\eta(0) = \Delta$. Following similar lines as above we get

$$C_{\mathbf{k}}^{BB^\dagger} \geq \frac{2Tm|\Delta + \eta(\mathbf{k})|}{k^2 n} \ . \tag{1.68}$$

The left hand side of this Bogolyubov inequality can be written in the form $C_{\mathbf{k}}^{BB^\dagger} = F_{\mathbf{k}} + R_{\mathbf{k}}$, where we introduced the Fourier transform of $F(\mathbf{r}_1 - \mathbf{r}_2) = \int\int d\mathbf{r}d\mathbf{r}' f(\mathbf{r}_1 - \mathbf{r})f(\mathbf{r}_2 - \mathbf{r}')\langle a_\downarrow^\dagger(\mathbf{r})a_\uparrow^\dagger(\mathbf{r}_1)a_\uparrow(\mathbf{r}_2)a_\downarrow(\mathbf{r}')\rangle$ and $R_{\mathbf{k}}$ is a regular function for small $\mathbf{k}$. Hence, one finally obtains

$$F_{\mathbf{k}} \geq -R_{\mathbf{k}} + \frac{2Tm|\Delta + \eta(\mathbf{k})|}{k^2 n} \ , \tag{1.69}$$

but $L^{-d}\sum_{\mathbf{k}\neq 0} F_{\mathbf{k}} < F(0) < \infty$. This is again the clear contradiction with Eq. (1.69) for one- and two-dimensional cases, unless $\Delta = 0$, *i.e.*, there is no nonzero superconducting order parameter for a system of interacting electrons in one and two space dimensions at any nonzero temperature.

We have to stress again that the Hohenberg theorem also does not exclude nonzero order parameter in the ground state, *i.e.*, quantum phase transitions.

Summarizing, for the most of low-dimensional electron systems with the finite radius of interaction between particles temperature fluctuations at $T \neq 0$ destroy possible ordering. This is why, the mean-field-like description, though very useful for systems in which an ordering can take place, cannot *a priori* be used. On the other hand, for most of low-dimensional multi-electron systems an interaction between particles is not weak, and, hence, one cannot use perturbative methods either. Moreover, singularities of the low-dimensional density of states enhance quantum fluctuations. All this, in fact, determines why when studying one- and two-dimensional electron systems it is important (and often, it is only possible) to use non-perturbative, optimally exact methods of the modern theoretical physics. Actually, in this book we want to introduce to the reader some very powerful non-perturbative methods of the modern theoretical physics, and to present a number of exact results for such low-dimensional electron systems.

Our book will mostly concentrate on one-dimensional correlated electron and quantum spin systems, because for two space dimensions one still has too small an amount of exact results for quantum many-body systems.

As for the references in this book: we shall introduce the relevant references at the end of each chapter. For example, the main description of thermodynamics and statistical mechanics can be found in [Landau and Lifshits (1980)]. The reader can find a description of "standard" electron systems in [Mahan (1990)], and read about magnetic systems in [Mattis (1965)]. The Fermi liquid description of correlation effects in standard three-dimensional metals can be found, *e.g.*, in [Pines and Nozières (1989)]. Our description of the Mermin–Wagner and Hohenberg theorems closely follows original papers [Mermin and Wagner (1966); Hohenberg (1967)]. Finally, the study of van Hove features in crystals was introduced in [van Hove (1952)].

Chapter 2

Quantum Spin-$\frac{1}{2}$ Chain with the Nearest-Neighbour Couplings

In this chapter we shall introduce the one-dimensional many-body quantum spin Hamiltonians and present some exact results for thermodynamic characteristics of the systems, described by those Hamiltonians. Such spin systems describe electron insulators in which charge degrees of freedom of electrons are frozen (*e.g.*, electrons are localized) and the only spin, magnetic excitations determine the states of electron systems. The simplest Hamiltonian of a spin system is the Zeeman Hamiltonian which describes the behaviour of spins in an external magnetic field. It shows how the degeneracy in the determination of the directions of spins is lifted by the magnetic field. However, one cannot describe the behaviour of most of magnetic insulators with only the Zeeman interaction. Generally speaking, there exists an interaction between spin degrees of motion of electrons in a crystal. It was concluded that the most important interaction, responsible for magnetic properties of multi-electron systems is the *exchange interaction*, which stems from the Coulomb interaction of electrons. The *Heisenberg model* is the seminal model of quantum mechanics. It was introduced to describe the exchange interaction of localized spin moments in insulators. It is commonly accepted that for most of the properties it is enough to consider exchange coupling between only nearest neighbours on the lattice. The reader already saw such a Hamiltonian in the previous chapter, when we considered the Mermin–Wagner theorem. In the case of a three-dimensional lattice, the Heisenberg model successfully describes ferromagnetic or antiferromagnetic ordering with the help of a mean-field-like approximation. However, for a one-dimensional situation, according to the Mermin–Wagner theorem, there is no magnetic ordering (at least for any nonzero temperature). This is why, it was necessary to use non-perturbative methods of theoretical physics to study thermodynamics of a one-dimensional quantum Heisenberg model. Historically, the spin-$\frac{1}{2}$ Heisenberg one-dimensional model (and its magnetically anisotropic versions, see below) was the first model of interacting quantum many-body system, for which an exact solution was obtained. The methods which were used for that purpose became very popular. Later, they were used to to treat many more complicated quantum correlated electron and spin models. This is why, we shall start our description of exact results of finite size effects for low-dimensional quantum models with this class of models. It turns out, that these results are interesting not only as "model" results to study important methods, but also were successfully used to describe low-temperature thermodynamics of some recently created quasi-one-dimensional magnetic compounds.

2.1 One-Dimensional Spin Hamiltonian

Let us start our consideration of quantum low-dimensional exactly solvable models with the Heisenberg model. The Heisenberg model, which we want to consider, describes the behaviour of spins $S = \frac{1}{2}$ in a one-dimensional chain, each of which interact with its nearest neighbours. For L spins in the chain the total number of states is 2^L, because the Hilbert space of each site (for definition we enumerate this site with the subscript j) of the line has two basis functions: one related to spin up (let us denote it by e_j^+), and the other one, related to spin down (e_j^-). There are four possible operators acting in each site: the unity operator, I_j and three operators of the projections of the site spin $S_j^{x,y,z}$ which constitute the SU(2) symmetry group. For $S = \frac{1}{2}$, one can also use the Pauli operators $\sigma_j^{x,y,z} = 2S_j^{x,y,z}$ with the well-known commutation relations. It is convenient to introduce the linear combinations of spin operators $S_j^{\pm} = S_j^x \pm iS_j^y$. The reader can check that the action of these operators on the basis functions is determined by the following relations $\sigma_j^+ e_j^+ = \sigma_j^- e_j^- = S_j^+ e_j^+ = S_j^- e_j^- = 0$, $\sigma_j^+ e_j^- = 2e_j^+$, $\sigma_j^- e_j^+ = 2e_j^-$, $\sigma_j^z e_j^{\pm} = \pm e_j^{\pm}$, $S_j^+ e_j^- = e_j^+$, $S_j^- e_j^+ = e_j^-$ and $S_j^z e_j^{\pm} = \pm\frac{1}{2} e_j^{\pm}$. The Hamiltonian of the quantum spin-$\frac{1}{2}$ chain with nearest-neighbour interactions can be written as

$$\begin{aligned}\mathcal{H}_{HI} &= \sum_{j=1}^{L-1} [J(S_j^x S_{j+1}^x + S_j^y S_{j+1}^y) + J_z S_j^z S_{j+1}^z] \\ &= \sum_{j=1}^{L-1} \left[\frac{J}{2} (S_j^+ S_{j+1}^- + S_j^- S_{j+1}^+) + J_z S_j^z S_{j+1}^z \right] . \end{aligned} \tag{2.1}$$

or as

$$\begin{aligned}\mathcal{H}_{HI} &= \frac{1}{4}\sum_{j=1}^{L-1}[J(\sigma_j^x\sigma_{j+1}^x + \sigma_j^y\sigma_{j+1}^y) + J_z\sigma_j^z\sigma_{j+1}^z] \\ &= \frac{1}{4}\sum_{j=1}^{L-1}\left[\frac{J}{2}\left(\sigma_j^+\sigma_{j+1}^- + \sigma_j^-\sigma_{j+1}^+\right) + J_z\sigma_j^z\sigma_{j+1}^z\right] . \qquad (2.2)\end{aligned}$$

Here exchange constants J and J_z define the interaction between the neighbouring sites of the spin chain. The case $J_z \neq J$ corresponds to the magnetically anisotropic situation of the uniaxial *magnetic anisotropy.* [Generally one can introduce 9 independent different exchange constants, but here we limit ourselves with the most known situation.] The case $J = 0$ describes the *Ising* chain. By the Heisenberg Hamiltonian people usually mean the magnetically isotropic case $J = J_z$. Finally, sometimes it is instructive to study the case $J_z = 0$ which is called the isotropic XY model (sometimes it is called XX0, XX, or planar model). In what follows we shall call the Hamiltonian Eq. (2.1) the *Heisenberg–Ising Hamiltonian.*

It is easy to show that the sign of J is irrelevant when one calculates the spectrum of eigenstates. For this purpose let us use the relations $\sigma_j^z\sigma_j^{x,y}\sigma_j^z = -\sigma_j^{x,y}$. Then one can apply to the Hamiltonian Eq. (2.1) the unitary transformation $U = \prod_j' \sigma_j^z$, where the prime means that the product extends over all odd sites of the chain. Naturally, this unitary transformation implies that we turn each odd spin down. Then it follows that $U\mathcal{H}_{HI}(J, J_z)U^\dagger = \mathcal{H}_{HI}(-J, J_z)$.

We can add to the Hamiltonian $\mathcal{H}_{HI}$ the Zeeman term, which describes the effect of an external magnetic field on a quantum spin chain. The Zeeman term commutes with the Heisenberg–Ising Hamiltonian Eq. (2.1), and, hence, it has the same eigenfunctions as $\mathcal{H}_{HI}$. The case $J_z < 0$ pertains to a ferromagnetic situation, while $J_z > 0$ determines an antiferromagnetic interaction between spins.

2.2 Ising Chain

Probably the simplest case is the Ising chain (let us consider periodic boundary conditions $S_{L+1}^z = S_1^z$). The Ising Hamiltonian was introduced by W. Lentz as a simplest model for a ferromagnet, and it is named due to E. Ising, who first solved it in one space dimension. For the Ising model one does not especially need to look for solutions of the stationary Schrödinger equation for the Hamiltonian, because local projections of spins, S_j^z, commute with the Ising Hamiltonian, and eigenfunctions are trivial. However, it is instructive to present the exact solution of statistical mechanics for an Ising chain, which introduces such an important value as the *transfer matrix*.

The partition function of the Ising chain can be written in a simple form

$$Z = \sum_{s_j} \exp\left\{-\beta \sum_{j=1}^{L} \left[J_z s_j s_{j+1} + \frac{1}{2} H(s_j + s_{j+1})\right]\right\}$$
$$= \sum_{s_j} \prod_{j=1}^{L} \exp\left\{-\beta \left[J_z s_j s_{j+1} + \frac{1}{2} H(s_j + s_{j+1})\right]\right\} \equiv \mathrm{Tr}(\hat{\tau})^L \ , \quad (2.3)$$

where s_j is the eigenvalue of S_j^z (equal to $\pm\frac{1}{2}$) and we introduced the transfer matrix as

$$\tau_{\mu,\nu} \equiv \exp\left\{-\beta \left[J_z s_\mu s_\nu + \frac{1}{2} H(s_\mu + s_\nu)\right]\right\} . \quad (2.4)$$

For $S = \frac{1}{2}$, each S_j^z has two possible values, so that $\hat{\tau}$ is 2×2 matrix. Moreover, the latter is symmetric in its indices. The eigenvalues of $\hat{\tau}$ are very easy to calculate, and, hence, one can calculate the trace of $\hat{\tau}^L$. Since $\hat{\tau}$ is symmetric and positive, it has positive eigenvalues. The trace is invariant under orthogonal transformations of $\hat{\tau}$. Therefore, it can be diagonalized by such a transformation. In the basis, in which $\hat{\tau}$ is diagonal, $\hat{\tau}^L$ is also diagonal. Suppose $\Lambda_0 > \Lambda_1$ are the eigenvalues of $\hat{\tau}$. Then it follows that in the thermodynamic limit

$$\lim_{L\to\infty} \frac{\ln(\mathrm{Tr}\hat{\tau})^L}{L} = \lim_{L\to\infty} \frac{\ln(\Lambda_0^L[1 - (\Lambda_1/\Lambda_0)^L])}{L} = \ln \Lambda_0 \ . \quad (2.5)$$

The larger eigenvalue of Eq. (2.4) is

$$\Lambda_0 = e^{-\beta J_z/4} \left[\cosh(\beta H/2) + \sqrt{\cosh^2(\beta H/2) - 1 + e^{\beta J_z}}\right] . \quad (2.6)$$

This is why, the Helmholtz free energy of an Ising chain per site in the thermodynamic limit is equal to

$$f = \frac{J_z}{4} - T \ln \left[\cosh(H/2T) + \sqrt{\cosh^2(H/2T) - 1 + e^{J_z/T}}\right] . \quad (2.7)$$

It is a smooth function of temperature, and, therefore, there is no phase transition in an Ising chain (except of the quantum phase transition at

$T = 0$). One can calculate the specific heat at $H = 0$:

$$c_{H=0} = \frac{J_z^2}{16T^2 \cosh^2(J_z/4T)} , \tag{2.8}$$

which shows that $c_H \to 0$ for $T \to 0$ and $T \to \infty$ and has a maximum as a function of temperature. The magnetization of the Ising ring per site m^z is calculated to be

$$m^z = \frac{\sinh(H/2T)}{2\sqrt{\cosh^2(H/2T) - 1 + e^{J_z/T}}} , \tag{2.9}$$

which implies that the Ising chain becomes magnetically ordered only at the infinite value of the external magnetic field for $T \neq 0$ and never becomes a ferromagnet, because $m^z \to 0$ for $H \to 0$ no matter what the value of T is. The magnetic susceptibility at zero H is calculated as $\chi_{H=0} = [4T\exp(|J_z|/2T)]^{-1}$, which tends to zero for high (as $1/T$) and low (exponentially) temperatures. The illustration of the temperature behaviour of the Ising chain is presented in Figs. 2.1 and 2.2. In the ground state (*i.e.*, at $T = 0$), the magnetization of the Ising chain is zero at zero value of the magnetic field and takes its nominal value (*i.e.*, $\pm\frac{1}{2}$ for positive and negative values of H, respectively) at any, even infinitesimally small value of $|H|$ for the ferromagnetic interaction. This suggests, naturally, that the point $H = 0$ (at $T = 0$) is the special point of the quantum phase transition for the ferromagnetic Ising chain.

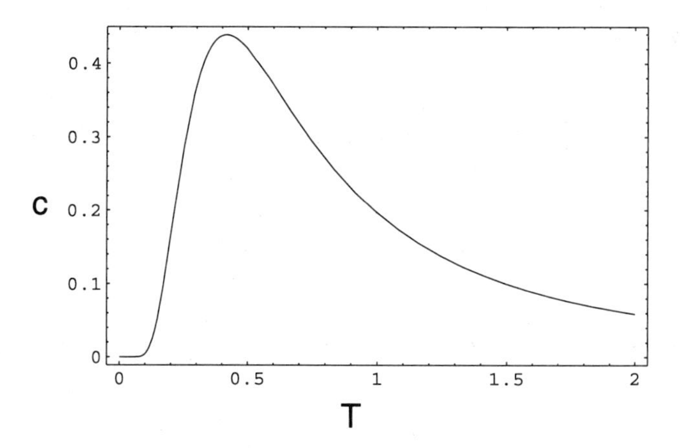

Fig. 2.1 The temperature dependence of the specific heat for the spin-$\frac{1}{2}$ Ising chain with $J_z = 2$ in a zero magnetic field.

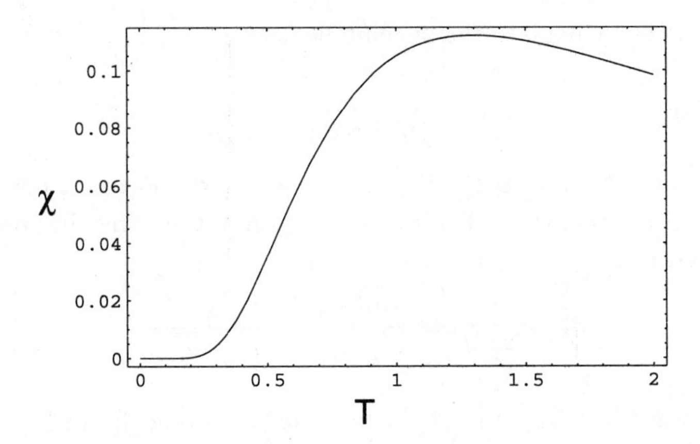

Fig. 2.2 The temperature dependence of the magnetic susceptibility for the spin-$\frac{1}{2}$ Ising chain with $J_z = 2$ in a zero magnetic field.

2.3 Isotropic XY Ring

Often it is also useful to change from spin operators to Fermi operators a_j and $a_j^\dagger$ (with the anticommutators $\{a_j, a_l^\dagger\} = \delta_{j,l}$ and $\{a_j, a_l\} = \{a_j^\dagger, a_l^\dagger\} = 0$) according to the well-known *Jordan–Wigner transformation*:

$$S_j^z = \frac{1}{2}(1 - 2a_j^\dagger a_j) \ , \ S_j^+ = \prod_{l=1}^{j-1}(1 - 2a_l^\dagger a_l)a_j \ ,$$
$$S_j^- = a_j^\dagger \prod_{l=1}^{j-1}(1 - 2a_l^\dagger a_l) \ , \tag{2.10}$$

(notice that $\nu_j \equiv \prod_{l=1}^{j-1}(1 - 2a_l^\dagger a_l) = \exp(i\pi \sum_{l=1}^{j} a_l^\dagger a_l))$. We point out that this transformation is non-local, *i.e.*, a spin operator in one site is described by Fermi operators of many sites, and reversible, *i.e.*, one can write Fermi operators in terms of spin operators. Suppose L is even. Then we observe that

$$(1 + \nu_j)\nu_j = (1 + \nu_j) \ , \ (1 - \nu_j)\nu_j = (\nu_j - 1) \ , \ \frac{1}{2}(1 + \nu_j)\frac{1}{2}(1 - \nu_j) = 0 \ ,$$
$$\frac{1}{2}(1 + \nu_j) + \frac{1}{2}(1 - \nu_j) = 1 \ , \ \frac{1}{4}(1 \pm \nu_j)^2 = \frac{1}{2}(1 \pm \nu_j) \ . \tag{2.11}$$

Actually these equalities imply that $\frac{1}{2}(1 + \nu_{L+1})$ is the projection of one of the halves of our total space and $\frac{1}{2}(1 - \nu_{L+1})$ is the complementary projection.

The Heisenberg–Ising Hamiltonian in an external magnetic field Eqs. (2.1) and (1.16) can be exactly re-written in terms of these spinless Fermi operators as

$$\begin{aligned}\mathcal{H}_{HI}+\mathcal{H}_{Z} &= \frac{J}{2}\left[\sum_{j=1}^{L-1}(a_j^{\dagger}a_{j+1}+a_{j+1}^{\dagger}a_j)-\nu_{L+1}a_L^{\dagger}a_1-\nu_{L+1}a_1^{\dagger}a_L\right]\\ &+\frac{J_z}{4}\left[\sum_{j=1}^{L}(1-4a_j^{\dagger}a_j)+4\sum_{j=1}^{L-1}a_j^{\dagger}a_ja_{j+1}^{\dagger}a_{j+1}+4a_L^{\dagger}a_La_1^{\dagger}a_1\right]\\ &-\frac{H}{2}\sum_{j=1}^{L}(1-2a_j^{\dagger}a_j)\ . \end{aligned} \tag{2.12}$$

Then one can resolve the total Hamiltonian into two parts by using projection operators

$$\begin{aligned}\mathcal{H}_{HI}+\mathcal{H}_{Z} &= \frac{1}{2}(1+\nu_L)[\mathcal{H}_{HI}+\mathcal{H}_{Z}]+\frac{1}{2}(1-\nu_L)[\mathcal{H}_{HI}+\mathcal{H}_{Z}]\\ &= \frac{1}{2}(1+\nu_L)\mathcal{H}^{+}+\frac{1}{2}(1-\nu_L)\mathcal{H}^{-}\ ,\end{aligned} \tag{2.13}$$

where

$$\begin{aligned}\mathcal{H}^{+} &= \frac{J}{2}\left[\sum_{j=1}^{L-1}(a_j^{\dagger}a_{j+1}+a_{j+1}^{\dagger}a_j)-a_L^{\dagger}a_1-a_1^{\dagger}a_L\right]\\ &+\frac{J_z}{4}\sum_{j=1}^{L}(1-4a_j^{\dagger}a_j+4a_j^{\dagger}a_ja_{j+1}^{\dagger}a_{j+1})-\frac{H}{2}\sum_{j=1}^{L}(1-2a_j^{\dagger}a_j)\end{aligned} \tag{2.14}$$

and

$$\begin{aligned}\mathcal{H}^{-} &= \frac{J}{2}\sum_{j=1}^{L-1}(a_j^{\dagger}a_{j+1}+a_{j+1}^{\dagger}a_j)+\frac{J_z}{4}\sum_{j=1}^{L}(1-4a_j^{\dagger}a_j+4a_j^{\dagger}a_ja_{j+1}^{\dagger}a_{j+1})\\ &-\frac{H}{2}\sum_{j=1}^{L}(1-2a_j^{\dagger}a_j)\ .\end{aligned} \tag{2.15}$$

We can find the eigenvalues for each of $\mathcal{H}^{\pm}$ separately and then take into account the effect of the factors $\frac{1}{2}(1\pm\nu_{L+1})$ by selecting half of the eigenvalues of $\mathcal{H}^{+}$ and half of those of $\mathcal{H}^{-}$. These two half-sets then constitute the full set of eigenvalues of the total quantum spin-$\frac{1}{2}$ chain Hamiltonian.

Now it is clear why the XY model is distinguished from others: for $J_z = 0$, the Hamiltonian of the spin-$\frac{1}{2}$ chain is exactly equal to the sum of Hamiltonians which are quadratic forms of the operators of free spinless fermions. These Hamiltonians can be obviously diagonalized by using the Fourier transform

$$\begin{aligned} a_k^\dagger &= (L)^{-1/2}\sum_{j=1}^{L} a_j^\dagger \exp\left[i\pi\left(\frac{kj}{L}-\frac{1}{4}\right)\right] , \\ a_k &= (L)^{-1/2}\sum_{j=1}^{L} a_j \exp\left[-i\pi\left(\frac{kj}{L}-\frac{1}{4}\right)\right] , \end{aligned} \tag{2.16}$$

after which the XY Hamiltonian can be written as

$$\begin{aligned} \mathcal{H}^+ = \frac{1}{2}\sum_{k=1}^{L/2}&[(J\cos[2\pi(2k-1)/L]+2H)(a_{2k-1}^\dagger a_{2k-1} \\ &+ a_{-2k+1}^\dagger a_{-2k+1}) - 2H] , \end{aligned} \tag{2.17}$$

and

$$\begin{aligned} \mathcal{H}^- &= \frac{1}{2}\sum_{k=1}^{(L/2)-1}[(J\cos[4\pi k/L]+2H)(a_{2k}^\dagger a_{2k} + a_{-2k}^\dagger a_{-2k}) - 2H] \\ &+ \frac{J+2H}{2}(a_0^\dagger a_0 + a_{k=L}^\dagger a_{k=L}) . \end{aligned} \tag{2.18}$$

Then it is not difficult to write down the partition function of the isotropic XY model, because it is just the sum of partition functions of several systems of free (non-interacting) spinless fermions

$$\begin{aligned} Z = 2^{L-1}\Bigg(&\prod_{k=1}^{L/2}\cosh^2\left[\frac{H+J\cos[2\pi(2k-1)/L]}{2T}\right] \\ &+ \cosh\left[\frac{H+J}{2T}\right]\cosh\left[\frac{H-J}{2T}\right]\prod_{k=1}^{(L/2)-1}\cosh^2\left[\frac{H+J\cos[4\pi k/L]}{2T}\right] \\ &- \sinh\left[\frac{H+J}{2T}\right]\sinh\left[\frac{H-J}{2T}\right]\prod_{k=1}^{(L/2)-1}\sinh^2\left[\frac{H+J\cos[4\pi k/L]}{2T}\right] \\ &+ \prod_{k=1}^{L/2}\sinh^2\left[\frac{H+J\cos[2\pi(2k-1)/L]}{2T}\right]\Bigg) . \end{aligned} \tag{2.19}$$

In the thermodynamic limit $L \to \infty$ the partition function is simplified to the expression

$$Z = \frac{2^L}{\pi} \int_0^\pi dk \cosh^2 \left[\frac{H - |J| \cos k}{2T} \right] . \tag{2.20}$$

The expression for the partition function according to the rules of statistical mechanics immediately yields the Helmholtz free energy per site of the isotropic XY chain

$$f = -\frac{T}{\pi} \int_0^\pi dk \ln \left[2 \cosh \left(\frac{H - |J| \cos k}{2T} \right) \right] . \tag{2.21}$$

It is easy to write down the magnetization per site of the isotropic XY model as

$$m^z = \frac{1}{2\pi} \int_0^\pi dk \tanh \left[\frac{H - |J| \cos k}{2T} \right] \tag{2.22}$$

and the magnetic susceptibility at $H = 0$ as

$$\chi_{H=0} = \frac{1}{4\pi T} \int_0^\pi \frac{dk}{\cosh^2 [|J| \cos k / 2T]} . \tag{2.23}$$

The internal energy per site at zero magnetic field is equal to

$$e = -\frac{|J|}{2\pi} \int_0^\pi \cos k dk \tanh [|J| \cos k / 2T] , \tag{2.24}$$

which produces the formula for the specific heat

$$c_{H=0} = \frac{J^2}{4\pi T^2} \int_0^\pi dk \frac{\cos^2 k}{\cosh^2 [|J| \cos k / 2T]} . \tag{2.25}$$

Figures 2.3 and 2.4 present results for the behaviour of the magnetic susceptibility and the low-temperature Sommerfeld coefficient of the specific heat, γ, for an isotropic XY chain for $H = 0$ and in a weak magnetic field $H = 0.1J$. Notice the difference in the low-temperature behaviour of these characteristics for the XY chain and for the Ising chain. The difference is due to the activation character of the spectrum of the Ising chain. It results in the exponentially small behaviour of χ and $c_{H=0}$ for the systems with gapped elementary excitations. The reader can also see that a weak magnetic field practically does not change the behaviour of thermodynamic characteristics of an isotropic XY chain.

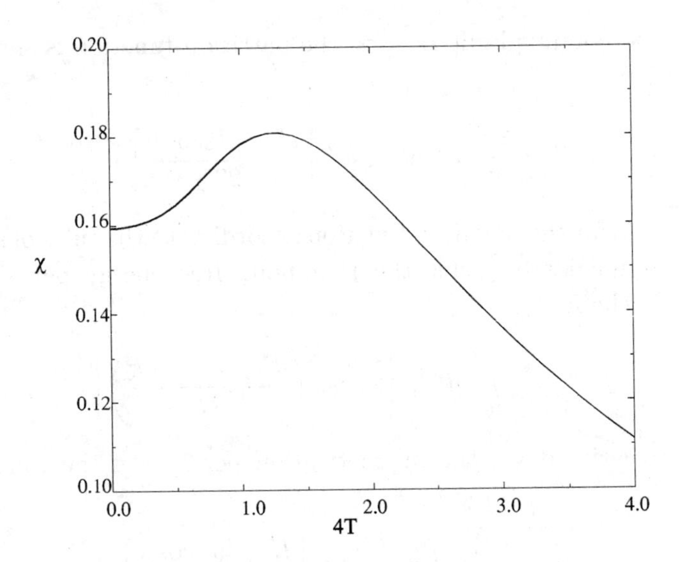

Fig. 2.3 The magnetic susceptibility of an isotropic XY spin-$\frac{1}{2}$ chain with $J = 2$. The solid line shows the results for $H = 0$, the dashed line — for $H = 0.2$. For this value of H the dashed and solid lines practically coincide.

It is interesting to study the behaviour of an isotropic XY model in the ground state, at $T = 0$. Remember that we deal with the fermionic system, and, therefore, it is necessary to determine its' Fermi sea for the ground state, *i.e.*, the ground state of free fermions pertains to the situation in which all possible states of the Hamiltonian with negative energies are filled and all states with positive energies are empty. This, naturally, depends on the value of an external magnetic field. For $H \geq |J|$ the ground state magnetization of the isotropic XY chain per site is equal to its nominal value $\frac{1}{2}$ (and for $H \leq -|J|$, it is $-\frac{1}{2}$). Obviously, in these domains of values of H the ground state magnetic susceptibility is zero. These phases are frequently referred to as ferromagnetic (or spin-saturated) phases. On the other hand, for $-|J| \leq H \leq |J|$ the ground state magnetization per site of an isotropic XY model is equal to $m^z = (1/\pi)\sin^{-1}(H/|J|)$, see Fig. 2.5, and the magnetic susceptibility reveals square-root singularities at $H_s = \pm|J|$.

The latter is a smooth function of the field, though, for $H \to 0$, in contrast with the previously considered behaviour of an Ising chain in the ground state. Nevertheless, this is again the manifestation of the quantum phase transition (of the *second kind*, according to the classification of P. Ehrenfest).

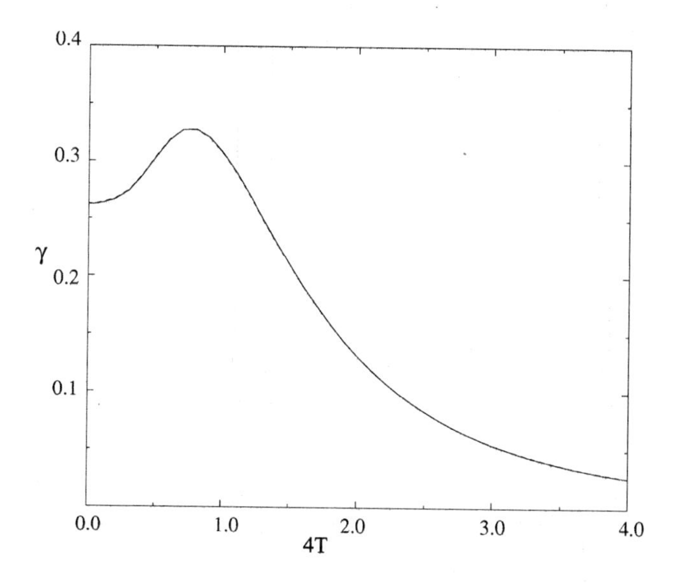

Fig. 2.4 The Sommerfeld coefficient $\gamma = c/T$ for the specific heat of an isotropic XY chain. The solid line shows the results for $H = 0$, the dashed line — for $H = 0.2$. For this value of H the dashed and solid lines practically coincide.

2.4 Ising Chain in a Transverse Magnetic Field

In one of the previous sections we considered the simplest case of an Ising ring in a magnetic field, parallel to the axis of the magnetic anisotropy of the spin–spin exchange coupling. In that case the z-projection of spin operator at each site commutes with the Hamiltonian, and, hence, there is no (non-trivial) spin dynamics. However, the interesting spin dynamics appears if the magnetic field is directed perpendicular to the axis of the magnetic anisotropy of the spin–spin interaction. Equally important is the fact that this case also permits us to obtain an exact solution, because, as is clear from the previous section, the introduction of a magnetic anisotropy in the xy plane of the XY chain (so-called anisotropic XY model) does not violate the important property of its Hamiltonian: it can also be exactly transformed to a quadratic form of Fermi spinless operators. This means that one can introduce the term $\sum_{j=1}^{L-1}(J_x S_j^x S_{j+1}^x + J_y S_j^y S_{j+1}^y)$ instead of $\sum_{j=1}^{L-1} J(S_j^x S_{j+1}^x + S_j^y S_{j+1}^y)$ in the Hamiltonian $\mathcal{H}_{HI}$. Then one has to change $2J \to (J_x + J_y)$ in formulae of the previous section and add to the

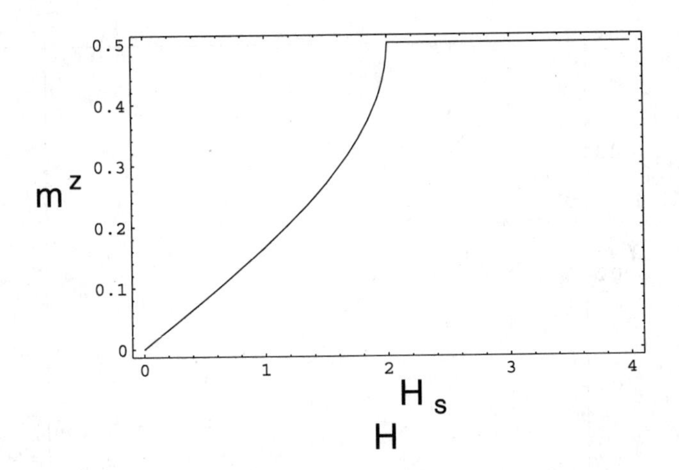

Fig. 2.5 The ground state magnetic moment of the isotropic XY spin-$\frac{1}{2}$ chain with $J = 2$.

Hamiltonians $\mathcal{H}^{\pm}$ the following terms:

$$\delta\mathcal{H}^{+} = \frac{J_x - J_y}{4}\left[\sum_{j=1}^{L-1}(a_j^{\dagger}a_{j+1}^{\dagger} - a_{j+1}a_j) - a_L^{\dagger}a_1^{\dagger} + a_1a_L\right]$$

$$= \frac{J_y - J_x}{2}\sum_{k=1}^{L/2}\sin[2\pi(2k-1)/L](a_{2k-1}^{\dagger}a_{-2k+1}^{\dagger} - a_{-2k+1}a_{2k-1})\ , \quad (2.26)$$

and

$$\delta\mathcal{H}^{-} = \frac{J_x - J_y}{4}\sum_{j=1}^{L}(a_j^{\dagger}a_{j+1}^{\dagger} - a_{j+1}a_j)$$

$$= \frac{J_y - J_x}{2}\sum_{k=1}^{(L/2)-1}\sin[4\pi k/L](a_{2k}^{\dagger}a_{-2k}^{\dagger} - a_{-2k}a_{2k})\ . \quad (2.27)$$

The reader can himself perform calculations, similar to those given above. The only difference is that for the diagonalization of the quadratic form of Fermi operators it is not enough in this case to use only Fourier transform; one also needs the usual *Bogolyubov transformation*:

$$a_k = u_kb_k + v_kb_{-k}^{\dagger}\ ,\ a_k^{\dagger} = u_kb_k^{\dagger} + v_kb_{-k}\ ,$$
$$u_{-k} = u_k\ ,\ v_{-k} = -v_k\ ,\ u_k^2 + v_k^2 = 1\ . \quad (2.28)$$

It has to be used in such a way that after the transformation the fermion Hamiltonian does not have the "anomalous" combinations $b_k^\dagger b_{-k}^\dagger$ and $b_{-k}b_k$. Then we obtain

$$2u_k^2 = 1 - \frac{(J_x + J_y)\cos[2\pi k/L]}{2\epsilon_k} , \; 2v_k^2 = 1 + \frac{(J_x + J_y)\cos[2\pi k/L]}{2\epsilon_k} ,$$
$$\epsilon_k = \frac{1}{2}\sqrt{J_x^2 + J_y^2 + 2J_xJ_y\cos[2\pi k/L] - 4H|J_x + J_y|\cos[2\pi k/L] + 4H^2} ,$$
$$\mathcal{H}_k = \epsilon_k(b_k^\dagger b_k + b_{-k}^\dagger b_{-k} - 1) . \tag{2.29}$$

After such a transformation we can write the expression for $\mathcal{H}^+ = (1/2)\sum_{k=1}^{L/2}(2\mathcal{H}_{2k-1} + (J_x + J_y)\cos[2\pi(2k-1)/L])$, and analogous one for $\mathcal{H}^-$ (different due to the boundary terms). Then, one can obtain the expression for the partition function. It coincides with Eq. (2.19), with the replacements

$$[H/2] + J\cos[2\pi(2k-1)/L] \to \epsilon_{2k-1} , \; [H/2] + J\cos[4\pi k/L] \to \epsilon_{2k} ,$$
$$[H/2] \pm J \to [H/2] \pm (J_x + J_y)/2 . \tag{2.30}$$

The special case of the Ising chain pertains to $J_x = 0$ (or $J_y = 0$). Then, in the thermodynamic limit, the Helmholtz free energy per site of the Ising chain in the perpendicular to the Ising axis magnetic field can be written as:

$$f = -\frac{T}{\pi}\int_0^\pi dk \ln[2\cosh(\sqrt{4H^2 + J^2 - 4H|J|\cos k}/4T)] \tag{2.31}$$

(we used $J_x = J$, or $J_y = J$) and the magnetization per site is equal to

$$m = \frac{1}{2\pi}\int_0^\pi dk \tanh\left[\frac{\sqrt{4H^2 + J^2 - 4H|J|\cos k}}{4T}\right] \times \frac{2H - |J|\cos k}{\sqrt{4H^2 + J^2 - 4H|J|\cos k}} . \tag{2.32}$$

The zero-field magnetic susceptibility is equal to

$$\chi_{H=0} = \frac{1}{8T}\left(\frac{1}{\cosh^2(J/4T)} + \frac{4T}{J}\tanh(J/4T)\right) . \tag{2.33}$$

It is important to point out that because $\sum_j S_j^z$ does not commute with the Hamiltonian in this case, there is no ferromagnetic phase (*i.e.*, the domain of values of the magnetic field, in which the magnetization becomes equal to its' nominal value, does not exist). The reader can see from Eq. (2.32) that

m tends to $\frac{1}{2}$ only in the infinite magnetic field. This is the consequence of the magnetic anisotropy in the system. In the ground state the magnetic susceptibility per site is equal to

$$\chi = \frac{J^2}{\pi}\int_0^{\pi} \frac{\sin^2 k dk}{(4H^2 + J^2 - 4H|J|\cos k)^{3/2}} , \tag{2.34}$$

which implies the logarithmic singularity (*i.e.*, weaker, than the square-root singularity of the ground state magnetic susceptibility of the isotropic XY chain) at $H_c = |J|/2$. Here we are again faced with a quantum phase transition. Such a quantum phase transition with a logarithmic singularity of the magnetic susceptibility is present for an anisotropic XY model, not only in the Ising limit.

2.5 Dimerized XY Chain

The other model which permits us to obtain an exact solution, is the *multimerized (multi-sublattice) XY chain.* The Hamiltonian of that model has the form:

$$\mathcal{H}_{mul} = \sum_j \left(\sum_{n=1}^{N-1} [J_n (S^x_{j,n} S^x_{j,n+1} + S^y_{j,n} S^y_{j,n+1})] \right.$$
$$\left. + J_N (S^x_{j,N} S^x_{j+1,1} + S^y_{j,N} S^y_{j+1,1}) - H \sum_{n=1}^{N} \mu_n S^z_{j,n} \right) , \tag{2.35}$$

where we take into account possible N different magnetic sublattices in a chain of spins $\frac{1}{2}$, each sitting in the j-th cell of a chain, $j = 1, \ldots, L$. Here J_n denote exchange constants, and μ_n denote effective magnetons. They distinguish magnetic sublattices in each elementary cell of the one-dimensional lattice from each other.

Using the generalized Jordan–Wigner transformation

$$\begin{aligned}
S^z_{j,n} &= \frac{1}{2}\sigma_{j,n} = \frac{1}{2} - a^\dagger_{j,n} a_{j,n} , \\
S^+_{j,1} &= \prod_{l<j} \prod_{n=1}^{N} \sigma^z_{l,n} a_{j,1} , \\
S^+_{j,m} &= \prod_{l<j} \prod_{n=1}^{N} \sigma^z_{l,n} \sigma^z_{j,1} \cdots \sigma^z_{j,m-1} a_{j,m} , \quad m = 2, \ldots, M , \\
S^-_{j,n} &= (S^+_{j,n})^+ , \quad n = 1, \ldots, M ,
\end{aligned} \tag{2.36}$$

where A^+ is the operator, Hermitian conjugated to A, the Hamiltonian Eq. (2.35) can be exactly re-written as a quadratic form of Fermi operators. Then, by using the Fourier transform $a_{k,n} = L^{-1/2} \sum_j a_{j,n} \exp(-ikj)$ (with $-\pi \le k \le \pi$), we get in the thermodynamic limit $L \to \infty$

$$\begin{aligned}\mathcal{H}_{mul} = & -\frac{LH}{2}\sum_{n=1}^{N}\mu_n + \sum_k \left[\sum_{n=1}^{N-1} \left[\mu_n H a_{k,n}^\dagger a_{k,n} \right.\right. \\ & \left. + \frac{J_n}{2}(a_{k,n}^\dagger a_{k,n+1} + a_{k,n+1}^\dagger a_{k,n}) \right] + \mu_N H a_{k,N}^\dagger a_{k,N} \\ & \left. + \frac{J_N}{2}(a_{k,N}^\dagger a_{k,1} e^{ik} + a_{k,1}^\dagger a_{k,N} e^{-ik}) \right] . \end{aligned} \tag{2.37}$$

This is a quadratic form of Fermi operators, which can be diagonalized by using a unitary transformation.

Let us consider the most important special case $N = 2$, *i.e.*, the *dimerized* spin-$\frac{1}{2}$ XY chain. Here one can say that the chain is *bond- and site-alternating*. Then the unitary transformation which diagonalizes the Hamiltonian equation (2.37) can be explicitly written as (b_{kn} destroys the fermion mode with the quasimomentum k, which belongs to the n-th branch of the spectrum)

$$a_{k1} = u_{11}(k)b_{k1} + u_{12}(k)b_{k2} \ , \ a_{k2} = u_{21}(k)b_{k1} + u_{22}(k)b_{k2} \ , \tag{2.38}$$

after which the Hamiltonian finally becomes

$$\begin{aligned}\mathcal{H}_{dim} &= -\frac{LH}{2}(\mu_1 + \mu_2) + \sum_k \sum_{n=1}^{2} \varepsilon_{k,n} b_{k,n}^\dagger b_{k,n} \\ &= \sum_k \sum_{n=1}^{2} \varepsilon_{k,n} [b_{k,n}^\dagger b_{k,n} - (1/2)] \ , \end{aligned} \tag{2.39}$$

where

$$\varepsilon_{k,1,2} = \frac{1}{2}[(\mu_1 + \mu_2)H \pm \sqrt{(\mu_1 - \mu_2)^2 H^2 + J_1^2 + J_2^2 + 2J_1 J_2 \cos k}] \ , \tag{2.40}$$

and the coefficients of the unitary transformations are

$$\begin{aligned} u_{11}(k) &= \left(\frac{\mu_2 H - \varepsilon_{k1}}{\varepsilon_{k2} - \varepsilon_{k1}}\right)^{\frac{1}{2}} e^{i\phi} \ , \ u_{12}(k) = \left(\frac{J_1 + J_2 e^{-ik}}{2(\mu_1 H - \varepsilon_{k2})}\right) u_{11}(k) e^{i(\psi - \phi)} \ , \\ u_{21}(k) &= \left(\frac{J_1 + J_2 e^{ik}}{2(\mu_2 H - \varepsilon_{k1})}\right) u_{11}(k) \ , \ u_{22}(k) = u_{11}(k) e^{i(\psi - \phi)} \ , \end{aligned} \tag{2.41}$$

where ϕ and ψ are arbitrary phases. Let us consider non-negative values of the magnetic field $H \geq 0$. We point out that in the thermodynamic limit $L \to \infty$ the first branch of the spectrum is positive for any value of the quasimomentum k. On the other hand, the second branch is negative for all k in the region $H < H_c$ and positive for all k for $H > H_s$, where

$$H_{c,s} = \frac{|J_1 \mp J_2|}{2\sqrt{\mu_1\mu_2}} . \tag{2.42}$$

In the region $H_c < H < H_s$, $\varepsilon_{k,2} > 0$ for $|k| > k_c$, and $\varepsilon_{k,2} < 0$ for $|k| < k_c$, where

$$k_c = \cos^{-1}\frac{2H^2 - H_c^2 - H_s^2}{H_s^2 - H_c^2} = \cos^{-1}\frac{\mu_1\mu_2 H^2 - J_1^2 - J_2^2}{2J_1J_2} . \tag{2.43}$$

Hence the spectrum of the system is gapless for $H_c < H < H_s$ and it has a gap (spin gap) for $H > H_s$ and $H < H_c$. The ground state of a dimerized XY spin chain is organized as follows. The first branch of excitations is non-occupied for any value of the magnetic field at $T = 0$. For small values of the magnetic field, $H < H_c$, the Dirac sea consists of all occupied states of fermions from the second branch. In this case excitations have the gap $G(H) = (1/2)[\sqrt{(\mu_1 - \mu_2)^2H^2 + (J_1 - J_2)^2} - (\mu_1 + \mu_2)H]$. It is equal to $|J_1 - J_2|/2$ for $H = 0$ and is closed at $H = H_c$. For intermediate values $H_c < H < H_s$ the Dirac sea consists of filled states of fermions from the second branch with $|k| < k_c$, and low-lying excitations are fermions from the second branch with $|k| > k_c$ and holes for $|k| < k_c$. Finally, in the region of $H_s < H$ the Dirac sea is empty and all excitations are activated.

The ground state magnetization of the dimerized spin chain for $H < H_c$ can be written as

$$\begin{aligned} m^z_{T=0} &= \sum_k \frac{(\mu_1 - \mu_2)^2 H}{2\sqrt{(\mu_1 - \mu_2)^2H^2 + J_1^2 + J_2^2 + 2J_1J_2\cos k}} \\ &= \frac{(\mu_1 - \mu_2)^2 H K(\kappa)}{\pi\sqrt{(\mu_1 - \mu_2)^2H^2 + (J_1 + J_2)^2}} , \end{aligned} \tag{2.44}$$

where $K(\kappa)$ is the complete elliptic integral of the first kind and $\kappa^2 = 4J_1J_2/[(\mu_1 - \mu_2)^2H^2 + (J_1 + J_2)^2]$. Obviously for $H \to 0$, there is no spontaneous magnetization. It turns out that $M^z = 0$ for $\mu_1 = \mu_2$ for any $H < H_c$. In this region of H absolute values of the z-projection of the total spin (mechanic moment) of each sublattice are equal to each other and compensate each other. For $\mu_1 \neq \mu_2$, there can be a weak magnetic moment in this phase. We can call this phase the "antiferromagnetic" one for $\mu_1 = \mu_2$, or "ferrimagnetic" one for the case of nonzero magnetization for $H < H_c$. Notice that for $J_1 = J_2 = J$, but for $\mu_1 \neq \mu_2$, the magnetic susceptibility has a logarithmic singularity at $H \to 0$, because

$$m^z_{T=0} = -\frac{(\mu_1 - \mu_2)^2 H}{\pi |J|} \ln \frac{|\mu_1 - \mu_2| H}{2|J|} . \tag{2.45}$$

In the region of large fields the ground state magnetization is

$$m^z_{T=0} = (\mu_1 + \mu_2)/2 , \tag{2.46}$$

i.e., the system is in the ferromagnetic phase, in which spins of each sublattice have their nominal values $\frac{1}{2}$. Finally, in the intermediate region of fields $H_c < H < H_s$ the ground state magnetization per site is equal to

$$m^z_{T=0} = \frac{(\mu_1 + \mu_2)}{2}\left(1 - \frac{k_c}{\pi}\right) + \frac{(\mu_1 - \mu_2)^2 H F(k_c/2, \kappa)}{\pi\sqrt{(\mu_1 - \mu_2)^2 H^2 + (J_1 + J_2)^2}} , \tag{2.47}$$

where $F(k_c/2, \kappa)$ is the incomplete elliptic integral of the first kind. At $H = H_{c,s}$ two second order quantum phase transitions take place. The magnetic susceptibility of the dimerized XY spin-$\frac{1}{2}$ chain reveals square root singularities in the ground state $\chi \sim 1/\sqrt{(H_s^2 - H^2)(H^2 - H_c^2)}$.

The partition function (and other thermodynamic characteristics) of the dimerized XY chain in the thermal equilibrium can be obtained straightforwardly. The Helmholtz free energy of the dimerized chain is equal to

$$f = -T \sum_k \sum_{n=1}^{2} \ln[2\cosh(\varepsilon_{k,n}/2T)] . \tag{2.48}$$

The magnetization is

$$\begin{aligned} m^z &= m^z_{T=0} + \frac{(\mu_1 + \mu_2)}{2} \sum_k \Bigg(\cosh \frac{(\mu_1 + \mu_2)H}{2T} \\ &\quad + \cosh \frac{\sqrt{(\mu_1 - \mu_2)^2 H^2 + J_1^2 + J_2^2 + 2J_1J_2\cos k}}{2T} \Bigg)^{-1} \\ &\quad \times \left[\sinh \frac{(\mu_1 + \mu_2)H}{2T} - \frac{(\mu_1 - \mu_2)H}{\sqrt{(\mu_1 - \mu_2)^2 H^2 + J_1^2 + J_2^2 + 2J_1J_2\cos k}} \right. \\ &\quad \left. \times \left(\cosh \left[\frac{(\mu_1 + \mu_2)H}{2T} \right] + e^{-\sqrt{(\mu_1 - \mu_2)^2 H^2 + J_1^2 + J_2^2 + 2J_1J_2\cos k}/2T} \right) \right] . \end{aligned} \tag{2.49}$$

The magnetic susceptibility can be written as

$$\chi = \sum_k \sum_{n=1}^{2} \left[\frac{1}{2} \frac{\partial^2 \varepsilon_{k,n}}{\partial H^2} \tanh \frac{\varepsilon_{k,n}}{2T} + \left[\frac{\partial \varepsilon_{k,n}}{\partial H} \right]^2 \frac{1}{4T \cosh^2(\varepsilon_{k,n}/2T)} \right]. \quad (2.50)$$

The specific heat can be calculated as

$$c = \sum_k \sum_{n=1}^{2} \frac{\varepsilon_{k,n}^2}{4T^2 \cosh^2(\varepsilon_{k,n}/2T)} . \quad (2.51)$$

It is instructive also to calculate the staggered magnetic susceptibility of the dimerized XY chain, *i.e.*, the response of the system to the alternating magnetic field

$$\chi_{st} = \frac{1}{2} \sum_k \sum_{n=1}^{2} \left[\frac{\partial^2 \varepsilon_{k,n}}{\partial(\mu_1 - \mu_2)^2 H^2} \tanh \frac{\varepsilon_{k,n}}{2T} + \left(\frac{\partial \varepsilon_{k,n}}{\partial(\mu_1 - \mu_2) H} \right)^2 \frac{1}{4T \cosh^2(\varepsilon_{k,n}/2T)} \right] . \quad (2.52)$$

Some analytic results can be written for the case $N = 4$, *i.e.*, *quadrimerized* XY chain. If $\mu_1 = \mu_2 = \mu_3 = \mu_4 = \mu$ there are four values of the magnetic field H, at which quantum phase transitions in the ground state take place:

$$\mu H_{c1,2,3,4} = \frac{1}{2\sqrt{2}} \left(J_1^2 + J_2^2 + J_3^2 + J_4^2 \pm \sqrt{[(J_1 + J_3)^2 + (J_2 \pm J_4)^2][(J_1 - J_3)^2 + (J_2 \mp J_4)^2]} \right)^{1/2} . \quad (2.53)$$

At these values of the magnetic field the ground state magnetic susceptibility reveals square root singularities, characteristic for quantum phase transitions of the second kind. Suppose that $J_1 J_2 J_3 J_4 > 0$. Then for H larger than the largest value of $H_{c1,2,3,4}$ the system at $T = 0$ is in the ferromagnetic phase with magnetic moments of each sublattice being equal to their nominal value $\mu/2$. For H smaller than the smallest value of $H_{c1,2,3,4}$ magnetic moments of sublattices compensate each other, and the ground state magnetization is zero. It turns our that in these phases, as well as for H being between two other critical values, the ground state magnetization of a multi-sublattice spin chain is not changed with the variation of an external magnetic field H, and the magnetic susceptibility is zero. It is used to refer to such a situation as about *magnetization plateaux.*

It is interesting to notice that for a N-sublattice XY spin-$\frac{1}{2}$ chain there exist N quantum phase transitions with respect to an external magnetic field directed along z axis. Moreover, it is important to emphasize that for even N there is a spin-gapped "antiferromagnetic" phase at low values of the magnetic field. On the other hand, there is no such a phase for N odd. Equally important, in the ground state of the N-sublattice model there are N phases with magnetization plateaux, in which the $T = 0$ magnetic susceptibility is equal to zero. As the reader will see below, similar conclusions can be made about the magnetic field behaviour of any "easy-plane" antiferromagnetic spin-$\frac{1}{2}$ chain.

Here it is worthwhile to present the theorem due to E. H. Lieb, T. Schultz and D. J. Mattis about the ground state and excitations of a Heisenberg–Ising spin-$\frac{1}{2}$ chain with an "easy-plane" magnetic anisotropy.

First, we can prove that for such an antiferromagnetic model the ground state is non-degenerate (and singlet) for even L. Let us use the same transformation which was used to prove that $\mathcal{H}_{HI}(J, J_z) = \mathcal{H}_{HI}(-J, J_z)$, *i.e.*, rotate all even (or odd) spins about z axis.

Consider, first, only states with $\sum_j S_j^z = 0$. A complete set of states is the set of configurations in which $L/2$ spins are up and $L/2$ are down. Denote these states by ϕ_a. Any eigenfunction of the Hamiltonian can be expanded as $\Psi = \sum_a c_a \phi_a$. The stationary Schrödinger equation in this representation for the Hamiltonian after the unitary transformation is

$$(E - J_z \epsilon_a) c_a = \frac{J}{2} \sum_{a'} c_{a'} \ , \tag{2.54}$$

where $\sum_j S_j^z S_{j+1}^z \phi_a = \epsilon_a \phi_a$. The Hamiltonian is real, this is why one can suppose that all c_a are also real. Then, assume that for some ground state Ψ_0 with the energy E_0 some $c_a = 0$ for $a = a_1, \ldots, a_r$. For these coefficients we obviously have $\sum_{a'} c_{a'} = 0$ for $a = a_1, \ldots, a_r$. In at least one of these equations (say the a_p-th) some of $c_a \neq 0$ (otherwise the Hamiltonian would break into blocks with no matrix elements connecting $\phi_{a_1}, \ldots, \phi_{a_r}$ with other configurations, which is impossible). Hence, there are nonzero c_a of both signs. Now consider the trial wave function $\Psi_0' = \sum_a |c_a| \phi_a$. On the one hand, it is not an eigenstate of $\mathcal{H}_{HI}$, because $|c_{a_p}| = 0$ but $\sum_{a'} |c_{a'}| \neq 0$, so that from the variational principle we have for its energy $E_0' > E_0$. On the other hand we have

$$E_0' = J_z \sum_a \epsilon_a c_a^2 - \frac{J}{2} \sum_{a,a'} |c_a||c_{a'}| \ , \tag{2.55}$$

and

$$E_0 = J_z \sum_a \epsilon_a c_a^2 - \frac{J}{2} \sum_{a,a'} c_a c_{a'} \ , \tag{2.56}$$

which imply that $E_0' \leq E_0$. This is in an obvious contradiction with the variational principle. Hence, for any ground state with $\sum_j S_j^z = 0$ all $c_a \neq 0$.

For any Ψ_0 to be a ground state one needs $E_0' = E_0$. This occurs, if and only if, all the terms $c_a c_{a'}$ are positive, *i.e.*, the coefficients of all configurations connected through J with each other should have the same sign. But each configuration is ultimately connected with every other through repeated application of the interaction with J. This is why, for every ground state with $\sum_j S_j^z = 0$ all c_a have the same sign. Then it is obvious that there can be only one ground state with $\sum_j S_j^z = 0$. Otherwise, several states would all have all positive coefficients and so could not be orthogonal to each other. The onset of another ground state (whatever its multiplicity) would imply that there is a second ground state with $\sum_j S_j^z = 0$, which is impossible, see above. This is why at least one ground state has $\sum_j S_j^z = 0$. This is, in fact, true for any dimensions and any *bipartite lattice* (*i.e.*, which can be decomposed into two equivalent sublattices and ferromagnetic interactions between spins of the same sublattice).

Next, consider the excited state

$$\Psi_k = \exp\left(ik \sum_n n S_n^z \right) \Psi_0 \equiv \mathcal{O}^k \Psi_0 \ . \tag{2.57}$$

Let us study the unitary operator T that displaces all the spins by one site cyclically $T S_j T^{-1} = S_{j+1}$ with periodic boundary conditions $S_{L+1} = S_1$. Observe that $T = \exp(iP)$, where P is the total momentum of a periodic spin chain. The reader can see that T obviously commutes with the Hamiltonian $[\mathcal{H}_{HI}, T] = 0$ (as well as $[\mathcal{H}_{HI}, P] = 0$). Then, if Ψ_0 is an eigenstate of $\mathcal{H}_{HI}$, so is $T\Psi_0$. The reader already knows that Ψ_0 is non-degenerate, then $T\Psi_0 = \exp(i\alpha)\Psi_0$. Thus

$$\langle \Psi_0 | \Psi_k \rangle = \langle \Psi_0 | \mathcal{O}^k | \Psi_0 \rangle = \langle \Psi_0 | T \mathcal{O}^k T^{-1} | \Psi_0 \rangle \ . \tag{2.58}$$

However, one has that

$$T\mathcal{O}^k T^{-1} = \mathcal{O}^k \exp(ikLS_1^z) \exp\left(-ik\sum_{n=1}^{L} S_n^z\right) . \tag{2.59}$$

Ψ_0 is a singlet. For even L we can choose $\exp(ikLS_1^z) = -1$, providing $k = 2\pi m/L$, where m is an odd integer. This means that $\langle\Psi_0|\Psi_k\rangle = -\langle\Psi_0|\Psi_k\rangle = 0$, *i.e.*, Ψ_k with $k = 2\pi m/L$ is orthogonal to Ψ_0. We can also calculate the energy of the state Ψ_k as $\langle\Psi_k|\mathcal{H}_{HI}|\Psi_k\rangle = \langle\Psi_0|\mathcal{O}^{k-1}\mathcal{H}_{HI}\mathcal{O}^k|\Psi_0\rangle$. Notice that

$$\begin{aligned}
\mathcal{O}^{k-1} S_j^x \mathcal{O}^k &= S_j^x \cos kj + S_j^y \sin kj \ , \\
\mathcal{O}^{k-1} S_j^y \mathcal{O}^k &= S_j^y \cos kj - S_j^x \sin kj \ , \\
\mathcal{O}^{k-1} S_j^z \mathcal{O}^k &= S_j^z \ ,
\end{aligned} \tag{2.60}$$

so that (taking into account periodic boundary conditions $S_{L+1} = S_1$ and that $k = 2\pi m/L$ for the boundary term $S_L S_1$)

$$\begin{aligned}
\langle\Psi_0|\mathcal{O}^{k-1}\mathcal{H}_{HI}\mathcal{O}^k|\Psi_0\rangle = \langle\Psi_0|\mathcal{H}_{HI} + J(\cos k - 1)\sum_{j=1}^{L}(S_j^x S_{j+1}^x + S_j^y S_{j+1}^y) \\
+ J\sin k \sum_{j=1}^{L}(S_j^x S_{j+1}^y - S_j^y S_{j+1}^x)|\Psi_0\rangle \ .
\end{aligned} \tag{2.61}$$

The reader knows that $\langle\Psi_0|\mathcal{H}_{HI}|\Psi_0\rangle = E_0$. Then it is easy to show that

$$\begin{aligned}
&(\cos k - 1)\langle\Psi_0|\sum_{j=1}^{L}(S_j^x S_{j+1}^x + S_j^y S_{j+1}^y)|\Psi_0\rangle \\
&\quad = -\frac{1}{2}\left[(2\pi/L)^2 - O(L^{-4})\right]\sum_{j=1}^{L}\langle\Psi_0|S_j^x S_{j+1}^x + S_j^y S_{j+1}^y|\Psi_0\rangle \\
&\quad \le [(2\pi/L)^2(L/2) + O(L^{-3})] \ ,
\end{aligned} \tag{2.62}$$

and

$$\begin{aligned}
&J\sin k\langle\Psi_0|\sum_{j=1}^{L}(S_j^x S_{j+1}^y - S_j^y S_{j+1}^x)|\Psi_0\rangle \\
&\quad = -i\sin k\langle\Psi_0|\left[\sum_n nS_n^z, \mathcal{H}_{HI}\right]|\Psi_0\rangle = 0 \ .
\end{aligned} \tag{2.63}$$

From these equations we conclude that for $k = 2\pi/L$ the energy of the state Ψ_k is

$$\langle\Psi_k|\mathcal{H}_{HI}|\Psi_k\rangle \leq E_0 + (2\pi^2 J/N) \ , \qquad (2.64)$$

i.e., there is no energy gap. This is why, an excited state for a periodic Heisenberg–Ising "easy-plane" antiferromagnetic chain with nearest neighbour interactions has vanishingly small excitation energy in the limit that the length of a chain is infinite, if the ground state is non-degenerate. One can obviously generalize this statement for any space dimension for bipartite lattices without spin frustration and for any half-integer values of a site spin (notice that one needs numbers of sites of the system in other than along the direction of the transformation T directions to be mutually prime with the number of magnetic sublattices along the T direction). In particular, the generalization of the Lieb–Schultz–Mattis theorem is often used for a description of the behaviour of *spin ladders*, *i.e.*, finite number of spin chains connected with each other (usually due to nearest-neighbour couplings between chains).

One can consider the generalization of this theorem for higher spins S and multi-sublattice spin chains in an external magnetic field parallel to the axis of a magnetic anisotropy, due to M. Oshikawa, M. Yamanaka and I. Affleck. In fact, they pointed out that the proof given above works for any ground state z-projection of an average spin moment per spin s^z (caused by a nonzero magnetic field H) except for values $N(S - s^z) = q$, where N is the number of magnetic sublattices (distinguished by non-equal exchange couplings, or by different effective magnetons, as we considered above) and q is integer. We see that $T\Psi_0$ is orthogonal to Ψ_0, except if $N(S - s^z) = q$, which follows from the action of the operator $\exp(ik[LS_1^z - ik\sum_{n=1}^{L} S_n^z])$. This implies that for any non-integer $N(S - s^z)$ there is a low-lying excited state with the energy $JO(1/L)$. For $S = \frac{1}{2}$ and $N = 1$, it is a trivial statement, naturally the reader knows that if the ground state moment per spin is equal to $\pm\frac{1}{2}$, we are in the ferromagnetic state (with gapped excitations; the latter will be shown in detail in the next chapter), and the points $s^z = \pm\frac{1}{2}$ are quantum critical points. From the above the reader can check that it is true for N-sublattice isotropic XY chain in a magnetic field. The situation with integer spins S, $N = 1$ and $s^z = 0$ is, in fact, related to the well-known conjecture due to F. D. M. Haldane (known as the *Haldane's hypothesis*), who was the first to emphasize that for Heisenberg antiferromagnetic integer-spin chains at $H = 0$, low-lying excitations have gaps, while for non-integer site spins there is no such a gap. The onset of magnetization plateaux, considered above, persists at least for an "easy-plane" magnetic anisotropy.

Summarizing, in this chapter we presented some simple exactly solvable models of quantum spin chains: Ising chains, isotropic and anisotropic XY chains, multimerized XY chains. Common features for the behaviours of all these models are the absence of $T \neq 0$ phase transitions, while quantum phase transitions in an external magnetic field are characteristic features for this class of systems. We presented thermodynamic characteristics for these models and pointed out similarities and differences in the temperature behaviours of magnetic susceptibility and specific heat. Finally, we presented the Lieb–Schultz–Mattis theorem and discussed the inset of magnetization plateaux in the ground state of (multimerized) quantum spin chains.

The interested reader can find the solution of the Ising chain in [Ising (1925)]. The Jordan–Wigner transformation is introduced in [Jordan and Wigner (1928)]. The solution of an isotropic XY chain for $H = 0$ is given in [Lieb, Schulz and Mattis (1961)]. We closely followed an appendix of that paper when discussing the Lieb–Schultz–Mattis theorem. The reader can find the generalization of that theorem for a nonzero magnetic field in [Yamanaka, Oshikawa and Affleck (1997)]. The solution of an isotropic and anisotropic XY model (including the case of an Ising chain in the perpendicular magnetic field) can be found in [Katsura (1962); Pikin and Tsukernik (1966); Pfeuty (1970)]. The generalized Jordan–Wigner transformation and an analysis of the behaviour of multi-sublattice spin chains is presented in [Zvyagin (1990a)]. The special important case of a dimerized XY chain can be found in [Kontorovich and Tsukernik (1967)]. The Haldane hypothesis was introduced in [Haldane (1983)].

Chapter 3

Co-ordinate Bethe Ansatz for a Heisenberg–Ising Ring

In this chapter we shall present the main ideas of the *co-ordinate Bethe ansatz* using as the basic model the simplest case of interacting spin-$\frac{1}{2}$ one-dimensional systems.

3.1 Bethe Ansatz

Now our goal is to find the eigenfunctions and eigenvalues of the Hamiltonian $\mathcal{H}_{HI}+\mathcal{H}_Z$ for the general case $J_z \neq 0$, $J \neq 0$. This, by now well-known method, is due to H. Bethe who proposed it first for the Heisenberg spin chain, is called the *Bethe's ansatz*.

The total spin (for $J_z = J$) , as well as the z-projection of the total spin $\sum_{j=1}^{L} S_j^z$, commute with the Hamiltonian $\mathcal{H}_{HI}+\mathcal{H}_Z$. This is why, we classify all states of the Hamiltonian by eigenvalues of the operator $\sum_{j=1}^{L} S_j^z$. It is convenient to choose the basis functions in the form

$$|x_1,\ldots,x_M\rangle \equiv e_1^+ \otimes \cdots \otimes e_{x_1}^- \otimes \cdots \otimes e_j^+ \otimes \cdots e_{x_M}^- \otimes \cdots \otimes e_L^+ \tag{3.1}$$

(here $\otimes$ denotes the tensor product), such that the values x_j in it determine M coordinates of sites with spins down (all other spins are directed up). [Naturally, we could choose the opposite basis with M spins up and $L-M$ spins down, and the results would be the same.] We suppose that $1 \leq x_1 < x_2 < \cdots < x_M \leq L$. Then the wave function can be written as

$$\Psi_M = \sum_{x_1<x_2<\cdots<x_M} a(x_1,\ldots,x_M)|x_1,\ldots,x_M\rangle \ , \tag{3.2}$$

where $a(x_1,\ldots,x_M)$ is the wave function in the co-ordinate representation. For this reason this method is often referred to as the co-ordinate Bethe ansatz. Then the action of the Hamiltonian on the wave function Eq. (3.2) is easy to get. The condition for Ψ_M to be the eigenfunction of $\mathcal{H}_{HI}+\mathcal{H}_Z$ (*i.e.*, to fulfill the stationary Schrödinger equation, $(\mathcal{H}_{HI}+\mathcal{H}_Z)\Psi_M = E\Psi_M$) is

$$Ea(x_1,\dots,x_M) = \left(-\frac{H(L-2M)}{2}+\frac{J_zL}{4}\right)a(x_1,\dots,x_M) + \frac{1}{2}\sum_{x_1',\cdots,x_M'}\left[Ja(x_1',\dots,x_M') - J_za(x_1,\dots,x_M)\right] , \quad (3.3)$$

where the set $x_1',\dots,x_M'$ differs from $x_1,\dots,x_M$ by an interchange of spins of some one pair of neighbours. Clearly, this set of finite difference equations is valid for any $\sum_{j=1}^{L} S_j^z = (L/2) - M$.

Let us now consider simple examples of the realization of Eq. (3.3). First, suppose that $M = 1$. In this case we have

$$Ea(x) = \frac{-2H(L-2)+J_zL}{4}a(x) - J_za(x) + \frac{J}{2}a(x-1) + \frac{J}{2}a(x+1) . \quad (3.4)$$

It has the well-known solution

$$a(x) = A\exp(ikx) , \; E = -\frac{H(L-2)}{2}+\frac{J_zL}{4} - (J_z - J\cos k) , \quad (3.5)$$

where an arbitrary constant A has to be determined from the normalization condition for the wave function and k is defined by boundary conditions. Let us consider periodic boundary conditions $a(x) = a(x+L)$. Then we have $\exp(ikL) = 1$, and, hence, $kL = 2\pi I$, where non-equal integers are $I = 0, \pm 1, \pm 2, \dots$.

For the case $M = 2$, we have to distinguish two situations. If $x_2 \neq x_1+1$, one has the equation

$$\begin{aligned} Ea(x_1,x_2) = &\frac{-2H(L-4)+J_zL}{4}a(x_1,x_2) - 2J_za(x_1,x_2) \\ &+\frac{J}{2}[a(x_1-1,x_2)+a(x_1+1,x_2) \\ &+a(x_1,x_2-1)+a(x_1,x_2-1)] , \end{aligned} \quad (3.6)$$

which general solution is

$$\begin{aligned} a(x_1,x_2) &= A_1e^{i(k_1x_1+k_2x_2)} + A_2e^{i(k_2x_1+k_1x_2)} , \\ E &= -\frac{H(L-4)}{2}+\frac{J_zL}{4} - 2J_z + J(\cos k_1+\cos k_2) , \end{aligned} \quad (3.7)$$

where $A_{1,2}$ are arbitrary constants.

On the other hand, when $x_2 = x_1 + 1$ (*i.e.*, down spins are situated at the nearest neighbour sites) one has

$$Ea(x, x+1) = \left(-\frac{H(L-4)}{2} + \frac{J_z L}{4}\right) a(x, x+1) - 2J_z a(x, x+1) + \frac{J}{2}[a(x-1, x+1) + a(x, x+2)] \ . \tag{3.8}$$

We can look for the solution of this equation in the form, similar to the previous case

$$a(x_1, x_2) = A_{12} e^{i(k_1 x_1 + k_2 x_2)} + A_{21} e^{i(k_2 x_1 + k_1 x_2)} \ , \quad E = -\frac{H(L-4)}{2} + \frac{J_z L}{4} - 2J_z + J(\cos k_1 + \cos k_2) \ , \tag{3.9}$$

but with the constraint on the amplitudes A_{12} and A_{21}

$$J_z \left(A_{12} e^{ik_2} + A_{21} e^{ik_1}\right) = \frac{1}{2}(A_{12} + A_{21}) \left(1 + e^{i(k_1 + k_2)}\right) \ . \tag{3.10}$$

The last formula implies

$$A_{21} = -A_{12} e^{i\theta(k_1, k_2)} \tag{3.11}$$

where A_{12} is determined from the normalization condition for the wave function and

$$\theta(k_1, k_2) = \tan^{-1}\left(\frac{J_z \sin \frac{k_1 - k_2}{2}}{J \cos \frac{k_1 + k_2}{2} - J_z \cos \frac{k_1 - k_2}{2}}\right) \ . \tag{3.12}$$

Since in the case with two down spins not nearest neighbours we used the same structure of the solution but with independent constants, one may suppose that the function Eq. (3.9) is the general eigenfunction of the Hamiltonian for the case $M = 2$ and any position of down spins. It turns out that the phase factor $\theta(k_1, k_2)$ does not depend on the value of an external magnetic field H, but the eigenvalue of the Hamiltonian does depend on it. It is also instructive to observe that for the isotropic XY model case $J_z = 0$, the phase $\theta(k_1, k_2) \equiv 0$. Actually, this is the manifestation of the fact, already known to the reader from the previous chapter, the Hamiltonian of the isotropic XY model can be exactly mapped onto the Hamiltonian of free (noninteracting) spinless fermions.

The periodic boundary condition $a(x_1, x_2) = a(x_2, L + x_1)$ (because we supposed that $x_1 < x_2$) implies two similar equations for the values k_1 and k_2:

$$e^{ik_1 L} = -e^{-i\theta(k_1,k_2)} \ , \ e^{ik_2 L} = -e^{-i\theta(k_2,k_1)} \ . \tag{3.13}$$

Taking the logarithm one can re-write these equations in the form

$$k_1 L = 2\pi I_1 - \theta(k_1, k_2) \ , \ k_2 L = 2\pi I_2 - \theta(k_2, k_1) \ , \tag{3.14}$$

with non-equal half-integers $I_{1,2} = \pm\frac{1}{2}, \pm\frac{3}{2}, \ldots$, because the logarithm is the multi-valued function.

The reader can directly check after some straightforward but lengthy calculations performed in a similar way that the cases $M = 3, 4$ can also be explicitly solved as for $M = 2$. This implies the general form of the eigenfunction in the co-ordinate representation as the superposition of the plane waves

$$a(x_1, \ldots, x_M) = \sum_P A_P \exp\left(i \sum_{j=1}^{M} k_{P_j} x_j\right) \ , \tag{3.15}$$

where P denotes a permutation of M indices $1, 2, \ldots, M$ (there are $M!$ terms here). This is nothing other than the famous Bethe ansatz! Amplitudes A_P are related to $A_{1,2,\ldots,M}$ as

$$A_P = \pm A_{1,2,\ldots,M} \exp[i \sum \theta(k_j, k_l)] \ , \tag{3.16}$$

where the summation is extended over all pairs of indices j, l obtained from the initial arrangement of them for A_P, which is necessary to interchange in order to get $A_{1,2,\ldots,M}$. The sign is determined by the parity of those permutations. The eigenvalue of the Hamiltonian $\mathcal{H}_{HI} + \mathcal{H}_Z$, which corresponds to the eigenfunction equation (3.15), is

$$E = -\frac{H(L - 2M)}{2} + \frac{LJ_z}{4} - \sum_{j=1}^{M} (J_z - J\cos k_j) \ . \tag{3.17}$$

Finally, the periodic boundary condition for the general case of any M can be written as

$$e^{ik_j L} = (-1)^{M-1} e^{-i\sum_{l=1, \ l\neq j}^{M} \theta(k_j, k_l)} \ , \ j = 1, \ldots, M \ , \tag{3.18}$$

which are well-known as the *Bethe ansatz equations.* Taking the logarithm we obtain

$$Lk_j = 2\pi I_j - \sum_{\substack{l=1,\\ l\neq j}}^{M} \theta(k_j, k_l) \tag{3.19}$$

with half-integers I_j for even M and integers I_j for odd M. For example, the convenient choice for these numbers is $I_j = (L+M+1)/2 \pmod 1$. One can naturally consider k_j as quasimomenta of eigenstates. It is clear that Bethe ansatz equations for essentially interacting system are very similar to standard quantization conditions for noninteracting particles in a one-dimensional box of length L.

These Bethe ansatz equations can be re-written in the following way. For the case $J = J_z < 0$ (*i.e.*, for the isotropic ferromagnetic Heisenberg chain) one can introduce the set of so-called *rapidities* $\{\lambda_j\}_{j=1}^{M}$ instead of quasimomenta $\{k_j\}_{j=1}^{M}$ as $\lambda_j = \frac{1}{2}\cot\frac{k_j}{2}$ (or, in other words, $k_j = 2\tan^{-1} 2\lambda_j = -i\ln[(2\lambda_j+i)/(2\lambda_j-i)]$). To have independent solutions of Eq. (3.18), one has to consider, *e.g.*, real k_j in the domain $0 \leq k_j \leq 2\pi$, which is related to the domain $-\infty \leq \lambda_j \leq \infty$. The goal of such an introduction is to re-write Bethe ansatz equations in a differential form

$$\left(\frac{\lambda_j + (i/2)}{\lambda_j - (i/2)}\right)^L = \prod_{\substack{l=1,\\ l\neq j}}^{M} \frac{\lambda_j - \lambda_l + i}{\lambda_j - \lambda_l - i} , \tag{3.20}$$

which, taking the logarithm, can be re-written as

$$2L\tan^{-1}(2\lambda_j) = 2\pi I_j + 2\sum_{\substack{l=1,\\ l\neq j}}^{M} \tan^{-1}(\lambda_j - \lambda_l) , \tag{3.21}$$

where I_j are (half)integers for M (even) odd, and the energy is

$$E = -\frac{HL}{2} + \frac{NJ}{4} + \sum_{j=1}^{M} [H - 2J(4\lambda_j^2 + 1)^{-1}] . \tag{3.22}$$

It turns out that for $J = J_z > 0$, one can introduce the set of rapidities as $\lambda_j = \frac{1}{2}\tan\frac{k_j}{2}$. The domain $-\infty \leq \lambda_j \leq \infty$ is now related to the one $-\pi \leq k_j \leq \pi$. The energy and the Bethe ansatz equations, however, formally have the same form as Eqs. (3.20) and (3.22).

For $J_z \neq J$ one can introduce the value $\cos\eta = J_z/J$ (real values of η are related to the "easy-plane" magnetic anisotropy $|J_z/J| \leq 1$, while the "easy-axis" magnetic anisotropy with $|J_z/J| \geq 1$ is described by imaginary values of η). The set of rapidities is now introduced *via* $k_j = -i\ln(\sin[\lambda_j + (\eta/2)]/\sin[\lambda_j - (\eta/2)])$. The Bethe ansatz equations and the energy can be written as

$$\left(\frac{\sin[\lambda_j + (\eta/2)]}{\sin[\lambda_j - (\eta/2)]}\right)^L = \prod_{\substack{l=1,\\ l\neq j}}^{M} \frac{\sin[\lambda_j - \lambda_l + \eta]}{\sin[\lambda_j - \lambda_l - \eta]} \tag{3.23}$$

and the energy

$$E = -\frac{HL}{2} + \frac{NJ_z}{4} + \sum_{j=1}^{M}\left(H - J_z + \frac{J\sin[\lambda_j + (\eta/2)]}{2\sin[\lambda_j - (\eta/2)]} + \frac{J\sin[\lambda_j - (\eta/2)]}{2\sin[\lambda_j + (\eta/2)]}\right). \tag{3.24}$$

For $J_z > J > 0$, it is convenient to use the parametrization $\eta \to \pi + i\eta$ and $\lambda_j \to (\pi - \lambda_j)/2$ within the domain $-\pi \leq \lambda_j \leq \pi$. On the other hand, for $0 > J > J_z$ the convenient parametrization is $\eta \to i\eta$ and $\lambda_j \to \lambda_j/2$ within $-\pi \leq \lambda_j \leq \pi$. In the "easy-plane" anisotropic situation $-J < J_z < J$ the convenient parametrization is $\eta \to \pi - \eta$ and $\lambda_j \to (\pi - i\lambda_j)/2$ within the domain $-\infty \leq \lambda_j \leq \infty$.

Notice that θ is a single-valued real analytic function of J, J_z, k_j, and k_l if the latter two are in the open interval given above, and $\theta(0,0) = 0$. These conditions uniquely define the branch of $\tan^{-1}$. Moreover, $\theta(-p,-q) = -\theta(p,q) = \theta(q,p)$. C. N. Yang and C. P. Yang proved (we shall not present the proof of those theorems and refer the reader to the original papers, where very transparent and elegant proofs are given) that

- For any $M \leq L/2$ and $0 \leq (J_z/J) < 1$, the Bethe ansatz equations have an unique solution in the interval for rapidities (momenta). Each k_j is an analytic function of J_z/J.
- This solution satisfies the condition $k_j = -k_{M-j+1}$, $j = 1,\ldots,M$.
- For $M \leq L/2$ and $(J_z/J) \leq 0$, the Bethe ansatz equations have a solution forming a continuous curve in the real $k_j \times (J_z/J)$ space with k_j in the above mentioned interval. The curve extends from $J_z = 0$ to all $(J_z/J) < 0$ and at each point on the curve one has $k_i \neq k_j$ and $k_j = -k_{M-j+1}$ for $j = 1,\ldots,M$.
- The ground state of the Heisenberg–Ising Hamiltonian for finite L and M is nondegenerate for any real J_z/J. The ground state energy is analytic in (J_z/J) for all real (J_z/J).

- For $J_z = 0$, the solution is unique.
- For any real $(J_z/J) < 1$ and for $M \le L/2$, the ground state is given by the Bethe eigenfunction with $k_j = -k_{M-j+1}$ for $j = 1, \ldots, M$.
- All k_j are analytic in (J_z/J) in an open strip containing the semi-infinite real axis $(J_z/J) < 1$.

Hence, the difficult problem of solving a stationary Schrödinger equation for a quantum interacting many-body system is reduced to the solution of a finite set of finite difference equations for rapidities. As we have seen, the solution of the Bethe ansatz equations parametrizes the eigenfunctions and eigenvalues of the considered quantum spin Hamiltonian.

3.2 Simple Solutions of the Bethe Ansatz Equations: Strings

It is instructive to consider possible solutions of Bethe ansatz equations. Let us for simplicity limit ourselves by the isotropic case $J_z = J$. For the simplest case $M = 1$, we already presented the solution in Eq. (3.5). The reader can see that only real $\lambda = (1/2)\cot 2\pi I/L$ realize the solution. It is not so for $M = 2$. Here the Bethe ansatz equations for $\lambda_{1,2}$ can be written explicitly as

$$\left(\frac{\lambda_1 + (i/2)}{\lambda_1 - (i/2)}\right)^L = \frac{\lambda_1 - \lambda_2 + i}{\lambda_1 - \lambda_2 - i}\,, \quad \left(\frac{\lambda_2 + (i/2)}{\lambda_2 - (i/2)}\right)^L = \frac{\lambda_2 - \lambda_1 + i}{\lambda_2 - \lambda_1 - i}\,. \tag{3.25}$$

We can replace one of these equations by

$$\left(\frac{\lambda_1 + (i/2)}{\lambda_1 - (i/2)} \times \frac{\lambda_2 + (i/2)}{\lambda_2 - (i/2)}\right)^L = 1\,. \tag{3.26}$$

The real $\lambda_{1,2}$ solutions (which in the thermodynamic limit $L \to \infty$ with the exponential accuracy in L decouple) have the same form as for the case $M = 1$. The energy of this solution is

$$\begin{aligned} E &= \frac{JL}{4} - \frac{H(L-4)}{2} - J(2 - \cos k_1 - \cos k_2) \\ &= \frac{JL}{4} - \frac{H(L-4)}{2} - 2J\left(1 - \cos\frac{k_1 + k_2}{2}\cos\frac{k_1 - k_2}{2}\right)\,. \end{aligned} \tag{3.27}$$

However, there exists a complex solution $\lambda_{1,2} = x_{1,2} + iy_{1,2}$. Suppose $y_1 > 0$. Then the first equation of Eqs. (3.25) in the limit $L \to \infty$ (*i.e.*, with the exponential accuracy in L) implies $x_1 = x_2 = x$, and $y_1 = y_2 + 1$. On the other hand, Eq. (3.26) yields $y_1 = \frac{1}{2}$ for any real x. Hence, the complex solution of the Bethe ansatz equations for $M = 2$ is $\lambda_{1,2} = x \pm i\frac{1}{2}$, which has the energy

$$E = \frac{JL}{4} - \frac{H(L-4)}{2} - J(x^2+1)^{-1}$$
$$= \frac{JL}{4} - \frac{H(L-4)}{2} - \frac{J}{2}[1 - \cos(k_1 + k_2)] \ . \quad (3.28)$$

This complex solution is usually referred to as the *bound state*, or the *string* of length 2. It exists only if there is an essential interaction in a system (*i.e.*, when $\theta(k_i, k_j) \neq 0$; there are no bound states for the case $J_z = 0$, in the isotropic XY model). From Eqs. (3.27) and (3.28), the reader can conclude that for a fixed total momentum $k_1 + k_2$ the values of the energy of a complex solution for $M = 2$ is lower than the energy of two real solutions in the ferromagnetic case $J_z < 0$, and it is higher than the energy of two real solutions in the antiferromagnetic situation $J_z > 0$.

The so-called *string hypothesis* is often used to study cases with arbitrary M. Define the string of length $2m + 1$ as

$$\lambda_m = \tilde{\lambda}_m + iy + O(e^{-\kappa L}) \ , \ y = -m, -m+1, \ldots, m-1, m \ , \quad (3.29)$$

where $\tilde{\lambda}_m$ is real (sometimes it is referred to as the *centre of mass* of the string), m is a positive integer or half-integer and $\kappa > 0$ (*i.e.*, strings are solutions to Bethe ansatz equations only in the thermodynamic limit, with the exponential in L accuracy). Then, real solutions of Bethe ansatz equations can be referred to as strings of length 1 (with $m = 0$) or *spinons*. Let us denote by μ_m the number of strings of length m. Then the total number of strings is equal to $\sum_m \mu_{2m+1}$ and the total number of down spins due to strings is

$$M = \sum_m (2m+1)\mu_{2m+1} \ . \quad (3.30)$$

μ_m, m and the total number of strings are often called a *configuration*. Introducing strings as $\lambda_j = \lambda_j^m + i[(m+1)/2 - \nu]$ with $\nu = 1, \ldots, m$ and summing Eq. (3.21) for a distinguished string of length m for the parameters λ_j occurring in it, one obtains

$$\theta_m(\lambda_j^m) = \frac{2\pi}{L} I_{j,m} + \frac{1}{L} \sum_{n,l \neq j,m} \theta_{mn}(\lambda_j^m - \lambda_l^n) \ , \quad (3.31)$$

where $\theta_m(x) = 2\tan^{-1}(2x/m)$,

$$\theta_{mn}(x) = (1-\delta_{m,n})\theta_{|m-n|}(x) + 2\theta_{|m-n|+2}(x) + \cdots + 2\theta_{m+n-2}(x) + \theta_{m+n}(x) \ , \tag{3.32}$$

and integers or half-integers $I_{j,m}$ appear because the logarithm is the multi-valued function. In what follows we shall show that it is often convenient to introduce *two* sets of these quantum numbers. Namely, the first set, $I_{j,m}$ which characterizes the strings present in the given configuration, parametrizes "quasiparticles" (often they are mentioned just as particles), and the second set, $I^{(h)}_{j,m}$, which characterizes the unoccupied vacancies of the given configuration of strings, parametrizes "quasiholes" (often mentioned just as holes).

The momentum of a string of length m for the isotropic case is

$$p_m = 2\cot^{-1}(2\lambda_j^m/m) \ , \tag{3.33}$$

and the energy is

$$E_m = E_0 + Hm + \frac{J}{2}\frac{d}{d\lambda_j^m}p_m = E_0 + Hm - \frac{2Jm}{4(\lambda_j^m)^2 + m^2} \ , \tag{3.34}$$

where $E_0 = L(J_z - 2H)/4$. The expression for the energy can be re-written in a convenient form

$$E = E_0 + \varepsilon_m(p_m) \ , \ \varepsilon_m(p_m) = mH - \frac{J}{m}(1-\cos p_m) \ . \tag{3.35}$$

The total energy with all possible strings can be written as

$$E = E_0 + \sum_m \sum_j \left[Hm - \frac{2Jm}{4(\lambda_j^m)^2 + m^2} \right] \ , \tag{3.36}$$

and the total magnetic moment of the system with all possible strings is equal to

$$M^z = \frac{L}{2} - \sum_{m=1}^{\infty} m\nu_m \ . \tag{3.37}$$

3.3 Thermodynamic Bethe Ansatz

The string hypothesis states that all solutions of Bethe ansatz equations can be written in the form of strings of all possible lengths.

Then the main goal is to find the solutions to the Bethe ansatz equations in the thermodynamic limit $L \to \infty$. In this limit it is convenient, following L. Hulthén, to introduce distribution functions for particles and holes, corresponding to strings of length m. We replace discrete quantum numbers by some continuous variable $I_{j,m} \to x_m$ and replace summation over j by integration over x_m. The *densities* of rapidities (sometimes referred to as "dressed densities", to emphasize the fact that they are different from the ones of noninteracting quasiparticles, like an isotropic XY model; the interaction "dresses" "bare" densities) $\rho_m(x)$ are introduced as $\rho_m(x) + \rho_m^{(h)}(x) = dx_m/d\lambda(x_m)$. Here we also introduced the density of holes $\rho_m^{(h)}(x)$ as the complementary to $\rho_m(x)$ function. Then, differentiating equations (3.31) with respect to real parts of λ_j and introducing continuous distributions of those real parts we obtain

$$a_m(\lambda) = \rho_m^{(h)}(\lambda) + \sum_{n=1}^{\infty} A_{mn} * \rho_n(\lambda) , \tag{3.38}$$

where $a_m(x) = 2m/[\pi(4x^2 + m^2)]$, the convolution, $A * B(x)$, means

$$A_{mn} * \rho_m(x) = \int_{-\infty}^{\infty} dy A_{mn}(x - y)\rho_m(y) \tag{3.39}$$

and

$$\begin{aligned} A_{mn}(x) = &\ a_{|m-n|}(x) + 2a_{(|m-n|+2)}(x) \\ &+ \cdots + 2a_{(m+n-2)}(x) + a_{m+n}(x) . \end{aligned} \tag{3.40}$$

Notice that here we introduced the term with $n = m$, unlike Eqs. (3.31), by using the identity

$$\lim_{|m-n|\to 0} \int_{-\infty}^{\infty} dy a_{|m-n|}(x - y)\rho_m(y) = \rho_m(x) . \tag{3.41}$$

Then the internal energy per site, $e = E/L$, and the magnetization $m^z = M^z/L$ per site are given as

$$\begin{aligned} e &= e_0 + \sum_{m=1}^{\infty} \int_{-\infty}^{\infty} d\lambda \varepsilon_m^{(0)}(\lambda)\rho_m(\lambda) , \\ m^z &= \frac{1}{2} - \sum_{m=1}^{\infty} m \int_{-\infty}^{\infty} d\lambda \rho_m(\lambda) , \end{aligned} \tag{3.42}$$

where we introduced the "bare" energy of the string of length m as $\varepsilon_m^{(0)}(\lambda) = mH - 2Jm/[(2\lambda)^2 + m^2] = mH - J\pi a_m(\lambda)$ and $e_0 = E_0/L$.

Then, following C. N. Yang and C. P. Yang, we can assume that the distribution function $\rho_m(\lambda)$ together with $\rho_m^{(h)}(\lambda)$ minimizes the Helmholtz free energy $F = E - T\mathcal{S}$. The change of the entropy for the given configuration of strings can be defined as

$$\begin{aligned} d\mathcal{S}_m &= \ln \frac{[L(\rho_m^{(h)}(\lambda) + \rho_m(\lambda))d\lambda]!}{[L\rho_m(\lambda)d\lambda]![L\rho_m^{(h)}(\lambda)d\lambda]!} \\ &\approx L[(\rho_m^{(h)}(\lambda) + \rho_m(\lambda)) \ln[\rho_m^{(h)}(\lambda) + \rho_m(\lambda)] \\ &\quad - \rho_m^{(h)}(\lambda) \ln \rho_m^{(h)}(\lambda) - \rho_m(\lambda) \ln \rho_m(\lambda)]d\lambda \ , \end{aligned} \tag{3.43}$$

where we used the Stirling formula $\ln x! \approx x \ln x$. Then the Helmholtz free energy per site of the Heisenberg–Ising chain in the thermodynamic limit can be written as

$$\begin{aligned} f = e_0 + \sum_{m=1}^{\infty} \int_{-\infty}^{\infty} d\lambda[\varepsilon_m^{(0)}(\lambda)\rho_m(\lambda) - T(\rho_m^{(h)}(\lambda) \\ + \rho_m(\lambda)) \ln(\rho_m^{(h)}(\lambda) + \rho_m(\lambda)) + T\rho_m^{(h)}(\lambda) \ln \rho_m^{(h)}(\lambda) \\ + T\rho_m(\lambda) \ln \rho_m(\lambda)] \ . \end{aligned} \tag{3.44}$$

Using the relation $\delta\rho_m^{(h)}(x) = -\sum_n A_{mn} * \delta\rho_n(x)$ one can write the variation of Eq. (3.44) with respect to $\delta\rho_n(\lambda)$ as

$$\begin{aligned} \delta f = \sum_m \int d\lambda \Big(\varepsilon_m^{(0)}(\lambda) - T \ln[1 + \eta_m(\lambda)] \\ - T \sum_n A_{nm} * \ln[1 + \eta_n^{-1}(\lambda)] \Big) \delta\rho_m(\lambda) \ , \end{aligned} \tag{3.45}$$

where we introduced the function $\eta_m(x) = \rho_m^{(h)}(x)/\rho_m(x) \equiv \exp[\varepsilon_m(\lambda)/T]$. The function $\varepsilon_m(\lambda)$ is often referred to as the *dressed energy*. Then $\delta f = 0$ implies

$$\varepsilon_m^{(0)}(\lambda) = T \ln[1 + \eta_m(\lambda)] - T \sum_n A_{nm} * \ln[1 + \eta_n^{-1}(\lambda)] \ , \tag{3.46}$$

which set of equations completes the set Eq. (3.38). Both two sets, first derived by M. Takahashi, M. Gaudin and M. Suzuki for a Heisenberg–Ising chain, are known as *thermodynamic (or thermal) Bethe ansatz equations.*

[We do not present here thermodynamic Bethe ansatz equations for a magnetically anisotropic situation, because they are more complicated than the isotropic case $J = J_z$ and refer the interested reader to the original works.] These equations can be re-written in the following useful form

$$\varepsilon_m(\lambda) = Ts(\lambda - \lambda') * \ln[1 + \eta_{m-1}(\lambda')][1 + \eta_{m+1}(\lambda')] - 2\pi\delta_{m,1}s(\lambda) ,$$
$$s(x) = 1/2\cosh(\pi x) , \ \lim_{m\to\infty} \frac{\varepsilon_m(\lambda)}{m} = H . \qquad (3.47)$$

Solving these equations for $\rho_m(\lambda)$ and $\eta_m(\lambda)$ one puts those solutions into Eq. (3.44) and it constitutes the exact (in the thermodynamic limit) Bethe ansatz solution for the quantum spin-$\frac{1}{2}$ Heisenberg chain. In fact, inserting the thermal equilibrium density functions into the expression for the Helmholtz free energy we obtain

$$f = e_0 + \frac{J}{2}[\psi(1/2) - \psi(1)] - T\int_{-\infty}^{\infty} d\lambda \frac{\ln(1 + \exp[\varepsilon_1(\lambda)/T])}{2\cosh(\pi\lambda)} , \qquad (3.48)$$

where $\psi(x)$ are digamma functions.

The results for the temperature behaviour of the Heisenberg chain for $H = 0$ and for a weak magnetic field $H = 0.1J$ for $J = 2$ are presented in Figs. 3.1 and 3.2. Notice the difference in the low-temperature behaviour of these characteristics for a Heisenberg chain and for an Ising chain, and the similarity with the results for an isotropic XY chain, *cf.* the previous chapter. Observe low-temperature logarithmic corrections for an isotropic Heisenberg chain in comparison with an XY chain, see the next section. The reader can see that a weak magnetic field practically does not change the temperature behaviour of the specific heat of a homogeneous Heisenberg chain, but removes logarithmic corrections in the susceptibility (the magnetic field reduces the symmetry of the system, naturally).

Thermodynamic Bethe ansatz equations are nonlinear integral equations. They can be solved analytically only in two limiting cases: high temperatures, $T \to \infty$, and low temperatures, $T \to 0$. For all intermediate temperatures one has to solve these two infinite (in the framework of the string hypothesis) sets of integral nonlinear equations numerically. Still, the computation problem is much simpler than numerically solving the stationary Schrödinger equation with the Heisenberg Hamiltonian. In the following chapters we shall present a more convenient way to study thermodynamics of quantum chains (in fact the results of Figs. 3.1 and 3.2 were obtained by using that method).

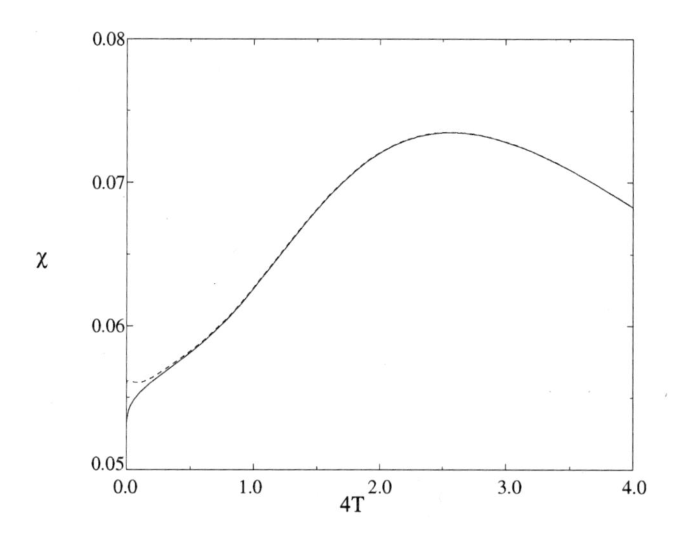

Fig. 3.1 The magnetic susceptibility of an isotropic Heisenberg antiferromagnetic spin-$\frac{1}{2}$ chain with $J = 2$. The solid line shows the results for $H = 0$, the dashed line — for $H = 0.2$.

In the limit of high temperatures $T \gg |J|$, but keeping the ratio H/T finite, one can consider $\varepsilon_m^{(0)}(\lambda) \approx mH$ (the term which does not depend on the function to be determined is often referred to as the *driving term* in the theory of integral equations). In this case driving terms do not depend on λ and we can solve Eqs. (3.46) and (3.38) exactly. The solutions are:

$$\eta_m = f^2(m) - 1 \;, f(m) = \frac{z^{m+1} - z^{-m-1}}{z - z^{-1}} \;, \; z = \exp(-H/2T) \;,$$
$$\rho_m(\lambda) = \frac{1}{f(1)f(m)}\left(\frac{a_m(\lambda)}{f(m-1)} - \frac{a_{m+2}(\lambda)}{f(m+1)}\right) \;. \tag{3.49}$$

These solutions permit us to find the magnetization of the quantum spin chain at high temperatures

$$m^z = \frac{1}{2} - \sum_{m=1}^{\infty} m \int_{-\infty}^{\infty} d\lambda \rho_m(\lambda) = \frac{1}{2} \tanh \frac{H}{2T} \;. \tag{3.50}$$

It is the magnetization of the free gas of quantum spins $\frac{1}{2}$ at high temperature in the nonzero magnetic field. This result is transparent, because we neglected the exchange interaction between spins in this limiting case. The magnetic susceptibility is equal to $\chi = (1/4T\cosh^2(H/2T))$. It is a smooth function of the temperature with $\chi \sim T^{-1}$ at high temperatures, as expected.

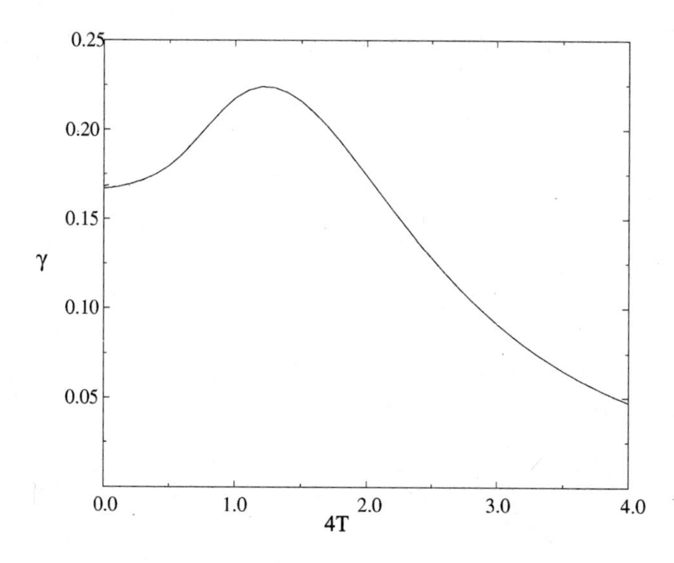

Fig. 3.2 The Sommerfeld coefficient for the specific heat of the isotropic Heisenberg antiferromagnetic spin-$\frac{1}{2}$ chain ($J = 2$). The solid line shows the results for $H = 0$, the dashed line — for $H = 0.2$. For this value of H the dashed line practically coincides with the solid one.

In Figs. 3.3–3.6, the temperature dependencies of magnetic susceptibilities and Sommerfeld coefficients of the specific heat ($\gamma = c/T$) at zero external magnetic field for Heisenberg–Ising spin-$\frac{1}{2}$ chains with an "easy-plane" magnetic anisotropy are presented. These figures manifest how the magnetic anisotropy changes the temperature behaviour of a Heisenberg–Ising chain. Notice, that for an isotropic ferromagnetic Heisenberg chain the magnetic susceptibility follows a Curie-like law.

3.4 The Ground State Behaviour

The other important limit which permits us to obtain an analytic solution of thermodynamic Bethe ansatz equations, is the limit of $T \to 0$. From Eqs. (3.46), we have

$$\varepsilon_m(\lambda) = (m-1)H + Ta_m * \ln(1 + \exp[\varepsilon_1(\lambda)/T]) \\ + T\sum_{n=2}^{\infty}(A_{m-1,n-1} - \delta_{m,n}\lim_{n\to 0} a_n) * \ln(1 + \exp[-\varepsilon_n(\lambda)/T]) \ , \tag{3.51}$$

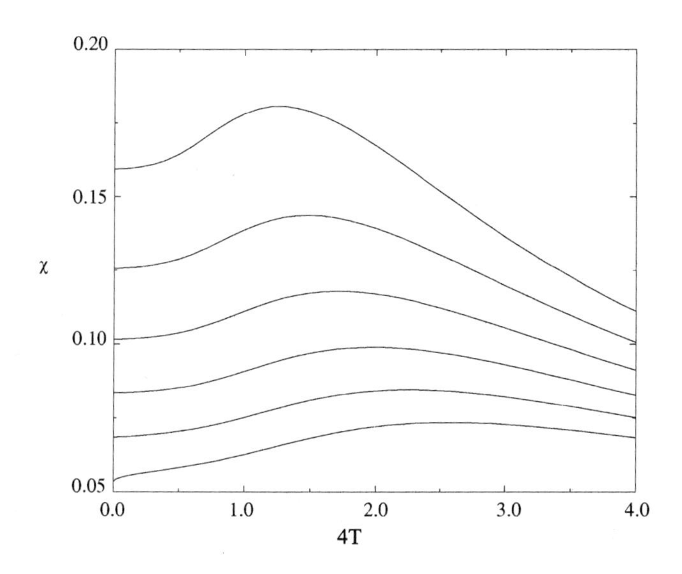

Fig. 3.3 The magnetic susceptibility at $H = 0$ of an antiferromagnetic Heisenberg–Ising spin-$\frac{1}{2}$ chain with $J = 2$ and an "easy-plane" magnetic anisotropy. The anisotropy $J_z/J = 0, 0.2, \ldots, 1$ for curves from top to bottom.

which implies that $\varepsilon_m(\lambda) > 0$ for $m = 2, 3, \ldots$ for the most interesting antiferromagnetic situation $J > 0$. [Notice that in the ferromagnetic situation $J < 0$, $\varepsilon_m(\lambda) > 0$ for all lengths of strings $m = 1, 2, 3, \ldots$, and the ground state energy is just e_0. This ferromagnetic state can also be referred to as the string of the infinite length for up spins.] Then one can introduce the positive and negative parts of $\varepsilon_1(\lambda)$ as: $\varepsilon_1^+(\lambda) = \varepsilon_1(\lambda)$ for $\varepsilon_1(\lambda) > 0$ and $\varepsilon_1^+(\lambda) = 0$ for $\varepsilon_1(\lambda) \leq 0$, and $\varepsilon_1^-(\lambda) = \varepsilon_1(\lambda)$ for $\varepsilon_1(\lambda) < 0$ and $\varepsilon_1^-(\lambda) = 0$ for $\varepsilon_1(\lambda) \geq 0$. By using these functions in the limit $T \to 0$, we obtain

$$\varepsilon_1(\lambda) + a_2 * \varepsilon_1^-(\lambda) = \varepsilon_1^{(0)}(\lambda) , \tag{3.52}$$

which can be re-written as (here we use the fact that the distribution of quantum numbers is symmetric with respect to zero)

$$\varepsilon_1(\lambda) + \int_{-B}^{B} d\lambda' \frac{\varepsilon_1^-(\lambda)}{\pi[(\lambda - \lambda')^2 + 1]} = H - \frac{J}{(2\lambda)^2 + 1} . \tag{3.53}$$

This is, in fact, the determination of the Dirac (Fermi) sea for quasiparticles (spinons) with dressed energies $\varepsilon_1(\lambda)$. The ground state pertains to the situation in which all states of spinons with negative energies are filled and all states with positive energies are empty. The limits of integration are determined from the natural conditions $\varepsilon_1(\pm B) = 0$. This means that $\pm B$

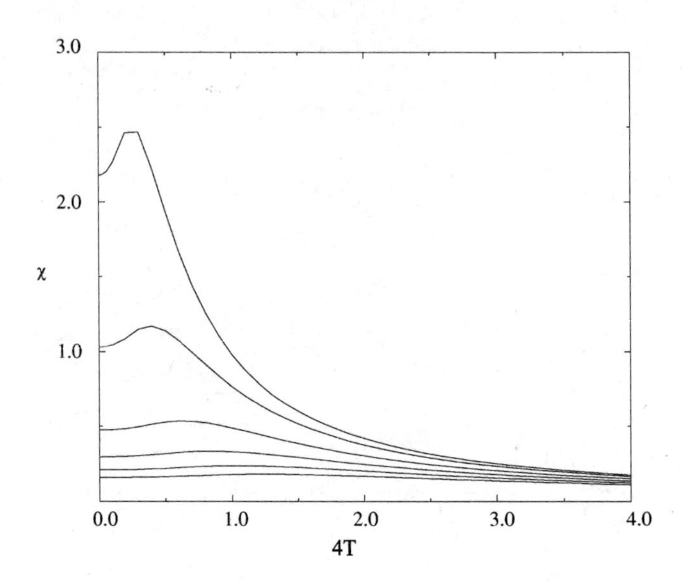

Fig. 3.4 The same as in Fig. 3.3, but for a ferromagnetic "easy-plane" Heisenberg–Ising spin-$\frac{1}{2}$ chain. The anisotropy $J_z/J = -0.9, -0.8, -0.6\dots, 0$ for curves from top to bottom.

can be considered as the Fermi points for quasiparticles (spinons). Then low-lying excitations for the antiferromagnetic Heisenberg spin-$\frac{1}{2}$ chain are quasiparticles with positive energies $\varepsilon_1(\lambda)$ and holes in the Fermi sea for quasiparticles with negative energies. It is important to emphasize that despite the symmetry of the wave function is *not* antisymmetric for permutation of two quasiparticles for any J_z and J, the statistical behaviour of quasiparticles is of the Fermi-Dirac type. As we shall show later, this property is the general property of Bethe ansatz solvable models. For the case of spin-$\frac{1}{2}$ systems this property has the natural origin, though, it is impossible to have more than one spin turn in each site of the lattice for spins $\frac{1}{2}$. The equations for densities stem from the equations for dressed energies. In the ground state, $T = 0$, we have

$$\rho_1(\lambda) + \rho_1^{(h)}(\lambda) = a_1(\lambda) - \int_{-B}^{B} d\lambda' a_2(\lambda - \lambda')\rho_1(\lambda') \ . \tag{3.54}$$

The ground state internal energy can be written as

$$e_{T=0} = e_0 + \int_{-B}^{B} d\lambda \frac{\varepsilon_1^-(\lambda)}{2\pi} = e_0 + \int_{-B}^{B} d\lambda \varepsilon_1^{(0)}(\lambda)\rho_1(\lambda) \tag{3.55}$$

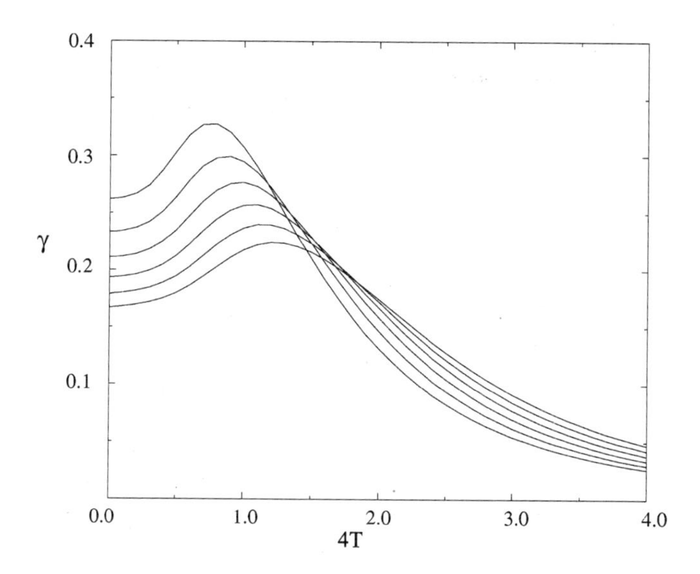

Fig. 3.5 The Sommerfeld coefficient of the specific heat at $H = 0$ for an antiferromagnetic Heisenberg spin-$\frac{1}{2}$ chain with $J = 2$ and an "easy-plane" magnetic anisotropy. The anisotropy $J_z/J = 0, 0.2, \ldots, 1$ for curves at high temperatures from bottom to top.

and the ground state magnetization is equal to

$$m^z = \frac{1}{2} - \int_{-B}^{B} d\lambda \rho_1(\lambda) . \tag{3.56}$$

Obviously, the value of the magnetic field H determines these limits of integration (*i.e.*, Fermi points). The reader can see that for the antiferromagnetic situation $J > 0$ and large values of the external magnetic field $H > 2J$ (or $H < -2J$) the system is in the ferromagnetic (spin-saturated) state and $B = 0$. Naturally, it means that in these regions of values of H the ground state energy is equal to e_0, the magnetization has its nominal values $\pm\frac{1}{2}$, and the magnetic susceptibility is zero. On the other hand, in zero magnetic field, for the antiferromagnetic situation at $H = 0$, we have $B = \infty$, *i.e.*, quantum numbers (rapidities) fill the total interval (in the thermodynamic limit $L \to \infty$). In the ferromagnetic situation, $J < 0$, the point of the quantum phase transition is $H = 0$: any infinitesimal magnetic field removes the two-fold degeneracy of a ferromagnetic chain and the magnetization of the latter becomes nominal.

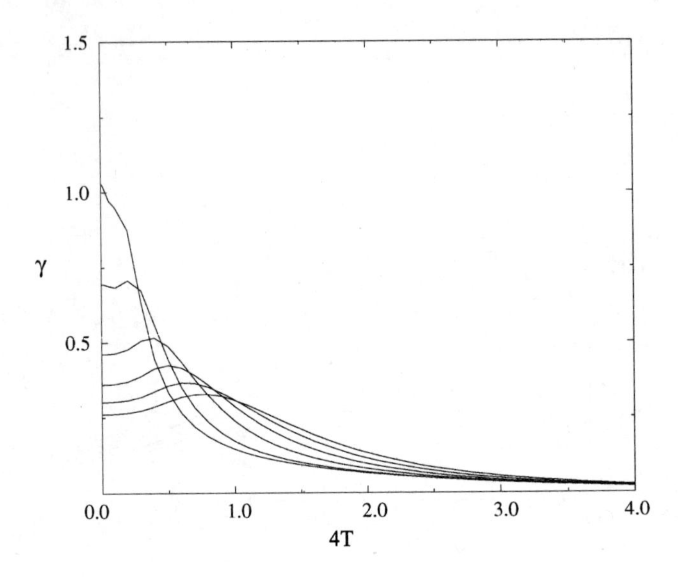

Fig. 3.6 The same as in Fig. 3.5, but for a ferromagnetic "easy-plane" Heisenberg–Ising spin-$\frac{1}{2}$ chain. The anisotropy $J_z/J = 0, -0.2, \ldots, -0.8, -0.9$ for curves at high temperatures from top to bottom.

3.5 Magnetic Field Behaviour in the Ground State: Wiener–Hopf Method

In this section we shall, for simplicity, consider the isotropic antiferromagnetic Heisenberg chain $J = J_z > 0$. For small B, *i.e.*, for $H - H_s \ll H_s$, where H_s is the critical value of the magnetic field at which the quantum phase transition takes place, we can evaluate the Helmholtz free energy by the direct iteration of Eq. (3.54). Thus, one gets the series

$$f = e_0 - \frac{\pi^2 J}{6}\left(\frac{M}{L}\right)^3 + H\left(\frac{M}{L}\right) + O\left(\left[\frac{M}{L}\right]^4\right) . \qquad (3.57)$$

The answer for the Helmholtz free energy for small values of the magnetic field can be also written analytically. Here we can apply the powerful *Wiener–Hopf method* of solutions of linear integral equations. We shall use this method in what follows very frequently, this is why, we want to present it here in detail.

First, to prepare for the use of the Wiener–Hopf method in our case, let us re-write the integral equation for dressed energies in the following way in the Fourier space

$$\varepsilon_m^+(\omega) = 2\pi m H\delta(\omega) - \frac{\pi J A_{1m}(\omega)}{2\cosh(\omega/2)} - A_{1m}(\omega)\varepsilon_m^-(\omega) , \tag{3.58}$$

where

$$A_{nm}(\omega) = \coth(\omega/2)[\exp[-|\omega||n-m|/2] - \exp[-|\omega|(n+m)/2]] . \tag{3.59}$$

Notice that for $H = 0$ it immediately follows $\varepsilon_m(\lambda) = 0$, for any $m > 1$ and $\varepsilon_1(\lambda) = -\pi J/2\cosh(\pi\lambda)$. Let us multiply Eq. (3.58) by $A_{11}(\omega)^{-1}$, where

$$A_{nm}(\omega)^{-1} = \delta_{n,m} - (\delta_{n,m+1} + \delta_{n,m-1})/2\cosh(\omega/2) . \tag{3.60}$$

Then we take the inverse Fourier transform which yields

$$\varepsilon_1(\lambda) = \frac{H}{2} - \frac{\pi J}{2\cosh(\pi\lambda)} + \left(\int_{-\infty}^{-B} + \int_B^{\infty}\right) d\lambda' J(\lambda-\lambda')\varepsilon_1(\lambda) , \tag{3.61}$$

where

$$J(x) = \int_{-\infty}^{\infty} \frac{d\omega}{2\pi} \frac{e^{-(|\omega|/2)-i\omega x}}{2\cosh(\omega/2)} . \tag{3.62}$$

Now we define $y(\lambda) = \varepsilon_1(\lambda+B)$, so that the Fermi point for dressed energies corresponds to $y(0)$. Then we re-write Eq. (3.61) as follows (here we use the identity $\varepsilon_1(\lambda) = \varepsilon_1(-\lambda)$)

$$\begin{aligned} y(\lambda) = \frac{H}{2} - \frac{\pi J}{2\cosh[\pi(\lambda+B)]} + \int_0^{\infty} d\lambda' J(\lambda-\lambda') y(\lambda') \\ + \int_0^{\infty} d\lambda J(\lambda+\lambda'+2B) y(\lambda') . \end{aligned} \tag{3.63}$$

If $H \ll J$, then B is very large and $J(\lambda+\lambda'+2B) \sim B^{-1}$. Hence, the last term in Eq. (3.63) is order B^{-1} smaller than the previous ones. We can, then, solve Eq. (3.63) iteratively $y(\lambda) = y_1(\lambda) + y_2(\lambda) + \ldots$, where

$$\begin{aligned} y_1(\lambda) &= \frac{H}{2} - \frac{\pi J}{2\cosh[\pi(\lambda+B)]} + \int_0^{\infty} d\lambda' J(\lambda-\lambda') y_1(\lambda') \\ y_2(\lambda) &= \int_0^{\infty} d\lambda' J(\lambda-\lambda') y_2(\lambda') + \int_0^{\infty} d\lambda J(\lambda+\lambda'+2B) y_1(\lambda') \end{aligned} \tag{3.64}$$

etc. These equations have the Wiener–Hopf structure and can be solved analytically.

Let us divide y into positive y^+ $(\lambda > 0)$ and negative y^- $(\lambda < 0)$ parts. It yields for the Fourier transform of the equation for y_1

$$\frac{y_1^+(\omega)}{A_{11}(\omega)} + y_1^-(\omega) = \pi H\delta(\omega) - \frac{\pi J \exp(i\omega B)}{2\cosh(\omega/2)} \,. \tag{3.65}$$

To apply the Wiener–Hopf method we re-write the kernel $(A_{11})^{-1}$ as a product of two functions, one, $G_1^+(\omega)$, being analytic in the upper half-plane, and the other one, $G_1^-(\omega)$, being analytic in the lower half-plane, where

$$G_1^+(\omega) = G_1^-(-\omega) = \frac{1}{\sqrt{2\pi}}\left(\frac{-i\omega+0}{2\pi e}\right)^{i\omega/2\pi}\Gamma[(1/2) - i(\omega/2\pi)] \,. \tag{3.66}$$

Observe that $G_1^{\pm}(\infty)$ is a constant. The Wiener–Hopf method uses the fact that from the analyticity of the functions $y_1^{\pm}(\omega)$ and $G_1^{\pm}(\omega)$ it follows that

$$y_1^+(\omega) = -\frac{q^+(\omega)}{G_1^+(\omega)} \,, \; y_1^-(\omega) = \frac{q^-(\omega)}{G_1^-(\omega)} \,, \tag{3.67}$$

where

$$q^{\pm}(\omega) = \frac{-iH}{\sqrt{2}(\omega \pm i0)} - \frac{iJ}{2}\int_{-\infty}^{\infty}\frac{d\omega'}{2\pi}\frac{\Gamma[(1/2)+i(\omega'/2\pi)]\Gamma[(1/2)-i(\omega'/2\pi)]e^{-i\omega' B}}{G_1^-(\omega')(\omega'-\omega \mp i0)} \,. \tag{3.68}$$

We are interested in the results for large positive B, hence, the contour of integration can be closed through the lower half-plane. Then, the value of the integral can be given as the sum of the residua of $\Gamma[(1/2) - i(\omega'/2\pi)]$. The leading term, *i.e.*, the pole closest to the real axes, yields the result (the next term is of the order of $\exp(-2\pi B)$ smaller)

$$y_1^+(\omega) = \frac{i}{\sqrt{2}G_1^+(\omega)}\left(\frac{H}{\omega + i0} - \frac{\pi^2 J\exp(-\pi B)}{\sqrt{2e}\Gamma(3/2)(\omega + i\pi)}\right) \,. \tag{3.69}$$

From the Fourier transform of the equation for y_2, it stems

$$y_2^+(\omega)G_1^+(\omega) + \frac{y_2^-(\omega)}{G_1^-(\omega)} = \frac{[1 - G_1^+(\omega)G_1^-(\omega)]\exp(-i2\pi B\omega)y_1^+(-\omega)}{G_1^-(\omega)} \,. \tag{3.70}$$

The analyticity of $y^{\pm}_{1,2}(\omega)$ and $G^{\pm}_1(\omega)$ implies

$$y^+_2(\omega)G^+_1(\omega) = -i\int_{-\infty}^{\infty}\frac{d\omega'}{2\pi}\frac{[1-G^+_1(\omega')G^-_1(\omega')]\exp(-i2\pi B\omega')y^+_1(-\omega')}{G^-_1(\omega')(\omega'-\omega-i0)}\,. \quad (3.71)$$

Again, we use the fact that $0<1\ll B$ and close the contour through the lower half-plane. In this half-plane only $G^+_1(\omega)$ has singularities. The leading singularity is the cut along the imaginary axis.

The parameter B is the function of the applied magnetic field H. It is determined from the condition $y(\lambda=0)=0$, which is equivalent to the condition $\lim_{\omega\to\infty}\omega y^+(\omega)=0$. By using the results for $y^+_{1,2}(\omega)$ we get

$$H\left[1+\frac{1}{2\pi B}-\frac{\ln(2\pi B)}{2(2\pi B)^2}+\cdots\right]=\frac{\pi^2 J\exp(-\pi B)}{\sqrt{2e}\Gamma(3/2)}\,. \quad (3.72)$$

Then the Helmholtz free energy can be written as

$$f(T=0,H\ll J)=e_0+\frac{J}{2}[\psi(1/2)-\psi(1)]-J\int_{-\infty}^{\infty}\frac{d\omega}{2\pi}\frac{e^{i\omega B}y^+(\omega)}{2\cosh(\omega/2)}\,. \quad (3.73)$$

The contour has to be closed through the upper half-plane. Then the value of the integral is given by the sum of the residua of the poles of $1/\cosh(\omega/2)$, any pole $\omega=i(2n+1)\pi$ yields the term $\sim(H/J)^{2n+2}$. The leading contribution arises from the closest to the real axis pole, $\omega=i\pi$, and it gives $-\exp(-\pi B)y^+(i\pi)\sim(H/J)^2$. Then the final answer for the Helmholtz free energy of the Heisenberg antiferromagnetic spin-$\frac{1}{2}$ chain in a weak magnetic field is

$$\begin{aligned}f(T=0,H\ll J)&=e_0+\frac{J}{2}[\psi(1/2)-\psi(1)]-\frac{H^2}{2\pi^2 J}\\&\times\left(1+\frac{1}{2\ln(AH/J)}-\frac{\ln|\ln(AH/J)|}{4\ln^2(AH/J)}+\cdots\right)\,, \end{aligned}\quad (3.74)$$

where $A=\sqrt{2e}\Gamma(3/2)/\pi^2$ is a constant. This implies the behaviour of the magnetic susceptibility

$$\chi=\frac{1}{\pi^2 J}\left(1+\frac{1}{2\ln(AH/J)}-\frac{\ln|\ln(AH/J)|}{4\ln^2(AH/J)}+\cdots\right)\,, \quad (3.75)$$

which is valid for even L. The antiferromagnetic spin-$\frac{1}{2}$ chain with the odd number of sites reveals the ground state and low-temperature behaviour, different from the above discussed. The reason for this difference is clear, there is a remnant spin-$\frac{1}{2}$, hence, the ground state is not a singlet at zero magnetic field. The ground state is degenerate for $H = 0$ with two different momenta $P = -\pi S(1 \pm L^{-1}) + \frac{1}{2}(1 \pm 1)$ mod 2π, where S is the eigenvalue of the operator of the z-projection of the total spin $\sum_{j=1}^{L} S_j^z$. For example, for the isotropic XY chain the ground state energy at $H = 0$ is

$$E = -\frac{J\cos(\pi/L)\cos(\pi S/L)}{L\sin(\pi/L)} . \tag{3.76}$$

Naturally, the remnant magnetization for $H \to 0$ for odd L implies the divergent ground state susceptibility, different from the above expression for even L.

To summarize, in this chapter we presented the co-ordinate Bethe ansatz for quantum spin-$\frac{1}{2}$ chains (Heisenberg–Ising chains) with periodic boundary conditions. We derived sets of transcendental equations for quantum numbers (rapidities), which parametrize eigenfunctions and eigenfunctions of stationary Schrödinger equations for these models. The difference in the behaviour of interacting systems and noninteracting ones (*e.g.*, XY chains) appears to be in the presence of bound states (complex solutions to Bethe anzatz equations), and in the distribution of rapidities, which depends on interactions, for real solutions. We considered the way of description of solutions of Bethe ansatz equations in the thermodynamic limit. In the framework of the string hypothesis thermodynamic Bethe ansatz integral equations are derived for (dresssed by interactions) densities of rapidities and dressed energies of all states. High-temperature solutions to those equations are presented. The transition to the ground state shows how the Fermi (Dirac) seas for these interacting models are organized. Finally, in the framework of the Wiener–Hopf method we analytically derived the Helmholtz free energy in the ground state as a function of an external magnetic field.

The method, presented in this chapter was pioneered in [Bethe (1931)]. The scheme of the solution of Bethe ansatz equations in the thermodynamic limit was given in [Hulthén (1938)]. Bethe ansatz equations for a Heisenberg–Ising chain in the ground state without external magnetic field were studied in [Orbach (1958)]. The ground state behaviour of a Heisenberg–Ising chain in an external magnetic field was studied in [Yang and Yang (1966a); Yang and Yang (1966b); Yang and Yang (1966c)]. Application of the Wiener–Hopf method for the ground state behaviour of a Heisenberg–Ising spin chain in a weak magnetic field was introduced there. In those papers the reader can also find the proofs of important theorems, which we presented above. Analysis of Bethe ansatz equations for Heisenberg–Ising chains can be also found in the books [Gaudin (1983); Izyumov and Skryabin (1990)]. Thermodynamic Bethe ansatz method was introduced in [Yang and Yang (1969)]. The introduction of the string hypothesis for Bethe ansatz-solvable models of condensed matter physics was reviewed in the excellent book [Takahashi (1999)].

Chapter 4

Correlated Electron Chains: Co-ordinate Bethe Ansatz

Quantum spin systems, considered in previous chapters, describe only spin dynamics of correlated electrons. However, it is interesting and important to also understand charge dynamics of correlated electron systems.

Usually there are two main energetical scales in the behaviour of electrons: the width of the band of itinerant electrons (and related to it characteristic velocity of electrons or the Fermi energy of electrons) and the strength of the Coulomb repulsion between electrons. If the former is much larger than the Coulomb repulsion, then electrons can be considered as a free lattice gas of itinerant electrons with Bloch-like wave functions. The weak interaction between electrons can be treated in the framework of perturbation theories. This kind of theory is well developed. The other limiting case, which is studied even better than the previous situation, is the atomic (localized) behaviour of electrons, where the effect of the Coulomb interaction is considered exactly, and the hopping of electrons between lattice sites can be considered perturbatively. In such a case wave functions of electrons are of Wannier-type rather than Bloch-like. However, the most interesting situation pertains to the case in which the energy of the hopping of electrons from site to site of the crystal lattice (which characteristic energy is the bandwidth of electrons) is of the same order as the strength of the repulsion between electrons. Here correlation effects and itinerant effects interfere with each other, which results in a reach behaviour of such systems: they can reveal metal-insulator phase transitions, heavy fermion behaviour, very special magnetic behaviour *etc.* However, the theoretical description of such a situation is very difficult. Why is it so? As we already mentioned, the well-developed methods of theoretical physics like perturbation theories cannot be applied in this region of parameters. On the other hand, in most of cases one cannot *a priori* state that there is any kind of ordering in these systems. Thus, the mean-field like methods, the other powerful approach of the theoretical physics, cannot also be applied in most cases. Then the only possibility is to study some approximate models of correlated electrons, which manifest both the itinerant and correlated nature of electrons, using exact methods. This program can be realized, for example, in one-dimensional models of correlated electrons.

4.1 Hubbard Chain

The *Hubbard model* was introduced (usually it is connected with the names of M. C. Gutzwiller, J. Hubbard and J. Kanamori) as a simple effective model for the treatment of correlation effects in metals. It is believed to provide a qualitative description of magnetic properties of correlated electron systems and possible metal-insulator transitions. It's Hamiltonian consists of the term, which describes the hopping of electrons between (usually neighbouring) sites of the lattice, and the term, which describes the Coulomb interaction between electrons in the simplest approximation: electrons with different spins can affect each other only locally, being at the same site of the lattice. Even for such great simplifications and its conceptual simplicity the Hubbard model does not permit us to obtain explicit results in any space dimension. The rare exclusion is the one-dimensional case, where the exact solution in the framework of the Bethe ansatz was obtained by E. H. Lieb and F. Y. Wu. In their study they used the *nested Bethe ansatz* scheme, discovered by M. Gaudin and C. N. Yang for a more simple continuous model of electrons with the local (so called δ-function) interaction.

The Hamiltonian of the one-dimensional Hubbard model can be written as:

$$\mathcal{H}_H = -t\sum_{j=1}^{L-1}\sum_{\sigma}(a_{j,\sigma}^{\dagger}a_{j+1,\sigma} + \text{H.c.}) + U\sum_{j=1}^{L} n_{j,\uparrow}n_{j,\downarrow} \ , \tag{4.1}$$

where $a_{j,\sigma}^{\dagger}$ ($a_{j,\sigma}$) creates (destroys) an electron with the spin projection $\sigma = \pm 1 \equiv \uparrow, \downarrow$ (it is used to simply explain that electrons can have up spins or down spins) at the lattice site j, $n_{j,\sigma} = a_{j,\sigma}^{\dagger}a_{j,\sigma}$, t is the hopping integral (in what follows we put it equal to unity, $t = 1$, *i.e.*, we shall measure all other energies in units of t) and U is the constant of the Hubbard interaction.

The Hilbert space of each site of the lattice realizes four possibilities: there can be an empty site without electrons, one electron sitting at the site (two possibilities because of two spins of electrons) and two electrons sitting at the same site (sometimes it is referred to as a local pair). This is why the total number of states for the lattice of L sites is 4^L. It turns out that the total numbers of electrons with spins up and down are conserved (the commutators of their operators with the Hubbard Hamiltonian are equal to zero $[\mathcal{H}_H, \sum_j n_{j,\sigma}] = 0$), and, therefore, one can classify all eigenstates with quantum numbers, related to those integrals of motion. In the Bethe ansatz scheme the convenient choice of these numbers is the total number of electrons, N, and the number of electrons with down spins, M. Let us consider the situation with $0 \leq N \leq L$ and $0 \leq M \leq N/2$. Other cases can be obviously obtained from this one by using unitary transformations (*e.g.*, by turning spins, or using a particle-hole transformation). We can consider the wave function

$$\Psi = \sum_{x_1 < x_2 < \cdots < x_N} \psi(x_1, \ldots, x_N, \sigma_1, \ldots, \sigma_N) a^\dagger_{x_1,\sigma_1} a^\dagger_{x_2,\sigma_2} \cdots a^\dagger_{x_N,\sigma_N} |0\rangle \ , \tag{4.2}$$

where the state $|0\rangle$ is taken such that $a_{j,\sigma}|0\rangle = 0$ for any j and σ. This wave function is very similar to the wave function of the Heisenberg–Ising chain, *cf.* the previous chapter. The difference is that now not only spin degrees of freedom can be spread through the lattice, but electrons themselves can move. The stationary Schrödinger equation for the wave function in the co-ordinate representation can be written as (we use here periodic boundary conditions)

$$\begin{aligned} E\psi(x_1, \ldots, x_N, \sigma_1, \ldots, \sigma_N) + \sum_j \psi(x_1, \ldots, x_j \pm 1, \ldots, x_N, \sigma_1, \ldots, \sigma_N) \\ + U \sum_{j<l} \delta_{x_j,x_l} \delta_{\sigma_j,-\sigma_l} \psi(x_1, \ldots, x_N, \sigma_1, \ldots, \sigma_N) = 0 \ . \end{aligned} \tag{4.3}$$

Let us again consider the simple cases to understand the situation. In the case $N = 1$, one has the equation

$$-\psi(x-1,\sigma) - \psi(x+1,\sigma) = E\psi(x,\sigma) \ , \tag{4.4}$$

which has the trivial solution (it is, naturally, doubly degenerate, because nothing depends on σ; this degeneracy can be removed when adding the Zeeman term $-(H/2)\sum_j (n_{j,\uparrow} - n_{j,\downarrow})$ to the Hubbard Hamiltonian, which, naturally, commutes with the latter, and, hence, it has the same eigenfunctions):

$$\begin{aligned}\psi(x,\sigma) &= A\exp(ikx) \ , \\ E &= -2\cos k \ ,\end{aligned} \tag{4.5}$$

where A is determined from the normalization condition, and the momentum k stems from the periodic boundary conditions $\exp(ikL) = 1$.

For the case $N = 2$, we again must distinguish two situations. If two electrons are situated at different sites, we have (there is no interaction in this case)

$$\begin{aligned}-\psi(x_1+1,x_2,\sigma_1,\sigma_2) - \psi(x_1-1,x_2,\sigma_1,\sigma_2) - \psi(x_1,x_2+1,\sigma_1,\sigma_2) \\ - \psi(x_1,x_2-1,\sigma_1,\sigma_2) = E\psi(x_1,x_2,\sigma_1,\sigma_2) \ ,\end{aligned} \tag{4.6}$$

which has the simple solution

$$\begin{aligned}\psi(x_1,x_2,\sigma_1,\sigma_2) &= A_{\sigma_1,\sigma_2}e^{i(k_1x_1+k_2x_2)} - A'_{\sigma_1,\sigma_2}e^{i(k_2x_1+k_1x_2)} \ , \\ E &= -2(\cos k_1 + \cos k_2) \ ,\end{aligned} \tag{4.7}$$

with arbitrary coefficients A, A' (to be determined from the normalization condition), and the values of $k_{1,2}$ are also determined from periodic boundary conditions $\exp(ik_{1,2}L) = 1$. Notice that the sign in the wave function is due to the antisymmetry of the wave function of two fermions.

A more interesting case is when $x_1 = x_2$. Now the Schrödinger equation for the wave function in the co-ordinate representation has the form

$$\begin{aligned}-\psi(x+1,x,\sigma_1,\sigma_2) - \psi(x-1,x,\sigma_1,\sigma_2) - \psi(x,x+1,\sigma_1,\sigma_2) \\ - \psi(x,x-1,\sigma_1,\sigma_2) + U\delta_{\sigma_1,-\sigma_2}\psi(x,x,\sigma_1,\sigma_2) = E\psi(x,x,\sigma_1,\sigma_2) \ .\end{aligned} \tag{4.8}$$

For equal spins of electrons $\sigma_1 = \sigma_2$, the solution coincides with the above solution for electrons in different sites. For $\sigma_1 \neq \sigma_2$, the interaction reveals itself: when electrons occupy the same site, they "feel" each other. This can be considered in terms of a scattering process, similar to the case of the Heisenberg–Ising chain, *i.e.*, the constants A become not independent, but connected to each other due to the interaction. This can be formally achieved by taking $A_{\sigma_1,\sigma_2}(k_1,k_2)$ dependent on the region $x_1 < x_2$, or $x_1 > x_2$:

$$\begin{aligned}\psi(x_1,x_2,\sigma_1,\sigma_2) = A_{\sigma_1,\sigma_2}(k_1,k_2)e^{i(k_1x_1+k_2x_2)} \\ - A_{\sigma_1,\sigma_2}(k_2,k_1)e^{i(k_2x_1+k_1x_2)} \ ,\end{aligned} \tag{4.9}$$

for $x_1 < x_2$ and

$$\psi(x_1, x_2, \sigma_1, \sigma_2) = A_{\sigma_2,\sigma_1}(k_1, k_2)e^{i(k_1x_1+k_2x_2)} - A_{\sigma_2,\sigma_1}(k_2, k_1)e^{i(k_2x_1+k_1x_2)} , \quad (4.10)$$

for $x_1 > x_2$. It can be written in the compact way as

$$\psi(x_1, x_2, \sigma_1, \sigma_2) = \sum_P \text{sign}(PQ)A_{\sigma_{Q_1},\sigma_{Q_2}}(k_{P_1}, k_{P_2})e^{i\sum_{j=1}^2 k_{P_j}x_{Q_j}} , \quad (4.11)$$

where $P = (P_1, P_2)$ is a permutation of the momenta labels 1,2, (*i.e.*, it is the element of the symmetric group S_2), and $Q = (Q_1, Q_2)$ is the permutation of the labels of co-ordinates (it is assumed that $x_{Q_1} \leq x_{Q_2}$). The case $x_1 = x_2$ requires the single-valuedness (continuity of the wave function),

$$A_{\sigma_1,\sigma_2}(k_1, k_2) - A_{\sigma_1,\sigma_2}(k_2, k_1) = A_{\sigma_2,\sigma_1}(k_1, k_2) - A_{\sigma_2,\sigma_1}(k_2, k_1) . \quad (4.12)$$

Then the wave function equation (4.11) is the eigenfunction of the Hubbard Hamiltonian in the co-ordinate representation, if the following equation holds:

$$A_{\sigma_1,\sigma_2}(k_2, k_1) = \frac{i(U/2)A_{\sigma_1,\sigma_2}(k_1, k_2) + (\sin k_1 - \sin k_2)A_{\sigma_2,\sigma_1}(k_1, k_2)}{\sin k_1 - \sin k_2 + i(U/2)} . \quad (4.13)$$

This condition can be re-written in a compact way as

$$A_{\sigma_2,\sigma_1}(k_2, k_1) = \sum_{\tau_1,\tau_2} S_{\sigma_2\tau_2}^{\sigma_1\tau_1}(k_1, k_2)A_{\tau_1,\tau_2}(k_1, k_2) , \quad (4.14)$$

where $S_{\sigma_2\tau_2}^{\sigma_1\tau_1}(k_1, k_2)$ is the *two-particle scattering matrix*:

$$S_{\sigma_2\tau_2}^{\sigma_1\tau_1}(k_1, k_2) = \frac{(\sin k_1 - \sin k_2)I_{\sigma_2\tau_2}^{\sigma_1\tau_1} + i(U/2)\Pi_{\sigma_2\tau_2}^{\sigma_1\tau_1}}{\sin k_1 - \sin k_2 + i(U/2)} , \quad (4.15)$$

where we introduced the identity operator $I_{\sigma_2\tau_2}^{\sigma_1\tau_1} = \delta_{\sigma_2\tau_2}\delta_{\sigma_1\tau_1}$ and the permutation operator $\Pi_{\sigma_2\tau_2}^{\sigma_1\tau_1} = \delta_{\sigma_1\tau_2}\delta_{\sigma_2\tau_1}$.

It is straightforward but quite tedious to generalize the above consideration to the N-electron problem. There are $N!$ possible arrangements of $x_1, \ldots, x_N$, and, hence, $N!$ space sectors to be matched at their common boundaries. In analogy with Eq. (4.11), we can write the eigenfunction of the Hubbard chain for the general N case as

$$\psi(x_1,\ldots,x_N,\sigma_1,\ldots,\sigma_N) = \sum_P \operatorname{sign}(PQ) A_{\sigma_{Q_1},\ldots,\sigma_{Q_N}}(k_{P_1},\ldots,k_{P_N}) \exp\left(i\sum_{j=1}^{N} k_{P_j}x_{Q_j}\right) . \quad (4.16)$$

Substituting it into the stationary Schrödinger equation we obtain the expression for the eigenvalue

$$E = \frac{U(L-2N)}{4} - 2\sum_{j=1}^{N} \cos k_j \ . \quad (4.17)$$

Using the condition of the single-valuedness of the wave function and solving the matching conditions, *i.e.*, the Schrödinger equation for the cases where two of the coordinates coincide, one gets

$$A_{\sigma_{Q'_1},\ldots,\sigma_{Q'_N}}(k_{P'_1},\ldots,k_{P'_N}) = \sum_{\tau_1,\tau_2} S^{\sigma_{Q_j}\tau_1}_{\sigma_{Q_{j+1}}\tau_2}(k_{P_j},k_{P_{j+1}}) A_{\sigma_{Q_1},\ldots,\sigma_{Q_{j-1}},\tau_1,\tau_2,\sigma_{Q_{j+2}},\ldots,\sigma_{Q_N}}(k_{P'_1},\ldots,k_{P'_N}) \ , \quad (4.18)$$

where Q and P are arbitrary permutations and $Q' = Q(j,j+1)$, $P' = P(j,j+1)$.

Periodic boundary conditions for $N=2$, *i.e.*,

$$\begin{aligned} \psi(L+1,x_2,\sigma_1,\sigma_2) &= \psi(1,x_2,\sigma_1,\sigma_2) \ , \\ \psi(0,x_2,\sigma_1,\sigma_2) &= \psi(L,x_2,\sigma_1,\sigma_2) \ , \\ \psi(x_1,L+1,\sigma_1,\sigma_2) &= \psi(x_1,1,\sigma_1,\sigma_2) \ , \\ \psi(x_1,0,\sigma_1,\sigma_2) &= \psi(x_1,L,\sigma_1,\sigma_2) \end{aligned} \quad (4.19)$$

imply the following equations

$$A_{\sigma_{Q_1},\sigma_{Q_2}}(k_{P_1},k_{P_2}) = e^{ik_{P_1}L} A_{\sigma_{Q_2},\sigma_{Q_1}}(k_{P_2},k_{P_1}) \ . \quad (4.20)$$

If we have two electrons with spins up (or with spins down), periodic boundary conditions imply $\exp(ik_{1,2}L) = 1$, with the wave function in the sector Q: $\psi(x_1,x_2,\sigma_1,\sigma_2) = \sum_P \operatorname{sign}(PQ)\exp(i\sum_{j=1}^{2} k_{P_j}x_{Q_j})$. If we have one electron with spin up and one electron with spin down the situation is more complicated, and one needs to introduce the nested Bethe ansatz.

The generalization for the N-electron case is straightforward. The periodic boundary conditions

$$\begin{aligned}
&\psi(x_1,\ldots,,x_{j-1},1,x_{j+1},\ldots,x_N,\sigma_1,\ldots,\sigma_N)\\
&\quad=\psi(x_1,\ldots,,x_{j-1},L+1,x_{j+1},\ldots,x_N,\sigma_1,\ldots,\sigma_N)\ ,\\
&\psi(x_1,\ldots,,x_{j-1},0,x_{j+1},\ldots,x_N,\sigma_1,\ldots,\sigma_N)\\
&\quad=\psi(x_1,\ldots,,x_{j-1},L,x_{j+1},\ldots,x_N,\sigma_1,\ldots,\sigma_N)\ ,
\end{aligned} \tag{4.21}$$

where $j=1,\ldots,N$, yield

$$\begin{aligned}
&A_{\sigma_{Q_1},\ldots,\sigma_{Q_N}}(k_{P_1},\ldots,k_{P_N})\\
&\quad=e^{ik_{P_1}L}A_{\sigma_{Q_2},\ldots,\sigma_{Q_N},\sigma_{Q_1}}(k_{P_2},\ldots,k_{P_N},k_{P_1})\ ,
\end{aligned} \tag{4.22}$$

where Q and P are arbitrary from S_N. These conditions can be re-written using two-particle scattering matrices as

$$e^{ik_jL}\xi=\sum_{\sigma'_1\ldots\sigma'_N}(T_j)^{\sigma_1\ldots\sigma_N}_{\sigma'_1\ldots\sigma'_N}(k_1,\ldots,k_N)\xi'\ , \tag{4.23}$$

where

$$\begin{aligned}
(T_j)^{\sigma_1\ldots\sigma_N}_{\sigma'_1\ldots\sigma'_N}(k_1,\ldots,k_N)&=\sum_{\tau_1\ldots\tau_{N-1}}(S^{\sigma_j\tau_1}_{\sigma_{j+1}\sigma'_{j+1}})^{-1}(k_j,k_{j+1})\\
&\times(S^{\tau_1\tau_2}_{\sigma_{j+2}\sigma'_{j+2}})^{-1}(k_j,k_{j+2})\cdots(S^{\tau_{j-2}\tau_{j-1}}_{\sigma_N\sigma'_N})^{-1}(k_j,k_N)\\
&\times(S^{\tau_{j-1}\tau_j}_{\sigma_1\sigma'_1})^{-1}(k_j,k_1)\cdots(S^{\tau_{N-1}\sigma'_j}_{\sigma_{j-1}\sigma'_{j-1}})^{-1}(k_j,k_{j-1})\ .
\end{aligned} \tag{4.24}$$

The vectors ξ,ξ' are composed of $N!$ coefficients $A_{\sigma_1,\ldots,\sigma_N}(k_1,\ldots,k_N)$ (which depend on P and Q). Including all coordinate permutations it is $N!\times N!\times N!$ such coefficients, which are not all independent but restricted by symmetries. It yields the solution for k_j, analogous to the case of N spins down for the Heisenberg–Ising chain, but one needs to distinguish $[N/2]+1$ cases (with $[A]$ being the integer part of A) corresponding to the possible values of M spins down.

We proceed further (with the help of the nested Bethe ansatz), introducing some auxiliary spin model on a one-dimensional ring of N sites. Every site allows two spin configurations, spin up and spin down, analogous to the ones for the Heisenberg–Ising chain. Let us introduce the following function

$$|k_{P_1},\ldots,k_{P_N}\rangle=\sum_{\sigma_1,\cdots,\sigma_N}A_{\sigma_1,\cdots,\sigma_N}(k_{P_1},\ldots,k_{P_N})|x_1,\ldots,x_M\rangle\ , \tag{4.25}$$

where $|x_1, \ldots, x_M\rangle$ denotes the wave function with M down spins at positions $x_1 < \cdots < x_M$, *cf.* the previous chapter devoted to the Heisenberg–Ising Hamiltonian of the spin-$\frac{1}{2}$ chain. Then the equality follows from the above definition of the two-particle scattering matrix

$$|k_{P_1}, k_{P_2}\rangle = Y_{1,2}(\sin k_{P_1}, \sin k_{P_2})|k_{P_2}, k_{P_1}\rangle \ , \tag{4.26}$$

where

$$Y_{1,2}(\lambda_1, \lambda_2) = \frac{(\lambda_1 - \lambda_2)I + i(U/2)\Pi_{12}}{\lambda_1 - \lambda_2 + i(U/2)} \tag{4.27}$$

with I and $\Pi_{12} = (I + 4\vec{S}_1\vec{S}_2)/2$ being the identity matrix and permutation operator for the Hilbert space of the auxiliary spin-$\frac{1}{2}$ model. The periodic boundary conditions Eq. (4.19) can be written as

$$|k_{P_1}, k_{P_2}\rangle = e^{ik_{P_1}L}\Pi_{12}|k_{P_2}, k_{P_1}\rangle \ , \tag{4.28}$$

or in the form

$$|k_{P_1}, k_{P_2}\rangle = e^{ik_{P_1}L}X_{12}(\sin k_{P_1}, \sin k_{P_2})|k_{P_1}, k_{P_2}\rangle \ , \tag{4.29}$$

where

$$X_{l,j}(\lambda_l, \lambda_j) = \Pi_{lj}Y_{l,j}(\lambda_l, \lambda_j) = \frac{(\lambda_l - \lambda_j)\Pi_{lj} + i(U/2)I}{\lambda_l - \lambda_j + i(U/2)} \ . \tag{4.30}$$

These equations can be straightforwardly generalized for the case of M spins down of N spins:

$$\begin{aligned}
&|k_{P_1}, \ldots, k_{P_N}\rangle = \exp(iLk_{P_1})\Pi_{12}\Pi_{23}\cdots\Pi_{N-1N}|k_{P_2}, \ldots, k_{P_N}, k_{P_1}\rangle \\
&= \exp(iLk_{P_1})\Pi_{12}\Pi_{23}\cdots\Pi_{N-1N}\prod_{m=0}^{N-2} Y_{N-m-1,N-m}(\sin k_{P_1}, \sin k_{P_{N-m}}) \\
&\times |k_{P_1}, \ldots, k_{P_N}\rangle = \exp(iLk_{P_1})X_{1,N}(\sin k_{P_1}, \sin k_{P_N}) \\
&\times X_{1,N-1}(\sin k_{P_1}, \sin k_{P_{N-1}})\cdots X_{1,2}(\sin k_{P_1}, \sin k_{P_2})|k_{P_1}, \ldots, k_{P_N}\rangle \ .
\end{aligned} \tag{4.31}$$

Now we need to distinguish $[N/2] + 1$ cases corresponding to the possible values of M.

Let us introduce the *monodromy matrix* (on the inhomogeneous lattice) as

$$T^{\sigma_1\ldots\sigma_N,\tau}_{\sigma'_1\ldots\sigma'_N,\tau'}(\lambda,\lambda^0_1,\ldots,\lambda^0_N)$$
$$= \sum_{\tau_1\ldots\tau_{N-1}} Y^{\tau\tau_1}_{\sigma_1\sigma'_1}(\lambda-\lambda^0_1)Y^{\tau_1\tau_2}_{\sigma_2\sigma'_2}(\lambda-\lambda^0_2)\cdots Y^{\tau_{N-1}\tau'}_{\sigma_N\sigma'_N}(\lambda-\lambda^0_N)\ , \quad (4.32)$$

where λ is a *spectral parameter*, the inhomogeneities are introduced via $\lambda^0_{1,\ldots,N}$, and

$$Y^{\sigma_1\tau_1}_{\sigma_2\tau_2}(x) = \frac{xI^{\sigma_1\tau_1}_{\sigma_2\tau_2} + i(U/2)\Pi^{\sigma_1\tau_1}_{\sigma_2\tau_2}}{x+i(U/2)}\ . \quad (4.33)$$

It turns out that (the summation over repeated indices is understood)

$$Y^{\sigma_1\sigma'_1}_{\sigma_2\sigma'_2}(x)Y^{\sigma'_1\sigma''_1}_{\sigma_3\sigma'_3}(x+y)Y^{\sigma'_2\sigma''_2}_{\sigma'_3\sigma''_3}(y) = Y^{\sigma_2\sigma'_2}_{\sigma_3\sigma'_3}(y)Y^{\sigma_1\sigma'_1}_{\sigma'_3\sigma''_3}(x+y)Y^{\sigma'_1\sigma''_1}_{\sigma'_2\sigma''_2}(x)\ , \quad (4.34)$$

which is well known as the famous *Yang–Baxter relation* for two-particle scattering matrices. The definition of the monodromy matrix can be re-written in a symbolic way by omitting spin indices as

$$T^{\tau}_{\tau'}(\lambda,\lambda^0_1,\ldots,\lambda^0_N) = Y_{01}(\lambda-\lambda^0_1)\cdots Y_{0N}(\lambda-\lambda^0_N)\ , \quad (4.35)$$

where the subscript 0 denotes the additional indices, which are summed over. Notice that $Y(x)Y(-x) = I$. With respect to the indices τ and τ' the monodromy matrix is 2×2 matrix. We define the trace of this 2×2 matrix as the transfer matrix on the inhomogeneous lattice, *i.e.*,

$$\hat{\tau}(\lambda,\lambda^0_1,\ldots,\lambda^0_N) = \mathrm{tr}T(\lambda,\lambda^0_1,\ldots,\lambda^0_N)$$
$$= \mathrm{tr}_0 Y_{01}(\lambda-\lambda^0_1)\cdots Y_{0N}(\lambda-\lambda^0_N)$$
$$\equiv \sum_{\tau_1\ldots\tau_{N-1}\tau} Y^{\tau\tau_1}_{\sigma_1\sigma'_1}(\lambda-\lambda^0_1)Y^{\tau_1\tau_2}_{\sigma_2\sigma'_2}(\lambda-\lambda^0_2)\cdots Y^{\tau_{N-1}\tau}_{\sigma_N\sigma'_N}(\lambda-\lambda^0_N)\ . \quad (4.36)$$

Let us denote the elements of the 2×2 monodromy matrix on the inhomogeneous lattice as

$$T^{\tau}_{\tau'}(\lambda,\lambda^0_1,\ldots,\lambda^0_N) = \begin{pmatrix} \hat{A} & \hat{B} \\ \hat{C} & \hat{D} \end{pmatrix}\ , \quad (4.37)$$

where the operators $\hat{A}$, $\hat{B}$, $\hat{C}$ and $\hat{D}$ in the matrix representation have indices $\sigma'_1,\ldots,\sigma'_N,\sigma_1,\ldots,\sigma_N$ and also depend on inhomogeneities $\lambda^0_{1,\ldots,N}$, but for simplicity we do not write that dependence explicitly. It follows from the definition that

$$\hat{\tau}(\lambda, \lambda_1^0, \ldots, \lambda_N^0) = \hat{A} + \hat{D} \ . \tag{4.38}$$

It is easy to show that

$$Y_{\tau_2\tau_2'}^{\tau_1\tau_1'}(\lambda - \lambda')T_{\tau_3}^{\tau_1'}(\lambda')T_{\tau_3'}^{\tau_2'}(\lambda) = T_{\tau_2'}^{\tau_2}(\lambda)T_{\tau_1'}^{\tau_1}(\lambda')Y_{\tau_2'\tau_3'}^{\tau_1'\tau_3}(\lambda - \lambda') \ , \tag{4.39}$$

which is the direct consequence of Eqs. (4.32) and (4.34). This equation is called the Yang–Baxter relation for monodromy matrices. Multiplying it from the left by $Y_{\tau_3'\tau_2}^{\tau_3\tau_1}(\lambda' - \lambda)$ and summing over the indices τ_1 and τ_2, we obtain

$$\hat{\tau}(\lambda')\hat{\tau}(\lambda) = \hat{\tau}(\lambda)\hat{\tau}(\lambda') \ , \tag{4.40}$$

where $\hat{\tau}(x)$ is the transfer matrix, which means that transfer matrices with different spectral parameters commute. It turns out that this result does not depend on inhomogeneities $\lambda_{1,\ldots,N}^0$.

The operators $\hat{A}$, $\hat{B}$, $\hat{C}$ and $\hat{D}$ obey the commutation relations, which stem from Eq. (4.39), some of which are relevant for the following consideration

$$\begin{aligned}
&[\hat{A}(x), \hat{A}(y)] = [\hat{D}(x), \hat{D}(y)] = [\hat{A}(x), \hat{D}(y)] = 0 \ , \\
&(x - y)\hat{A}(x)\hat{B}(y) = [x - y + i(U/2)]\hat{B}(y)\hat{A}(x) - i(U/2)\hat{B}(x)\hat{A}(y) \ , \\
&(y - x)\hat{D}(x)\hat{B}(y) = [y - x + i(U/2)]\hat{B}(y)\hat{D}(x) - i(U/2)\hat{B}(x)\hat{D}(y) \ , \\
&[\hat{B}(x), \hat{B}(y)] = [\hat{C}(x), \hat{C}(y)] = 0 \ .
\end{aligned} \tag{4.41}$$

Let us denote the state with no spins down as $|0\rangle$ (it is often referred to as the *mathematical vacuum*). Then the action of some matrix $Y_{0j}(x)$ (where the index j denotes the position of this matrix in the inhomogeneous lattice) on this vacuum state can be symbolically written as

$$Y_{0j}(x)|0\rangle = \frac{1}{x + i(U/2)} \begin{pmatrix} x + i(U/2) & iUS_j^- \\ 0 & x \end{pmatrix} |0\rangle \ . \tag{4.42}$$

It is important to emphasize that the lower left element of this matrix is zero, this is why such a form is usually called as *triangular matrix* form. From the definition of the monodromy matrix the operators $\hat{A}$, $\hat{B}$, $\hat{C}$ and $\hat{D}$ can be obtained by successive multiplications of such matrices Y for each site of the lattice of N sites. Then one can see that

$$\hat{C}|0\rangle = 0 \ , \tag{4.43}$$

as the consequence of the properties of the triangular matrices. The action of the operators $\hat{A}$ and $\hat{D}$ on the mathematical vacuum state is diagonal:

$$\hat{A}(\lambda)|0\rangle = |0\rangle ,$$
$$\hat{D}(\lambda)|0\rangle = \prod_{j=1}^{N} \frac{\lambda_j^0 - \lambda}{\lambda_j^0 - \lambda + i(U/2)}|0\rangle . \tag{4.44}$$

The operator $\hat{B}$ plays the role of a "spin-lowering" operator. We can consider the state with M down spins as a result of action of M operators $\hat{B}$:

$$|\lambda_1, \ldots, \lambda_M\rangle = \prod_{\beta=1}^{M} \hat{B}(\lambda_\beta)|0\rangle . \tag{4.45}$$

Let us act with the operator $\hat{\tau} = \hat{A} + \hat{D}$ on the state Eq. (4.45) using the commutation relations Eq. (4.41). We get

$$\hat{\tau}(\lambda)|\lambda_1, \ldots, \lambda_M\rangle = \Lambda(\lambda, \lambda_1^0, \ldots, \lambda_N^0, \lambda_1, \ldots, \lambda_M)|\lambda_1, \ldots, \lambda_M\rangle$$
$$+ \sum_{\gamma=1}^{M} \Lambda_\gamma(\lambda, \lambda_1^0, \ldots, \lambda_N^0, \lambda_1, \ldots, \lambda_M) \prod_{\substack{\beta=1,\\ \beta\neq\gamma}}^{M} \hat{B}(\lambda_\beta)B(\lambda)|0\rangle , \tag{4.46}$$

where

$$\Lambda(\lambda, \lambda_1^0, \ldots, \lambda_N^0, \lambda_1, \ldots, \lambda_M) = \prod_{\beta=1}^{M} \frac{\lambda - \lambda_\beta + i(U/2)}{\lambda - \lambda_\beta}$$
$$+ \prod_{j=1}^{N} \frac{\lambda_j^0 - \lambda}{\lambda_j^0 - \lambda + i(U/2)} \prod_{\beta=1}^{M} \frac{\lambda - \lambda_\beta - i(U/2)}{\lambda - \lambda_\beta} \tag{4.47}$$

and

$$\Lambda_\gamma(\lambda, \lambda_1^0, \ldots, \lambda_N^0, \lambda_1, \ldots, \lambda_M)$$
$$= \frac{i(U/2)}{\lambda - \lambda_\gamma}\Bigg(- \prod_{\substack{\beta=1\\ \beta\neq\gamma}}^{M} \frac{\lambda_\gamma - \lambda_\beta + i(U/2)}{\lambda_\gamma - \lambda_\beta}$$
$$+ \prod_{j=1}^{N} \frac{\lambda_j^0 - \lambda_\gamma}{\lambda_j^0 - \lambda_\gamma + i(U/2)} \prod_{\substack{\beta=1\\ \beta\neq\gamma}}^{M} \frac{\lambda_\gamma - \lambda_\beta - i(U/2)}{\lambda_\gamma - \lambda_\beta}\Bigg) . \tag{4.48}$$

The state $|\lambda_1, \ldots, \lambda_M\rangle$ is the eigenstate of the transfer matrix if $\Lambda_\gamma = 0$. It is true, if

$$\prod_{j=1}^{N} \frac{\lambda_j^0 - \lambda_\gamma + i(U/2)}{\lambda_j^0 - \lambda_\gamma} = \prod_{\substack{\beta=1 \\ \beta\neq\gamma}}^{M} \frac{\lambda_\gamma - \lambda_\beta - i(U/2)}{\lambda_\gamma - \lambda_\beta + i(U/2)} , \tag{4.49}$$

which holds for any $\gamma = 1, \ldots, M$.

It is convenient to choose $P = Q = 1$ (the identity) in Eq. (4.23) and restrict ourselves to the spin subspace only. The conditions Eq. (4.49) guarantee that operators T_j commute with each other, hence, they can be diagonalized simultaneously. It is easy to show, by using the properties of Y matrices, that

$$T_j(k_1, \ldots, k_N) = \hat{\tau}(\lambda, \lambda_1^0 = \sin k_1, \ldots, \lambda_N^0 = \sin k_N)|_{\lambda=\sin k_j} , \tag{4.50}$$

which implies

$$e^{ik_jL} = \Lambda(\lambda, \lambda_1^0 = \sin k_1, \ldots, \lambda_N^0 = \sin k_N, \lambda_1, \ldots, \lambda_M)|_{\lambda=\sin k_j} . \tag{4.51}$$

It is convenient to shift $\lambda_\gamma \to \lambda_\gamma + i(U/4)$. Then the conditions for the sets k_j $(j = 1, \ldots, N)$ and λ_γ $(\gamma = 1, \ldots, M)$ are

$$\begin{aligned} \exp(ik_jL) &= \prod_{\beta=1}^{M} \frac{\sin k_j - \lambda_\beta + i(U/4)}{\sin k_j - \lambda_\beta - i(U/4)} , \\ \prod_{j=1}^{N} \frac{\lambda_\gamma - \sin k_j + i(U/4)}{\lambda_\gamma - \sin k_j - i(U/4)} &= \prod_{\substack{\beta=1 \\ \beta\neq\gamma}}^{M} \frac{\lambda_\gamma - \lambda_\beta + i(U/2)}{\lambda_\gamma - \lambda_\beta - i(U/2)} , \end{aligned} \tag{4.52}$$

which are nothing other than the famous Bethe ansatz equations for the Hubbard chain, first obtained by E. H. Lieb and F. Y. Wu. These Bethe ansatz equations for charge (k_j) and spin (λ_γ) rapidities are quantization conditions (similar to simple quantization conditions for noninteracting particles). One has to solve these equations, and then put the solution(s) into Eq. (4.17) to obtain the eigenvalues of the stationary Schrödinger equation for the Hubbard Hamiltonian in one space dimension for arbitrary number of electrons N and number of electrons with spins down M. It is also instructive to write down the expression for the eigenfunctions (Π are permutations from S_M)

$$\psi(x_1, \ldots, x_N, \sigma_1, \ldots, \sigma_N)$$
$$= \sum_P \mathrm{sign}(PQ) A_{\sigma_{Q_1}, \ldots, \sigma_{Q_N}}(k_{P_1}, \ldots, k_{P_N}) \exp(i \sum_{j=1}^{2} k_{P_j} x_{Q_j}) ,$$
$$A_{\sigma_{Q_1}, \ldots, \sigma_{Q_N}}(k_{P_1}, \ldots, k_{P_N}) = \sum_{1 \le y_1 < \cdots < y_M \le N} \sum_{\Pi}$$
$$\times \prod_{1 \le j \le m \le M} \frac{\lambda_{\Pi_l} - \lambda_{\Pi_m} - i(U/2)}{\lambda_{\Pi_l} - \lambda_{\Pi_m}} \prod_{t=1}^{M} \frac{1}{\lambda_{\Pi_t} - \sin k_{\Pi_{y_t}} + i(U/4)}$$
$$\times \prod_{s=1}^{y_t - 1} \frac{\lambda_{\Pi_t} - \sin k_{\Pi_s} - i(U/4)}{\lambda_{\Pi_t} - \sin k_{\Pi_s} + i(U/4)} , \tag{4.53}$$

where y_j are the positions of down spins in the sequence $\sigma_1, \ldots, \sigma_N$.

Observe that the structure of the Bethe ansatz equations (4.52) does not depend on whether one has repulsive or attractive Hubbard interaction. For the attraction one has to replace $U \to -U$ there.

The limit of small k_j of the Hubbard model describes the continuum gas of electrons with the δ-function interaction of the strength $U/2$ with the Hamiltonian, here presented in the first-quantized form

$$\mathcal{H}_\delta = -\sum_{j=1}^{N} \left(\frac{\partial^2}{\partial x_j^2} \right) + (U/2) \sum_{j<l} \delta(x_j - x_l) , \tag{4.54}$$

in which the spin degrees of freedom are not present explicitly: they are incorporated *via* the symmetry of the wave function. The Bethe ansatz solution to this problem is given by

$$\exp(ik_j L) = \prod_{\beta=1}^{M} \frac{k_j - \lambda_\beta + i(U/4)}{k_j - \lambda_\beta - i(U/4)} ,$$
$$\prod_{j=1}^{N} \frac{\lambda_\gamma - k_j + i(U/4)}{\lambda_\gamma - k_j - i(U/4)} = \prod_{\substack{\beta=1 \\ \beta \ne \gamma}}^{M} \frac{\lambda_\gamma - \lambda_\beta + i(U/2)}{\lambda_\gamma - \lambda_\beta - i(U/2)} , \tag{4.55}$$

where $j = 1, \ldots, N$ and $\gamma = 1, \ldots, M$, and the energy

$$E = \sum_{j=1}^{N} k_j^2 + \text{const} . \tag{4.56}$$

Now we can study thermodynamic properties of a Hubbard chain. For this purpose we shall use the string hypothesis, already known to the reader from the previous chapter. In the thermodynamic limit $L, N, M \to \infty$ with N/L and M/L kept fixed we can consider three main classes of solutions of Eq. (4.52). The first class consists of $N - 2M^*$ real charge rapidities k_j, which correspond to unbound electron excitations with densities $\rho(k)$, densities of holes $\rho_h(k)$ and dressed energies $\varepsilon(k) = T\ln[\rho_h(k)/\rho(k)] = T\ln\xi(k)$. The second class represents complex charge rapidities describing spin-singlet pairs (bound states) of electrons or bound states of them with $\sin k^l_{\alpha,n} = \lambda^{l'}_{\alpha,n} \pm i(U/4)$, where $\lambda^{l'}_{\alpha,n} = \lambda'_{\alpha,n} + i(n+1-2l)(U/4)$ $(l = 1, \ldots, n)$ are parts of the string of length $(n-1)$ with $n = 1, \ldots, \infty$. $\lambda'_{\alpha,n}$ are real and characterize the centre of motion of the bound state of n pairs and $\alpha = 1, \ldots, M'_n$ label the strings. Notice that $M^* = \sum_{n=1}^{\infty} nM'_n$. These excitations have densities $\sigma'_n(\lambda)$, densities of holes $\sigma'_{nh}(\lambda)$, and dressed energies $\psi_n(\lambda) = T\ln[\sigma'_{nh}(\lambda)/\sigma'_n(\lambda)] = T\ln\kappa_n(\lambda)$. These two classes of solutions are different from the solutions of the Heisenberg chain, presented in the previous chapter, because they describe the propagation of charge degrees of freedom of electrons. The third class, however, is already familiar to the reader. This class consists of M_n spin strings (bound states) of length $(n-1)$ of the form $\lambda^l_{\alpha,n} = \lambda_{\alpha,n} + i(n+1-2l)(U/4)$ $(l = 1, \ldots, n)$ with real $\lambda_{\alpha,n}$ and $\alpha = 1, \ldots, M_n$. Naturally, because M is the number of down spins, we have $M = M^* + \sum_{n=1}^{\infty} nM_n = \sum_{n=1}^{\infty} n(M'_n + M_n)$. These excitations carry only spin and no charge and have densities $\sigma_n(\lambda)$, densities of holes $\sigma_{nh}(\lambda)$, and dressed energies $\phi_n(\lambda) = T\ln[\sigma_{nh}(\lambda)/\sigma_n(\lambda)] = T\ln\eta_n(\lambda)$. We remark that the case of the continuum gas of electrons with the δ-function coupling has no bound states between pairs.

By using straightforward but tedious procedures, similar to the case of the Heisenberg spin chain, we obtain the thermodynamic Bethe ansatz equations for densities

$$
\begin{aligned}
\rho(k) + \rho_h(k) &= \frac{1}{2\pi} + \cos k \sum_{n=1}^{\infty} a_{nU/4}(\sin k - \lambda) * [\sigma_n(\lambda) + \sigma'_n(\lambda)] \, , \\
\sigma_{nh}(\lambda) &= a_{nU/4}(\lambda - \sin k) * \rho(k) - \sum_{m=1}^{\infty} A_{nm}(\lambda - \lambda') * \sigma_m(\lambda') \, , \\
\sigma'_{nh}(\lambda) &= \frac{1}{\pi}\mathrm{Re}\frac{1}{\sqrt{1 - [\lambda - in(U/4)]^2}} \\
&\quad - \sum_{m=1}^{\infty} A_{nm}(\lambda - \lambda') * \sigma'_m(\lambda') - a_{nU/4}(\lambda - \sin k) * \rho(k) \, .
\end{aligned}
\tag{4.57}
$$

The thermodynamic Bethe ansatz equations for dressed energies have the form

$$\begin{aligned}
\varepsilon(k) &= -\left(2\cos k + \frac{H}{2} + \mu\right) + T\sum_{n=1}^{\infty} a_{nU/4}(\sin k - \lambda) * \ln\frac{1+\kappa_n^{-1}(\lambda)}{1+\eta_n^{-1}(\lambda)} , \\
T\ln[1+\eta_n(\lambda)] &= nH - T\cos k \; a_{nU/4}(\lambda - \sin k) * \ln[1+\xi^{-1}(k)] \\
&\quad + T\sum_{m=1}^{\infty} A_{nm}(\lambda-\lambda') * \ln[1+\eta_m^{-1}(\lambda')] , \\
T\ln[1+\kappa_n(\lambda)] &= -4\,\mathrm{Re}\sqrt{1-[\lambda - in(U/4)]^2} - 2n\mu \\
T\ln[1+\eta_n(\lambda)] &= -\; T\cos k \; a_{nU/4}(\lambda - \sin k) * \ln[1+\xi^{-1}(k)] \\
&\quad + T\sum_{m=1}^{\infty} A_{nm}(\lambda-\lambda') * \ln[1+\kappa_m^{-1}(\lambda')] .
\end{aligned} \tag{4.58}$$

Here $*$ denotes convolution, $a_{Un/4}(x) = (nU/4)/\pi[x^2 + (nU/4)^2]$, the Fourier transform of $A_{nm}(x)$ is $\coth(|\omega U|/8)[\exp(-|n-m||\omega U|/8) + \exp(-(n-m)|\omega U|/8)]$, H is the external magnetic field, μ is the chemical potential, and T is the temperature. The internal energy, the number of electrons and the magnetization per site are given by

$$\begin{aligned}
e &= -2\int_{-\pi}^{\pi} dk \cos k\rho(k) - 4\sum_{n=1}^{\infty}\mathrm{Re}\int_{-\infty}^{\infty} d\lambda\sqrt{1-[\lambda - in(U/4)]^2}\sigma_n'(\lambda) , \\
\tfrac{N}{L} &= \int_{-\pi}^{\pi} dk\rho(k) + 2\sum_{n=1}^{\infty} n\int_{-\infty}^{\infty} d\lambda\sigma_n'(\lambda) , \\
m^z &= \frac{1}{2}\int_{-\pi}^{\pi} dk\rho(k) - \sum_{n=1}^{\infty} n\int_{-\infty}^{\infty} d\lambda\sigma_n(\lambda) .
\end{aligned} \tag{4.59}$$

The Helmholtz free energy of the Hubbard chain per site is equal to

$$\begin{aligned}
f &= -T\sum_{n=1}^{\infty}\int_{-\infty}^{\infty}\frac{d\lambda}{\pi}\mathrm{Re}\frac{1}{\sqrt{1-[\lambda - in(U/4)]^2}}\ln[1+\kappa_n^{-1}(\lambda)] \\
&\quad - T\int_{-\pi}^{\pi}\frac{dk}{2\pi}\ln[1+\xi^{-1}(k)] = e_0 - \mu - T\int_{-\infty}^{\infty} d\lambda\sigma_0(\lambda)\ln[1+\kappa_1(\lambda)] \\
&\quad - T\int_{-\pi}^{\pi} dk\rho_0(k)\ln[1+\xi(k)] ,
\end{aligned} \tag{4.60}$$

where the ground state (internal) energy per site and densities for $H = 0$ and half-filled band with $\mu = U/2$ are

$$e_0 = -4\int_0^{\infty} d\omega \frac{J_0(\omega)J_1(\omega)}{\omega[1+\exp(\omega U/2)]} ,$$
$$\rho_0(k) = \frac{1}{2\pi} + \cos k \int_{-\infty}^{\infty} d\lambda \sigma_0(\lambda) a_{U/4}(\lambda - \sin k) , \tag{4.61}$$
$$\sigma_0(\lambda) = \int_0^{\infty} \frac{d\omega}{2\pi} \frac{J_0(\omega)\cos(\omega\lambda)}{\cosh(\omega U/4)} ,$$

where $J_{0,1}(x)$ are Bessel functions.

It is important to notice that the thermodynamic Bethe ansatz equations for densities of the attractive Hubbard model follow from Eq. (4.57) with the change of the signs of the second term in the right hand side of the first equation. On the other hand, the thermodynamic Bethe ansatz equations for dressed energies of the attractive Hubbard model follow from Eq. (4.58) with the change of the signs of $\ln(1+\xi^{-1})$ and the sign of the driving term for dressed energies of pair excitations ($2n\mu$ keeps the same sign though). One can see that these differences reflect the simple change $k \to \pi - k$. This simple difference is clear from the observation that the transformation related to that change $a_{j,\sigma} \to (-1)^j a_{j,\sigma}$, $a^{\dagger}_{j,\sigma} \to (-1)^j a^{\dagger}_{j,\sigma}$ reverses the sign of hopping terms in the Hubbard Hamiltonian, but leaves the other properties invariant. Then the Hamiltonian with negative U is just the Hamiltonian for the repulsive Hubbard chain but with the total negative sign. This means that the eigenstates are the same, but the energies differ by their signs. This is why, for the attractive Hubbard chain one has to replace $\xi(k) \to \xi^{-1}(k)$ in the last term of Eq. (4.60), to add the term $-U/2$ to e_0 (which now pertains to the case $\mu = -U/2$), and to change the sign of the second term for the expression for $\rho_0(k)$.

As for the spin-$\frac{1}{2}$ Heisenberg chain, considered in the previous chapter, thermodynamic Bethe ansatz equations for a Hubbard chain have analytic solutions for high and low temperatures.

For high T we consider the limit $T \to \infty$, but with U/T, H/T and μ/T kept finite. In this limit the terms, which depend on k and λ in driving terms can be neglected, and ξ, η_n and κ_n are constants. There is no movement of excitations from site to site in this limit. Then the solutions of the thermodynamic Bethe ansatz equations are

$$\xi = \frac{w + w^{-1}}{z + z^{-1}} , \ \kappa_n = f^2(n) - 1 , \ \eta_n = g^2(n) - 1 , \tag{4.62}$$

where $z = \exp(-H/2T)$, $w = \exp[(2\mu - U)/2T]$, and

$$f(n) = \frac{w^{n+1} - w^{-n-1}}{w - w^{-1}} \ , \ g(n) = \frac{z^{n+1} - z^{-n-1}}{z - z^{-1}} \ . \tag{4.63}$$

The solutions for densities are

$$\begin{aligned}
\rho &= \frac{1}{1+\xi}\left(\frac{1}{2\pi} + \frac{1}{\pi}\frac{\cos k}{(z+z^{-1})(w+w^{-1})}\mathrm{Re}\frac{1}{\sqrt{1-[\sin k - i(U/2)]^2}}\right) , \\
\sigma'_n &= \frac{1}{\pi}\frac{1}{z+z^{-1}+w+w^{-1}}\left(\frac{1}{f(n-1)f(n)}\mathrm{Re}\frac{1}{\sqrt{1-[\lambda - in(U/4)]^2}}\right. \\
&\quad \left. - \frac{1}{f(n+1)f(n)}\mathrm{Re}\frac{1}{\sqrt{1-[\lambda - i(n+2)(U/4)]^2}}\right) , \\
\sigma_n &= \frac{1}{\pi}\frac{1}{z+z^{-1}+w+w^{-1}}\left(\frac{1}{g(n-1)g(n)}\mathrm{Re}\frac{1}{\sqrt{1-[\lambda - in(U/4)]^2}}\right. \\
&\quad \left. - \frac{1}{g(n+1)g(n)}\mathrm{Re}\frac{1}{\sqrt{1-[\lambda - i(n+2)(U/4)]^2}}\right) .
\end{aligned} \tag{4.64}$$

This is why, the numbers of electrons with spins up and down per site are

$$\begin{aligned}
\frac{(N-M)}{L} &= (1 + \exp[(2U - 2\mu - H)/2T])^{-1} , \\
\frac{M}{L} &= (1 + \exp[(2U - 2\mu + H)/2T])^{-1} .
\end{aligned} \tag{4.65}$$

These are obvious Fermi distribution functions. The total number of electrons and the magnetization per site are

$$\begin{aligned}
\frac{N}{L} &= \frac{\cosh(H/2T) + \exp[(2\mu - U)/2T]}{\cosh(H/2T) + \cosh[(2\mu - U)/2T]} , \\
m^z &= \frac{1}{2}\frac{\sinh(H/2T)}{\cosh(H/2T) + \cosh[(2\mu - U)/2T]} .
\end{aligned} \tag{4.66}$$

Let us consider the behaviour of the magnetic susceptibility of the Hubbard chain at high temperatures. It is also possible to consider the charge stiffness (charge susceptibility), defined as $\chi_c = -(\partial^2 f/\partial\mu^2) = (\partial(N/L)/\partial\mu)$. They are

$$\chi_c = \frac{1}{T}\frac{1+\exp[(2\mu-U)/2T]\cosh(H/2T)}{(\cosh(H/2T)+\cosh[(2\mu-U)/2T])^2} ,$$
$$\chi = \frac{1}{4T}\frac{1+\cosh[(2\mu-U)/2T]}{(\cosh(H/2T)+\cosh[(2\mu-U)/2T])^2} . \qquad (4.67)$$

These expressions manifest that at high temperatures the Hubbard chain has no phase transitions and the behaviour of its characteristics is smooth with T, μ and H.

For low T we again, as in the previous chapter, separate dressed energies into their positive and negative parts, so that the terms with "+" superscript are positive (empty states in the ground state) and those with "−" superscript are negative (those which form Dirac seas). Re-writing Eq. (4.58), we obtain

$$\ln\kappa_n(\lambda) = \frac{1}{U\cosh(2\pi(\lambda-\lambda')/U)} * \ln([1+\kappa_{n-1}(\lambda')][1+\kappa_{n+1}(\lambda')]) ,$$
$$\ln\eta_n(\lambda) = \frac{1}{U\cosh(2\pi(\lambda-\lambda')/U)} * \ln([1+\eta_{n-1}(\lambda')][1+\eta_{n+1}(\lambda')]) ,$$

for $n \geq 2$, hence $\psi_n > 0$ and $\phi_n > 0$ for $n \geq 2$, and these excitations have no Dirac seas. In the limit of low temperatures it is important to distinguish the sign of the Hubbard coupling U. First, let us consider the repulsive case $U > 0$. The $T = 0$ equation for ψ_1 has the form

$$\psi_1(\lambda) = U - 2\mu - a_{U/2}(\lambda-\lambda') * \psi_1^-(\lambda') - \cos k\, a_{U/4}(\lambda - \sin k) * \varepsilon^+(k) . \qquad (4.68)$$

Since for $N \leq L$ we have $2\mu \leq U$, then $\psi_1^- = 0$ for any λ. Then the integral ground state equations for dressed energies for the repulsive Hubbard chain can be written as

$$\phi_1(\lambda) + \int_{-B}^{B} d\lambda' a_{U/2}(\lambda'-\lambda)\phi_1(\lambda')$$
$$= H + \int_{-Q}^{Q} dk \cos k a_{U/4}(\lambda - \sin k)\varepsilon(k) , \qquad (4.69)$$
$$\varepsilon(k) = -2\cos k - \mu - \frac{H}{2} + \int_{-B}^{B} d\lambda a_{U/4}(\lambda - \sin k)\phi_1(\lambda) ,$$

and the equations for densities are

$$\sigma_1(\lambda) + \sigma_{1h}(\lambda) + \int_{-B}^{B} d\lambda' a_{U/2}(\lambda' - \lambda)\sigma_1(\lambda')$$
$$= \int_{-Q}^{Q} dk a_{U/4}(\lambda - \sin k)\rho(k) , \tag{4.70}$$
$$\rho(k) + \rho_h(k) = \frac{1}{2\pi} + \cos k \int_{-B}^{B} d\lambda a_{U/4}(\lambda - \sin k)\sigma_1(\lambda) ,$$

where the Fermi points for unbound electrons and spin strings of the length 1 (they are often called spinons, as for the Heisenberg–Ising chain) are related to the values of the chemical potential and magnetic field and determined from the conditions $\varepsilon(\pm Q) = 0$ and $\phi_1(\pm B) = 0$.

Let us consider the internal energy of the Hubbard chain as the function of the number of electrons. From the symmetry we have

$$\begin{aligned} E(N-M, M, U) &= -(L-N)U + E(L-N+M, L-M, U) \\ &= (N-M)U + E(N-M, L-M, -U) \\ &= MU + E(L-N+M, M, -U) , \end{aligned} \tag{4.71}$$

where $N - M$ is the number of electrons with spins up. The chemical potentials for adding or removing an electron can be defined as

$$\begin{aligned} \mu_+ &= E(N-M, M+1, U) - E(N-M, M, U) , \\ \mu_- &= E(N-M, M, U) - E(N-M-1, M, U) . \end{aligned} \tag{4.72}$$

Notice that for the half-filled band, unless $N - M = M = N/2$, the chemical potential depends on the spin of the electron added or removed, and $\mu_\pm$ are related to opposite directions of spins. Since a metal (*i.e.*, a conductor) has to have a Fermi surface, then necessarily it follows that $\mu_+ = \mu_-$. On the other hand, the insulator has to have an excitation gap when changing the number of electrons, so $\mu_+ > \mu_-$. One can see that $\mu_+ = U - \mu_-$ for $N = L$, while $\mu_+ = \mu_-$ if the band is not half-filled for $U > 0$. Hence, the repulsive Hubbard chain is a metal (conductor) for $N < L$, and for the half-filled situation it is an insulator unless $U = 0$. The gap $\mu_+ - \mu_-$ in zero magnetic field asymptotically approaches $U - 4$ for large U and it is proportional to $\exp(-2\pi U)$ for small U (*i.e.*, nonanalytically vanishes as $U \to 0$). We can consider this metal-insulator quantum critical behaviour from the viewpoint of the commensurability of the backward scattering of electrons. If the Fermi vector $4k_F$ is equal to the Brillouin zone, *i.e.*, it is commensurate with the reciprocal lattice vector, then the spin-flip backward scattering may give rise to a gap at the Fermi level in the charge excitation spectrum (notice that one has the electron-hole symmetry). We would like to emphasize here that in the continuum version of the Hubbard chain (the electron gas with the δ-function repulsion) there is no metal-insulator transition.

Let us now consider the half-filled band $N = L$. In this case we have $Q = \pi$ and $\rho_h(k) = 0$ and $\rho(k) = 1/2\pi$. The charged unbound electron excitations have a gap, and we have from Eqs. (4.69) and (4.70)

$$\phi_1(\lambda) + \int_{-B}^{B} d\lambda' a_{U/2}(\lambda' - \lambda)\phi_1(\lambda') = H - \int_{-\pi}^{\pi} dk \cos^2 k a_{U/4}(\lambda - \sin k) \ ,$$

$$\sigma_1(\lambda) + \sigma_{1h}(\lambda) + \int_{-B}^{B} d\lambda' a_{U/2}(\lambda' - \lambda)\sigma_1(\lambda')$$
$$= \int_{-\pi}^{\pi} \frac{dk}{2\pi} a_{U/4}(\lambda - \sin k). \tag{4.73}$$

The ground state energy and the magnetization are

$$\begin{aligned} e_0 &= -2\int_{-\pi}^{\pi} dk \cos^2 k \int_{-B}^{B} d\lambda a_{U/4}(\lambda - \sin k)\sigma_1(\lambda) \ , \\ m^z &= \frac{1}{2} - \int_{-B}^{B} d\lambda \sigma_1(\lambda) \ . \end{aligned} \tag{4.74}$$

It is interesting to note that if U is large the k dependence in driving terms and the ground state energy can be neglected ($\sin k \to 0$) and the resulting integral equations and expressions for the energy and magnetization coincide with those for the spin-$\frac{1}{2}$ antiferromagnetic Heisenberg chain, *cf.* the previous chapter. Hence, in the limit of large U for the half-filling we can use already known to us results to describe the quantum phase transition in the external magnetic field. Moreover, for any U such a transition also takes place, but at the critical field

$$H_s = \frac{16}{U + \sqrt{16 + U^2}} \ , \tag{4.75}$$

which pertains to the case $B = 0$. At this value of the magnetic field all spins of electrons are polarized along the field direction, and the magnetic susceptibility diverges as $\chi \sim 1/\sqrt{H_s - H}$. In zero magnetic field $B = \infty$

and we can solve the integral equations analytically with the ground state energy and densities given above in Eq. (4.61) and the magnetic susceptibility and the low temperature specific heat per site are equal to

$$\chi = = \frac{I_0(2\pi/U)}{8\pi I_1(2\pi/U)} ,$$
$$c = \frac{\pi I_0(2\pi/U)}{6I_1(2\pi/U)}T + \dots , \tag{4.76}$$

where $I_{0,1}(x)$ are modified Bessel functions. One can see that the following (Wilson) relation for the Sommerfeld coefficient γ of the specific heat holds:

$$\gamma = \frac{4\pi^2\chi}{3} . \tag{4.77}$$

For $U \to 0$, we have $H_s = 4$ and the magnetization behaves as $m^z = (1/2) - (1/\pi)\sin^{-1}\sqrt{1-(H/4)^2}$. For $U \to \infty$ we have $H_s \to 0$, and, hence, the divergent magnetic susceptibility at $H = 0$ (*i.e.*, an infinitesimal magnetic field transfers the infinite-U repulsive Hubbard chain into the spin-polarized, ferromagnetic ground state). Notice that at the values for which van Hove singularities of empty one-dimensional Dirac seas of low-lying excitations take place, see below, the low-temperature specific heat is proportional to $\sqrt{T}$.

Usually it is convenient to relate both of these characteristics with the velocity of low-lying excitations (spinons in this case) taken at the Fermi point

$$v_\sigma^F = (2\pi\sigma_1(\lambda))^{-1}\frac{\partial\phi_1(\lambda)}{\partial\lambda}|_{\lambda=B} \tag{4.78}$$

as $\gamma = \pi/3v_\sigma^F$ and $\chi = 1/4\pi v_\sigma^F$ (at half-filling). It turns out that $v_\sigma^F = 0$ at $H = H_s$ (*i.e.*, at $B = 0$), and the susceptibility diverges. Naturally, the Fermi velocity of unbound electron excitations

$$v_\rho^F = (2\pi\rho(k))^{-1}\frac{\partial\varepsilon(k)}{\partial k}|_{k=Q} \tag{4.79}$$

is equal to zero at half filling.

Considering the charge stiffness the reader can better understand why we wrote about a metal-insulator transition at half-filling as about the quantum phase transition: the ground state charge stiffness of the repulsive Hubbard chain diverges at half filling as $\chi_c = 1/4\pi v_\rho^F$ (similar to the magnetic susceptibility at H_s). In the metallic situation, where both $v_\rho^F \neq 0$

and $v_\sigma^F \neq 0$, one has the general formula

$$\gamma = \frac{\pi}{3}\left(\frac{1}{v_\rho^F} + \frac{1}{v_\sigma^F}\right) . \tag{4.80}$$

Naturally, this formula is valid except for situations with the van Hove singularities.

Some other analytic results can be obtained for the metallic case $N < L$ for $H = 0$. Here ϕ_1 and σ_1 can be eliminated by the Fourier transformation and we have

$$\begin{aligned} \varepsilon(k) &= -2\cos k - \mu + \int_{-Q}^{Q} dk' \cos k' A(\sin k' - \sin k)\phi_1(\lambda)\varepsilon(k') , \\ \rho(k) + \rho_h(k) &= \frac{1}{2\pi} + \cos k \int_{-Q}^{Q} dk' A(\sin k' - \sin k)\rho(k') , \end{aligned} \tag{4.81}$$

where $A(x) = \mathrm{Re}(\psi[1 + i(x/U)] - \psi[(1/2) + i(x/U)])/\pi U$, and $\psi(x)$ is a digamma function. The number of electrons per site is given by $N/L = \int_{-Q}^{Q} dk\rho(k)$. The chemical potential is, by its definition, obtained from the condition $\varepsilon(\pm Q) = 0$. It increases with increasing U and the number of electrons. For $U = 0$, we have $e_0 = -(4/\pi)\sin(\pi N/2L)$ and $\mu = -2\cos(\pi N/2L)$, while for $U \to \infty$ (where it is forbidden for two electrons to occupy the same site) we get $e_0 = -(2/\pi)\sin(\pi N/L)$ and $\mu = -2\cos(\pi N/L)$. The difference is transparent, because the effective lattice size becomes twice as small for these effective "spinless fermions". The magnetic susceptibility, charge stiffness and Sommerfeld coefficient are related to Fermi velocities of spinons and charged excitations, which are

$$\begin{aligned} v_\sigma^F &= -\frac{\int_{-Q}^{Q} dk \cos k\varepsilon(k)\exp(2\pi\sin k/U)}{U\int_{-Q}^{Q} dk\rho(k)\exp(2\pi\sin k/U)} , \\ v_\rho^F &= (2\pi\rho(k))^{-1}\frac{\partial\varepsilon(k)}{\partial k}|_{k=Q} . \end{aligned} \tag{4.82}$$

The quantum phase transition to the spin-polarized state at $T = 0$ takes place at

$$H_s = -\int_{-Q}^{Q} dk \cos k\varepsilon(k)a_{U/4}(\sin k) \tag{4.83}$$

with the square-root singularity of the magnetic susceptibility. As we see, this square-root singularity of the magnetic susceptibility is characteristic for any one-dimensional system with SU(2) spin symmetry. It is related to the van Hove singularity of the one-dimensional empty Dirac sea of spinons (low-lying spin excitations). The critical field H_s is equal to $4[1 - \cos(\pi N/2L)]$ for $U = 0$, and, for given N decreases monotonically with increasing U (vanishing at $U \to \infty$).

Now we turn to the attractive Hubbard chain, $U < 0$. Here one can see that $\phi_n > 0$ for any n for $T \to 0$. Then the ground state equations for dressed energies and densities have the form

$$\begin{aligned}
&\psi_1(\lambda) + \int_{-Q}^{Q} d\lambda' a_{U/2}(\lambda' - \lambda)\psi_1^-(\lambda') \\
&\quad = -4\mathrm{Re}\sqrt{1 - [\lambda + i(U/4)]^2} - 2\mu - \int_{-B}^{B} dk \cos k a_{U/4}(\lambda - \sin k)\varepsilon^-(k) , \\
&\varepsilon(k) = -2\cos k - \mu - \frac{H}{2} - \int_{-Q}^{Q} d\lambda a_{U/4}(\lambda - \sin k)\psi_1^-(\lambda) ,
\end{aligned} \tag{4.84}$$

and

$$\begin{aligned}
&\sigma_1'(\lambda) + \sigma_{1h}'(\lambda) + \int_{-Q}^{Q} d\lambda' a_{U/2}(\lambda' - \lambda)\sigma_1'(\lambda') \\
&\quad = \frac{1}{\pi}\mathrm{Re}\frac{1}{\sqrt{1 - [\lambda + i(U/4)]^2}} - \int_{-B}^{B} dk a_{U/4}(\lambda - \sin k)\rho(k) , \\
&\rho(k) + \rho_h(k) = \frac{1}{2\pi} - \cos k \int_{-Q}^{Q} d\lambda a_{U/4}(\lambda - \sin k)\sigma_1'(\lambda) ,
\end{aligned} \tag{4.85}$$

where the Fermi points for unbound electrons and spin-singlet pairs are related to the values of the chemical potential and magnetic field and determined from the conditions $\varepsilon(\pm B) = 0$ and $\psi_1(\pm Q) = 0$ (do not confuse with the repulsive case: now charge excitations are connected with quantum numbers λ and Q, while spin is carried by excitations with quantum numbers k and B).

At $H = 0$, the magnetization is zero and, hence, $B = 0$. All electrons are bound in pairs, and unbound electron excitations have a spin gap. The chemical potential for the empty band is equal to $-2\sqrt{1 + (U/4)^2}$, while for the half-filled band it is $-U/2$. μ monotonically decreases with increasing $|U|$ (as U^2 for N small and as U for $N \to L$). It requires a magnetic field larger than the critical one to have unbound electron excitations. The critical field needed to overcome the binding energy of a spin-singlet pair is equal to

$$H_c = -4 - 2\mu - 2\int_{-Q}^{Q} d\lambda a_{U/4}(\lambda)\psi_1(\lambda) \ , \tag{4.86}$$

where $\psi_1(\lambda)$ is determined from the first of Eq. (4.84) with $B = 0$. Naturally, H_c vanishes at $U = 0$ and grows with increasing $|U|$. The ground state magnetization of an attractive Hubbard chain is zero for $H \leq H_c$ and it is proportional to $\sqrt{H - H_c}$ for values of the magnetic field slightly above H_c. Hence, at H_c the ground state magnetic susceptibility has a square-root divergence which signals a quantum phase transition. It is the consequence of the van Hove singularity of the empty one-dimensional band of unbound electron excitations. For larger values of the magnetic field, the ground state magnetization of an attractive Hubbard chain saturates at H_s at which all spin-singlet pairs are broken up, *i.e.*, at $Q = 0$. Again, at H_s we have the quantum phase transition, related to the van Hove one-dimensional singularity of the empty band of pairs and all spins of electrons are polarized by the magnetic field. The magnetic susceptibility diverges at H_s and H_c (it is zero for $H > H_s$ and $H < H_c$) and it is finite for fields just below H_s (except for the half-filled case $N = L$).

The situation for the ground state of an attractive Hubbard chain is reminiscent of the one for type-II superconductors. Namely, there are two critical values of the external magnetic field. For $H < H_c$ only Cooper-like singlet pairs are low-lying excitations, while unbound electron excitations are gapped. In the intermediate phase, $H_c \leq H \leq H_s$ both pairs and unbound electron excitations are gapless. Finally, at $H > H_s$ all excitations have gaps. However, there is a drastic difference: in a one-dimensional attractive Hubbard chain pairs are not coherent even at $T = 0$ (and, moreover, for $T \neq 0$, remember the Hohenberg theorem), and there is no spontaneous superconductive ordering in that model.

The magnetic susceptibility, charge stiffness and Sommerfeld coefficient for an attractive Hubbard model are again related to Fermi velocities of pairs and unbound electron excitations

$$v_{\sigma'}^F = (2\pi\sigma_1'(\lambda))^{-1}\frac{\partial\psi_1(\lambda)}{\partial\lambda}|_{\lambda=Q} \ , \ v_\rho^F = (2\pi\rho(k))^{-1}\frac{\partial\varepsilon(k)}{\partial k}|_{k=B} \tag{4.87}$$

as $\chi = 1/4\pi v_\rho^F$, $\chi_c = 1/4\pi v_{\sigma'}^F$ (for $H \leq H_c$), and $\gamma = (4\pi^2/3)[(v_\rho^F)^{-1} + (v_{\sigma'}^F)^{-1}]$. At $H = 0$, the Sommerfeld coefficient of the attractive Hubbard chain is

$$\gamma = \frac{\pi I_0(2\pi/U)}{6I_1(2\pi/U)} \ . \tag{4.88}$$

Naturally, at the points of van Hove singularities of empty bands of low-lying excitations the low-temperature specific heat is proportional to $\sqrt{T}$.

At finite but low temperatures the magnetic susceptibility is exponentially small for $H < H_c$ and $H > H_s$. At $H = H_c$ or H_s the magnetic susceptibility displays the $\sqrt{T}$ feature corresponding to the van Hove singularity of empty bands. For $H_c < H < H_s$, on the other hand, the magnetic susceptibility is finite as $T \to 0$.

The spin gap implies that the limits $U \to 0$ and $T \to 0$ cannot be interchanged for a Hubbard chain.

It is important also to emphasize that except for $U = 0$, there are no other critical values of the Hubbard coupling constant for spin-$\frac{1}{2}$ electrons (it is not so for the so-called degenerate SU$(2S + 1)$-symmetric Hubbard chain with electrons carrying arbitrary spin S).

4.2 *t-J* Chain

Another important model of correlated electron systems, which possesses an exact Bethe ansatz solution is the *t-J model* (it is integrable with some restrictions on the values of coupling constants, see below). It became popular when it was realized that in the limit of strong repulsion the Hubbard model with $U > 0$ reduces to it (for the antiferromagnetic situation). The strong on-site repulsions limit site occupations to at most one electron. States with double occupation of a site are energetically unfavourable and can be projected out. Hence, in the t-J model there are only three states per site: one empty state and two states with an electron, either with spin up, or down, and the total number of states in the Hilbert space of the t-J chain of the length L is 3^L. However, virtual transitions to states with doubly occupied sites give rise to an exchange and direct interaction between electrons on nearest neighbour sites. It turns out that a t-J model is important as it is, without direct relation to the large U limit of a repulsive Hubbard model. It is frequently invoked as a model for strongly correlated electrons, in particular it is popular for the description of high-T_c cuprate superconductors and heavy fermion systems.

The Hamiltonian of the one-dimensional t-J model can be written as:

$$\begin{aligned}\mathcal{H}_{tJ} = \sum_{j=1}^{L-1} [-t \sum_{\sigma} \mathcal{P}(a_{j,\sigma}^{\dagger} a_{j+1,\sigma} + \text{H.c.})\mathcal{P} \\ + J\vec{S}_j\vec{S}_{j+1} + V\sum n_{j,\sigma}n_{j+1,\sigma}] \ ,\end{aligned} \tag{4.89}$$

where $a_{j,\sigma}^\dagger$ $(a_{j,\sigma})$ creates (destroys) an electron with the spin projection $\sigma = \pm 1 \equiv \uparrow, \downarrow$ in the lattice site j, $n_{j,\sigma} = a_{j,\sigma}^\dagger a_{j,\sigma}$, $\mathcal{P} = (1-n_{j,-\sigma})(1-n_{j+1,-\sigma})$ is the projection operator which excludes double occupation of each site, $\vec{S}_j = a_{j,\sigma}^\dagger \vec{S}_{\sigma,\sigma'} a_{j,\sigma'}$ is the operator of the spin of the electron in the lattice site j $(S_j^z = (1/2)(n_{j,\uparrow} - n_{j,\downarrow})$, $S_j^+ = a_{j,\uparrow}^\dagger a_{j,\downarrow}$, and $S_j^- = a_{j,\downarrow}^\dagger a_{j,\uparrow})$, t is the hopping integral (in what follows we put it equal to unity, $t = 1$, *i.e.*, we shall measure all other energies in units of t), J is the exchange constant and V is the coupling constant of the nearest-neighbour interactions, respectively. This Hamiltonian can be expressed in terms of *Hubbard operators* $X_j^{ab} = |a_j\rangle\langle b_j|$, where a, b denote states with one electron with spin up or down and empty state (we can define them as $\uparrow, \downarrow, 0$, respectively), supplemented with the local constraint $X_j^{\uparrow\uparrow} + X_j^{\downarrow\downarrow} + X_j^{00} = 1$. These Hubbard operators satisfy the following relations

$$X_j^{ab} X_l^{cd} \pm X_l^{cd} X_j^{ab} = \delta_{j,l}(\delta_{b,c} X_j^{ad} \pm \delta_{a,d} X_j^{cb}) \ . \tag{4.90}$$

The Hamiltonian equation (4.89) can be re-written as

$$\mathcal{H}_{tJ} = -t \sum_{j=1}^{L-1} \Bigg[\sum_{\sigma} (X_j^{0\sigma} X_{j+1}^{\sigma 0} + X_{j+1}^{0\sigma} X_j^{\sigma 0}) + J \sum_{\sigma\sigma'} X_j^{\sigma\sigma'} X_{j+1}^{\sigma'\sigma} + V X_j^{00} X_{j+1}^{00} \Bigg] \ . \tag{4.91}$$

It turns out that the one-dimensional t-J Hamiltonian is exactly solvable by the Bethe's ansatz only for $J = \pm 2t = \pm 2$, and $V = -J/4$, or $V = 3J/4$. The model with these values of parameters used to be called a *supersymmetric* model. We shall explain this definition in the following chapter. Using the same method as in the previous section for a Hubbard chain it is not difficult to find that the two-particle scattering matrix of a supersymmetric t-J model for $V = -J/4$ is equal to

$$S_{\sigma_2\tau_2}^{\sigma_1\tau_1}(k_1, k_2) = \frac{1}{2}\Big([1 + \exp(-2i\psi_{k_1,k_2})] I_{\sigma_2\tau_2}^{\sigma_1\tau_1} + [1 - \exp(-2i\psi_{k_1,k_2})] \Pi_{\sigma_2\tau_2}^{\sigma_1\tau_1} \Big) \ , \tag{4.92}$$

where $I_{\sigma_2\tau_2}^{\sigma_1\tau_1} = \delta_{\sigma_2\tau_2}\delta_{\sigma_1\tau_1}$ is the identity operator, $\Pi_{\sigma_2\tau_2}^{\sigma_1\tau_1} = \delta_{\sigma_1\tau_2}\delta_{\sigma_2\tau_1}$ is the permutation operator, and

$$\cot \psi_{k_1 k_2} = \frac{J}{2} \frac{\cot(k_1/2) - \cot(k_2/2)}{[1 - (J/2)] \cot(k_1/2) \cot(k_2/2) - [1 + (J/2)]} \ . \tag{4.93}$$

Equation (4.92) can be re-written as

$$S^{\sigma_1\tau_1}_{\sigma_2\tau_2}(k_1,k_2) = \frac{(p_1-p_2)I^{\sigma_1\tau_1}_{\sigma_2\tau_2} + i(J/2)\Pi^{\sigma_1\tau_1}_{\sigma_2\tau_2}}{p_1-p_2+i(J/2)} , \tag{4.94}$$

where $p_{1,2} = (1/2)\cot(k_{1,2}/2)$ for $J=2$, and $p_{1,2} = (1/2)\tan(k_{1,2}/2)$ for $J=-2$. On the other hand, for $V=3J/4$, we have

$$S^{\sigma_1\tau_1}_{\sigma_2\tau_2}(k_1,k_2) = \frac{-(p_1-p_2)I^{\sigma_1\tau_1}_{\sigma_2\tau_2} - i(J/2)\Pi^{\sigma_1\tau_1}_{\sigma_2\tau_2}}{p_1-p_2-i(J/2)} . \tag{4.95}$$

These equations have the same form as the two-particle scattering matrix of a Hubbard model, Eq. (4.15), up to the re-formulation of charge rapidities and, hence, we can use the results of the previous analysis and after straightforward, but tedious calculations we have for the sets p_j ($j=1,\ldots,N$, where N is the number of electrons) and λ_γ ($\gamma = 1,\ldots,M$, where M is the number of down spin electrons) for $V=-J/4$

$$\begin{aligned} \left(\frac{p_j+i(1/2)}{p_j-i(1/2)}\right)^L &= \prod_{\beta=1}^{M} \frac{p_j-\lambda_\beta+i(1/2)}{p_j-\lambda_\beta-i(1/2)} , \\ \prod_{j=1}^{N} \frac{\lambda_\gamma-p_j+i(1/2)}{\lambda_\gamma-p_j-i(1/2)} &= \prod_{\substack{\beta=1\\ \beta\neq\gamma}}^{M} \frac{\lambda_\gamma-\lambda_\beta+i}{\lambda_\gamma-\lambda_\beta-i} . \end{aligned} \tag{4.96}$$

Please pay attention that the Bethe ansatz equations do not depend on the sign of J. The energy, contrary, depends on J

$$E = -J\sum_{j=1}^{N}\left(1-\frac{2}{4p_j^2+1}\right) . \tag{4.97}$$

The magnetic moment is equal to $S^z = (N/2) - M$. On the other hand, for $V=3J/4$, the Bethe ansatz equations are

$$\begin{aligned} \left(\frac{p_j+i(1/2)}{p_j-i(1/2)}\right)^L &= \prod_{\substack{l=1\\ l\neq j}}^{N} \frac{p_j-p_l+i}{p_j-p_l-i} \prod_{\beta=1}^{M} \frac{p_j-\lambda_\beta-i(1/2)}{p_j-\lambda_\beta+i(1/2)} , \\ \prod_{j=1}^{N} \frac{\lambda_\gamma-p_j+i(1/2)}{\lambda_\gamma-p_j-i(1/2)} &= \prod_{\substack{\beta=1\\ \beta\neq\gamma}}^{M} \frac{\lambda_\gamma-\lambda_\beta+i}{\lambda_\gamma-\lambda_\beta-i} , \end{aligned} \tag{4.98}$$

and the expression for the energy coincides with Eq. (4.97) taken with the opposite total sign. It turns out that the Bethe ansatz equations of the supersymmetric t-J model with $V = 3J/4$ are equivalent to the ones of the SU(3)-symmetric spin-1 Hamiltonian

$$\mathcal{H}_{SU(3)} = \frac{J}{2}\sum_{j=1}^{L}[\vec{S}_j\vec{S}_{j+1} + (\vec{S}_j\vec{S}_{j+1})^2] , \tag{4.99}$$

with $N \to N_2 + N_3$ and $M \to N_3$ ($N_1 + N_2 + N_3 = L$), where $N_{1,2,3}$ are the number of spins with the z-projections up, zero and down.

It is important to point out that for a supersymmetric t-J model one can introduce a magnetic anisotropy, and this introduction does not violate the exact integrability. Namely, the Hamiltonian

$$\mathcal{H}_{an} = -\sum_{j=1}^{L-1}\Bigg[\sum_{\sigma}\mathcal{P}(a_{j,\sigma}^{\dagger}a_{j+1,\sigma} + \text{H.c.})\mathcal{P} - \frac{J}{2}\sum_{\sigma,\sigma'}(e^{\eta\text{sign}(\sigma'-\sigma)}n_{j,\sigma}n_{j+1,\sigma'} - a_{j,\sigma}^{\dagger}a_{j,\sigma'}a_{j+1,\sigma'}^{\dagger}a_{j+1,\sigma})\Bigg] , \tag{4.100}$$

where η is the parameter of the "easy-plane" magnetic anisotropy (with $J = \pm 2$) is integrable by the Bethe's ansatz. Naturally, for $\eta = 0$, this Hamiltonian coincides with Eq. (4.89). The Bethe ansatz equations (we keep the same notations for rapidities) are

$$\begin{aligned}\left(\frac{\sin[p_j + i(\eta/2)]}{\sin[p_j - i(\eta/2)]}\right)^L &= \prod_{\beta=1}^{M}\frac{\sin[p_j - \lambda_\beta + i(\eta/2)]}{\sin[p_j - \lambda_\beta - i(\eta/2)]} ,\\ \prod_{j=1}^{N}\frac{\sin[\lambda_\gamma - p_j + i(\eta/2)]}{\sin[\lambda_\gamma - p_j - i(\eta/2)]} &= \prod_{\substack{\beta=1\\ \beta\neq\gamma}}^{M}\frac{\sin[\lambda_\gamma - \lambda_\beta + i\eta]}{\sin[\lambda_\gamma - \lambda_\beta - i\eta]} ,\end{aligned} \tag{4.101}$$

with the energy

$$E = -JN\cosh\eta + J\sinh^2\eta\sum_{j=1}^{N}\frac{1}{\cosh\eta - \cos 2p_j} \tag{4.102}$$

and the magnetic moment $S^z = (N/2) - M$. It turns out that while the Bethe ansatz eigenstates equation (4.102) are real for the "easy-plane" magnetic anisotropy $\eta \to i\eta$, the Hamiltonian equation (4.100) is non-Hermitian for such a choice of the anisotropy, and, hence, we shall not consider that case in what follows.

Now let us study the thermodynamic properties of the supersymmetric t-J chain using the string hypothesis. Let us start with the case $V = -J/4$.

In the thermodynamic limit $L, N, M \to \infty$ with N/L and M/L kept fixed we can consider three main classes of solutions of Eq. (4.96). The first class consists of $N - 2M^*$ real charge rapidities p_j, which correspond to unbound electron excitations with densities $\rho(p)$, densities of holes $\rho_h(p)$ and dressed energies $\varepsilon(p) = T\ln[\rho_h(p)/\rho(p)] = T\ln\xi(p)$. The second class represents complex charge rapidities describing spin-singlet pairs (bound states) of electrons with $p_\alpha^\pm = \lambda'_\alpha \pm i(1/2)$, where λ'_α are real and characterize the centre of motion of pairs and $\alpha = 1, \ldots, M'$. These excitations have densities $\sigma'(\lambda)$, densities of holes $\sigma'_h(\lambda)$, and dressed energies $\psi(\lambda) = T\ln[\sigma'_h(\lambda)/\sigma'(\lambda)] = T\ln\kappa(\lambda)$. It turns out that the supersymmetric t-J chain has no bound states between pairs. The third class consists of M_n spin strings (bound states) of length $(n-1)$ of the form $\lambda^l_{\alpha,n} = \lambda_{\alpha,n} + i(n+1-2l)/2$ $(l = 1, \ldots, n)$ with real $\lambda_{\alpha,n}$ and $\alpha = 1, \ldots, M_n$. Naturally, because M is the number of down spins, we have $M = M^* + \sum_{n=1}^\infty nM_n$. These excitations carry only spin and no charge and have densities $\sigma_n(\lambda)$, densities of holes $\sigma_{nh}(\lambda)$, and dressed energies $\phi_n(\lambda) = T\ln[\sigma_{nh}(\lambda)/\sigma_n(\lambda)] = T\ln\eta_n(\lambda)$.

Then, by using straightforward procedures, similar to the case of a Hubbard chain, we get thermodynamic Bethe ansatz equations for densities

$$\begin{aligned} \rho(p) + \rho_h(p) &= a_1(p) - \sum_{n=1}^\infty a_n(p-\lambda) * \sigma_n(\lambda) - a_1(p-\lambda) * \sigma'(\lambda) , \\ \sigma_{nh}(\lambda) &= a_n(\lambda - p) * \rho(p) - \sum_{m=1}^\infty A_{nm}(\lambda - \lambda') * \sigma_m(\lambda') , \\ \sigma'_h(\lambda) &= a_2(p) - a_2(\lambda-\lambda') * \sigma'(\lambda') - a_1(\lambda - p) * \rho(p) . \end{aligned} \tag{4.103}$$

The thermodynamic Bethe ansatz equations for dressed energies have the form

$$\begin{aligned} \varepsilon(p) &= -\left(J[1 - \pi a_1(p)] + \frac{H}{2} + \mu\right) \\ &\quad + Ta_1(p-\lambda) * \ln[1 + \kappa^{-1}(\lambda)] \\ &\quad - T\sum_{n=1}^\infty a_1(p-\lambda) * \ln[1 + \eta_n^{-1}(\lambda)] , \\ T\ln[1 + \eta_n(\lambda)] &= nH + Ta_n(\lambda - p) * \ln[1 + \xi^{-1}(p)] \\ &\quad + T\sum_{m=1}^\infty A_{nm}(\lambda - \lambda') * \ln[1 + \eta_m^{-1}(\lambda')] , \end{aligned}$$

$$\psi(\lambda) = -\left(J[2 - \pi a_2(\lambda)] + 2\mu\right) + Ta_1(\lambda - p) * \ln[1 + \xi^{-1}(p)]$$
$$+ Ta_2(\lambda - \lambda') * \ln[1 + \kappa^{-1}(\lambda')] \,. \tag{4.104}$$

Here $*$ denotes convolution, $a_n(x) = (n/2)/\pi[x^2 + (n/2)^2]$, and the Fourier transform of $A_{nm}(x)$ is $\coth(|\omega|/4)[\exp(-|n - m||\omega|/4) + \exp(-(n - m)|\omega|/4)]$. The internal energy, the number of electrons and the magnetization per site of the supersymmetric t-J chain for $V = -J/4$ are given by

$$e = -J\frac{N}{L} + J\pi \int_{-\infty}^{\infty} dp a_1(p)\rho(p) + J\pi \int_{-\infty}^{\infty} d\lambda a_2(\lambda)\sigma'(\lambda) \,,$$
$$\frac{N}{L} = \int_{-\infty}^{\infty} dp\rho(p) + 2\int_{-\infty}^{\infty} d\lambda\sigma'(\lambda) \,, \tag{4.105}$$
$$m^z = \frac{1}{2}\int_{-\infty}^{\infty} dp\rho(p) - \sum_{n=1}^{\infty} n \int_{-\infty}^{\infty} d\lambda\sigma_n(\lambda) \,.$$

The Helmholtz free energy of the supersymmetric t-J chain for $V = -J/4$ per site is equal to

$$f = -T\int_{-\infty}^{\infty} d\lambda a_2(\lambda)\ln[1 + \kappa^{-1}(\lambda)]$$
$$-T\int_{-\infty}^{\infty} dp a_1(p)\ln[1 + \xi^{-1}(p)] = \psi(0) - \mu - J \,. \tag{4.106}$$

At high temperatures we consider the limit $T \to \infty$, keeping H/T and μ/T finite. In this limit the terms, which depend on p and λ in driving terms can be neglected, and ξ, η_n and κ are constants. There is no movement of excitations from site to site in this limit. The solutions of the thermodynamic Bethe ansatz equations are

$$\xi = \sqrt{\frac{1+\kappa}{1+\eta_1}} \,, \quad \eta_n = \left[\frac{\sinh\left(\frac{nH}{2T} + x\right)}{\sinh\left(\frac{H}{2T}\right)}\right]^2 - 1 \,,$$

$$(1+\kappa)^{3/2}\kappa^2 = \exp(-2\mu/T)[\sqrt{1+\kappa} + \sqrt{1+\eta_1}] \,, \tag{4.107}$$

$$\exp(x) = \exp(H/T)\sqrt{\frac{1 + \exp[-(H + 2\mu)/2T]}{1 + \exp[(H - 2\mu)/2T]}} \,.$$

The Helmholtz free energy per site at high temperatures is

$$f = -T\ln[1 + 2\exp(\mu/T)\cosh(H/2T)] , \tag{4.108}$$

which describes three degrees of freedom per site (a hole and a free spin-$\frac{1}{2}$). The total number of electrons and the magnetization per site are

$$\begin{aligned} \frac{N}{L} &= \frac{2\cosh(H/2T)}{2\cosh(H/2T) + \exp(-\mu/T)} , \\ m^z &= \frac{\sinh(H/2T)}{2\cosh(H/2T) + \exp(-\mu/T)} . \end{aligned} \tag{4.109}$$

The charge and magnetic susceptibilities of the t-J chain at high temperatures are

$$\begin{aligned} \chi_c &= \frac{1}{T}\frac{2\exp(-\mu/2T)\cosh(H/2T)}{[2\cosh(H/2T) + \exp(-\mu/T)]^2} , \\ \chi &= \frac{1}{T}\frac{1 + \exp(-\mu/T)\cosh(H/2T)}{[2\cosh(H/2T) + \exp(-\mu/T)]^2} . \end{aligned} \tag{4.110}$$

These expressions manifest that at high temperatures the supersymmetric t-J chain with $V = -J/4$ has no phase transitions and the behaviour of its characteristics is smooth with T, μ and H.

For low T we again, as in the previous chapter, separate dressed energies into their positive and negative parts, so that the terms with "+" superscript are positive (empty states in the ground state) and those with "−" superscript are negative (those which form Dirac seas). By using similar procedures as in the previous section for a Hubbard chain, the reader can see that dressed energies are positive, $\phi_n > 0$, for any n, hence, these excitations have no Dirac seas. Then the ground state equations for dressed energies and densities have the form

$$\begin{aligned} &\psi_1(\lambda) + \int d\lambda' a_2(\lambda' - \lambda)\psi_1^-(\lambda') \\ &\quad = -2J + \pi J a_2(\lambda) - 2\mu - \int dp a_1(\lambda - p)\varepsilon^-(p) , \end{aligned} \tag{4.111}$$

$$\varepsilon(p) = -J + \pi J a_1(p) - \mu - \frac{H}{2} - \int d\lambda a_1(\lambda - p)\psi_1^-(\lambda) ,$$

and

$$\sigma_1'(\lambda) + \sigma_{1h}'(\lambda) + \int d\lambda' a_2(\lambda' - \lambda)\sigma_1'(\lambda')$$
$$= a_2(\lambda) - \int dp a_1(\lambda - p)\rho(p) \,, \tag{4.112}$$
$$\rho(p) + \rho_h(p) = a_1(p) - \int d\lambda a_1(\lambda - p)\sigma_1'(\lambda) \,,$$

where the Fermi points for unbound electrons and spin-singlet pairs are related to the values of the chemical potential and magnetic field and determined from the conditions $\varepsilon(\pm B) = 0$ and $\psi_1(\pm Q) = 0$. These equations are similar to the ones for the attractive Hubbard chain.

For the ferromagnetic coupling $J = -2$, two dressed energies $\varepsilon(p)$ and $\psi(\lambda)$ have their minima at $p = \lambda = 0$. Then the ground state energy is minimum if $\psi(0)$ is maximum, *i.e.*, when the band of pairs is empty, $Q = 0$, (no spin-paired electrons, which is clear for the ferromagnetic case). All electrons occupy states of the band $\varepsilon(p)$, with

$$\varepsilon(p) = 2 - \frac{4}{4p^2 + 1} - \mu - \frac{H}{2} \,. \tag{4.113}$$

Hence, all states with $|p| < B$ are occupied (the Dirac sea for unbound electrons), where $B^2 = (4 + 2\mu + H)/4(4 - 2\mu - H)$.

For the antiferromagnetic situation $J = 2$, the dressed energies are decreasing functions of $|p|$ and $|\lambda|$. Hence, the Dirac seas, and, therefore, the integrations in Eqs. (4.111) and (4.112) are in the intervals $|p| \geq B$ and $|\lambda| \geq Q$.

The case with one electron per site $N = L$, since the double occupancy is excluded, pertains to the insulator. The hole density σ_h' vanishes, $Q = 0$, and eliminating σ' by using the Fourier transform we get

$$\rho(p) + \rho_h(p) = G_0(p) + \int_{|p'| \geq B} dp' G_1(p' - p)\rho(p') \,, \tag{4.114}$$

where $G_n(x) = \int_{-\infty}^{\infty} d\omega \exp[-i\omega x - (n|\omega|/2)]/4\pi \cosh(\omega/2)$. In the absence of the magnetic field we have $\rho = 0$ (and $\sigma'(\lambda) = G_0(\lambda)$). This is why, the ground state energy is equal to

$$e_0 = -2 \ln 2 \,. \tag{4.115}$$

The ground state is singlet. There is no gap for spin-carrying excitations (unlike an attractive Hubbard chain). The magnetic susceptibility is finite.

It coincides with the magnetic susceptibility of a Heisenberg spin-$\frac{1}{2}$ antiferromagnetic chain, described in the previous chapter, with $J = 2$. Any weak magnetic field begins to polarize electrons linearly with small logarithmic corrections. At the critical value H_s (related to $B = 0$), the magnetic field totally polarizes all the spins of electrons, and the system undergoes the quantum phase transition into the spin-polarized (ferromagnetic) phase. It is important to point out that, despite the ground state Bethe ansatz equations for a supersymmetric antiferromagnetic t-J chain with $V = -1/2$ are similar to the ones for an attractive Hubbard chain, there is no gap for unbound electron excitations for a t-J model, and, hence there is no phase with only paired electrons (which implies $H_c = 0$).

For the metallic case $N < L$, the integral equation for density of unbound electrons can be written as

$$\rho(p) + \rho_h(p) = G_0(p) + \int_{|p'|\geq B} dp' G_1(p' - p)\rho(p') + \int_{-Q}^{Q} d\lambda G_0(p - \lambda)\sigma_h'(\lambda) \ . \tag{4.116}$$

Actually the holes of pairs determine the metallic behaviour of the chain. The magnetization is zero for $H \to 0$. Notice, that even for a small magnetic field the Fermi point B is much larger than any given Q. Then the last term in the right hand side can be considered as small perturbation and for small H (large B) we can solve this equation using the Wiener–Hopf method, described the previous chapter. The $H = 0$ magnetic susceptibility ($\chi = 1/4\pi v_\rho^F$, where v_ρ^F is the Fermi velocity of unbound electron excitations, defined as for the case of a Hubbard chain), is

$$\chi = \frac{1}{4\pi^2} \frac{1 + \int_{-Q}^{Q} d\lambda \sigma_h'(\lambda) \exp(\pi\lambda)}{1 + (1/2\pi)\int_{-Q}^{Q} d\lambda \psi(\lambda) \exp(\pi\lambda)} \ , \tag{4.117}$$

where σ_h and ψ are solutions for zero magnetic field. The zero-field magnetic susceptibility diverges for $N \to 0$ as the consequence of the van Hove singularity of the empty band of pairs.

At $H = 0$, the magnetization is zero and, hence, $B = \infty$. All electrons are bound in pairs (we assumed the even number of electrons), and there are no unbound electron excitations. The equation for the density of pairs in this case is

$$\sigma(\lambda) + \sigma_h(\lambda) = G_0(\lambda) + \int_{-Q}^{Q} d\lambda' G_1(\lambda' - \lambda)\sigma_h(\lambda') \ . \tag{4.118}$$

If the band is almost half-filled (Q small) we have $\sigma'(\lambda) = [2-(N/L)]G_1(\lambda)$ for $|\lambda| \geq Q$ and zero elsewhere, where the number of electrons per site is $(N/L) = 1 - (2Q/\pi)\ln 2$. On the other hand, if the band is almost empty (Q large) we have for the number of electrons per site $(N/L) = (\pi Q)^{-1} + (1/2)\ln(Q)(\pi Q)^{-2} + \ldots$. The Fermi point can be related to the value of the chemical potential $\mu = Q^{-2} - 2$. The charge stiffness is inverse proportional to the Fermi velocity of pairs. It is, naturally, divergent at $N \to 0$ (for empty band, where $\mu = -2$), because the Fermi velocity of pairs goes to zero in this limit.

In the metallic situation, where both $v_\rho^F \neq 0$ and $v_{\sigma'}^F \neq 0$, except of van Hove singularities (where the specific heat is proportional to $\sqrt{T}$) one has the general formula for the Sommerfeld coefficient of the low-temperature specific heat

$$\gamma = \frac{\pi}{3}\left(\frac{1}{v_\rho^F} + \frac{1}{v_{\sigma'}^F}\right) . \tag{4.119}$$

At finite but low temperatures the magnetic susceptibility is exponentially small for $H > H_s$. At $H = H_s$, the magnetic susceptibility displays the $\sqrt{T}$ feature corresponding to the van Hove singularity of empty bands. For $H < H_s$, on the other hand, the magnetic susceptibility is finite as $T \to 0$.

For the system with $V = 3J/4$, the thermal equilibrium properties are classified according to the string hypothesis as strings of the length $(n-1)$ of both p_j and λ_α, introduced as

$$\begin{aligned} p_{j,n}^\nu = p_{j,n} + \frac{1}{2}i\nu \ , \ p_{\alpha,n}^\nu = \lambda_{\alpha,n} + \frac{1}{2}i\nu \ , \\ \nu = -(n-1), -(n-3), \ldots, (n-1), \quad n = 1, \ldots, \infty \ , \end{aligned} \tag{4.120}$$

where $p_{j,n}$ and $\lambda_{\alpha,n}$ are real and related to the momentum of the centre of mass of the bound state. Notice, that real rapidities p_j and λ_j pertain to $n = 1$. Dressed energies of string solutions can be denoted by $\varepsilon_n^{(l)} = T\ln\eta_n^{(l)}$, where $l = 1$ pertains to p-strings and $l = 2$ corresponds to λ-strings. Thermodynamic Bethe ansatz equations for dressed energies can be written as

$$\begin{aligned} \ln\eta_n^{(l)} &= -\frac{\pi J}{T}\delta_{n,1}F_m + F_2 * \ln\left[\frac{(1+\eta_{n+1}^{(l)})(1+\eta_{n-1}^{(l)})}{(1+\eta_n^{(m)})}\right] \\ &\quad + F_1 * \ln\left[\frac{(1+\eta_{n+1}^{(m)})(1+\eta_{n-1}^{(m)})}{(1+\eta_n^{(l)})}\right] , \end{aligned} \tag{4.121}$$

where $l, m = 1, 2$ $(l \neq m)$, $\eta_0^{(l)} = 0$ and

$$F_m(x) = \frac{1}{\sqrt{3}[2\cosh(2\pi x/3) - (-1)^m]} . \tag{4.122}$$

Equation (4.121) (which are similar in structure to the second form of thermodynamic Bethe ansatz equations for a Heisenberg chain, *cf.* the previous chapter) have to be complemented by the asymptotic conditions

$$\lim_{n\to\infty} \frac{1}{n} \ln \eta_n^{(1)} = \frac{2J - 2\mu - H}{2T} , \quad \lim_{n\to\infty} \frac{1}{n} \ln \eta_n^{(2)} = \frac{H}{T} . \tag{4.123}$$

The Helmholtz free energy per site of the supersymmetric t-J chain with $V = 3J/4$ is equal to

$$\begin{aligned} f = -\frac{J}{3} - T \int_{-\infty}^{\infty} dp F_2(p) \ln[1 + \eta_1^{(1)}(p)] \\ - T \int_{-\infty}^{\infty} d\lambda F_1(\lambda) \ln[1 + \eta_1^{(2)}(\lambda)] . \end{aligned} \tag{4.124}$$

The ground state pertains to the Dirac seas fillings of two low-lying excitations (*i.e.*, to solutions of Bethe ansatz equations with negative energies), namely, unbound electron excitations and spinons. The ground state Bethe ansatz equations for dressed energies are

$$\begin{aligned} &\varepsilon_1^{(1)}(p) + \int_{-Q}^{Q} dp' a_2(p - p') \varepsilon_1^{(1)}(p') \\ &\quad = J - \pi J a_1(p) - \mu - \frac{H}{2} + \int_{-B}^{B} d\lambda a_1(p - \lambda) \varepsilon_1^{(2)}(\lambda) , \\ &\varepsilon_1^{(2)}(\lambda) + \int_{-B}^{B} d\lambda' a_2(\lambda - \lambda') \varepsilon_1^{(2)}(\lambda') = H + \int_{-Q}^{Q} dp a_1(p - \lambda) \varepsilon_1^{(1)}(p) , \end{aligned} \tag{4.125}$$

and the equations for densities are

$$\begin{aligned} &\rho_1(p) + \rho_{1h}(p) + \int_{-Q}^{Q} dp' a_2(p - p') \rho_1(p') \\ &\quad = a_1(p) + \int_{-B}^{B} d\lambda a_1(p - \lambda) \sigma_1(\lambda) , \\ &\sigma_1(\lambda) + \sigma_{1h}(\lambda) + \int_{-B}^{B} d\lambda' a_2(\lambda - \lambda') \sigma_1(\lambda') \\ &\quad = \int_{-Q}^{Q} dp a_1(p - \lambda) \rho_1(p) . \end{aligned} \tag{4.126}$$

The internal energy, the number of electrons and the magnetization per site in the ground state are

$$
\begin{aligned}
e_0 &= J \int_{-Q}^{Q} dp[1 - \pi a_1(p)]\rho_1(p) , \\
\frac{N}{L} &= \int_{-Q}^{Q} dp\rho_1(p) , \\
m^z &= \frac{N}{2L} - \int_{-Q}^{Q} d\lambda\sigma_1(\lambda) .
\end{aligned}
\tag{4.127}
$$

The ferromagnetic situation $J = -2$ is trivial, as for the case $V = -J/4$. For the most interesting antiferromagnetic situation $J = 2$ in the absence of a magnetic field we have $B = \infty$ and the Dirac sea of spinons is totally filled, $\sigma_{1h} = 0$. Then the dressed energy and density of spinons can be eliminated *via* a Fourier transformation. Hence, in zero field the ground state problem reduces to the solution of Fredholm integral equations of the second kind

$$
\begin{aligned}
&\varepsilon_1^{(1)}(p) + \int_{-Q}^{Q} dp' G_3(p - p')\varepsilon_1^{(1)}(p') = 2 - 2\pi a_1(p) - \mu , \\
&\varepsilon_1^{(2)}(\lambda) = \int_{-Q}^{Q} dp G_0(p - \lambda)\varepsilon_1^{(1)}(p) , \\
&\rho_1(p) + \rho_{1h}(p) + \int_{-Q}^{Q} dp' G_3(p - p')\rho_1(p') = a_1(p) , \\
&\sigma_1(\lambda) = \int_{-Q}^{Q} dp G_0(p - \lambda)\rho_1(p) ,
\end{aligned}
\tag{4.128}
$$

where $G_n(x)$ is the Fourier transform of $\exp(-n|\omega|/2)/2\cosh(\omega/2)$. We can solve these equations analytically for small Q (which pertains to the small number of electrons in the system) and large Q. At small Q the dominant feature is the van Hove singularity of the empty band for unbound electron excitations, and we have $\mu = -2 + 16Q^2$, $(N/L) = (1/\pi)\sqrt{2+\mu}$ and $E_0 = -2N$. On the other hand, for $Q \to \infty$ we obtain $\mu = 2$, $(N/L) = 2/3$ (it is the maximum band filling for this model, which corresponds to the equal number of electrons with spins up, down, and empty sites; it is the point of higher symmetry) and $e_0 = (2/3)[\psi(1/3) - \psi(1) + 2]$, where $\psi(x)$ is a digamma function. For large, but finite Q the Wiener–Hopf-like solution of the integral equations gives $Q \sim |\ln(2-\mu)|$, *e.g.*, $(N/L) = (2/3) - (2-\mu)/3\pi$.

Why does the solution not exist for $N > 2L/3$? The Bethe ansatz is formulated for the mathematical vacuum state with all spins pointing upward. All states with non-negative magnetization can be generated by reversing spins. The construction, however, breaks down if more than half of the spins are reversed, because the wave functions (plane waves for the Bethe ansatz) are no longer linearly independent. States of negative magnetization are created by flipping states from the mathematical vacuum with all spins pointing downward. A similar situation arises here for band fillings larger than 2/3. Then empty states are in the minority as compared to the electrons with spins up and down, and a different representation of the mathematical vacuum state has to be used. If we assume that the number of electrons with spins up is larger or equal to the number of electrons with spins down the mathematical vacuum state consists of the filled band with all L electron spins pointing upward, rather than the mathematical vacuum corresponding to the absence of electrons, which we used.

The charge stiffness is the monotonically decreasing function of N, which becomes zero at $N = 2L/3$ and it is divergent at small N as $1/Q$, as the consequence of the van Hove singularity of the empty band of charged unbound electron excitations. It is inverse proportional to the Fermi velocity of those charged low-lying excitations. The magnetic susceptibility is equal to

$$\chi = -\frac{\int_{-Q}^{Q} dp \exp(-\pi p)\rho_1(p)}{2\pi \int_{-Q}^{Q} dp \exp(-\pi p)\varepsilon_1^{(1)}(p)} \ . \tag{4.129}$$

Again, except for van Hove singularities (where the specific heat is proportional to $\sqrt{T}$) one has the general formula for the Sommerfeld coefficient of the low-temperature specific heat

$$\gamma = \frac{\pi}{3}\left(\frac{1}{v_{\rho_1}^F} + \frac{1}{v_{\sigma_1}^F}\right) , \tag{4.130}$$

and the magnetic susceptibility is $\chi = (1/4\pi)[(v_{\sigma_1}^F)^{-1} + (v_{\rho_1}^F)^{-1}]$. The latter is divergent at $N \to 0$, where $v_{\sigma_1}^F \sim Q^2$, and monotonically decreases with increasing N.

At high temperatures for $N = 2L/3$ (where there is no charge dynamics) for $H = 0$, the Helmholtz free energy per site is

$$f = -T \ln 3 \ . \tag{4.131}$$

This result is simple to understand, if one takes into account that the Hamiltonian of the supersymmetric t-J model with $V = 3J/4$ is equivalent to the Hamiltonian of the spin-1 system. For the finite ratios H/T and μ/T the free energy at high temperatures is the smooth function of μ, H and T, *i.e.*, there are no phase transitions in this model at high T.

Now we want to consider the supersymmetric antiferromagnetic t-J chain with an "easy-axis" magnetic anisotropy and $V = -1/2$. This class of models reveals the most interesting properties in the ground state, and, hence, we restrict ourselves here with only $T = 0$ and low temperatures. The ground state of the system is given by $N - 2M$ unbound electron states (with real charge rapidities p_j) and M singlet Cooper-like bound states for which charge rapidities are complex conjugated pairs. It follows from Eq. (4.101) that they are related to spin rapidities λ_β, such that (to exponential accuracy e^{-L}) $p_\alpha^\pm = \lambda_\beta \pm i\frac{\eta}{2}$. Inserting real charge rapidities p_j and pair solutions (characterized by λ_α) into Eq. (4.101) and taking the logarithm of resulting equations we obtain

$$
\begin{aligned}
L\Theta\left[p_j, \eta/2\right] &= 2\pi I_j + \sum_{\alpha=1}^{M} \Theta\left[p_j - \lambda_\alpha, \eta/2\right] \ , \ j = 1, ..., N - 2M \ , \\
L\Theta\left[\lambda_\alpha, \eta\right] &= 2\pi J_\alpha + \sum_{j=1}^{N-2M} \Theta\left[\lambda_\alpha - p_j, \eta/2\right] \\
&\quad + \sum_{\substack{\beta=1 \\ \beta\neq\alpha}}^{M} \Theta\left[\lambda_\alpha - \lambda_\beta, \eta\right] \ , \ \alpha = 1, ..., M \ ,
\end{aligned}
\tag{4.132}
$$

where $\Theta[x, \eta] = 2\tan^{-1}(\tan x \coth \eta)$. The quantum numbers I_j and J_α arise because the logarithm is a multivalued function. Quantum numbers completely determine the solutions for the ground state and elementary excitations. The ground state energy of the system is

$$
\begin{aligned}
E_0 = -2 &\sum_{j=1}^{N-2M} \frac{1 - \cos(2p_j)\cosh(\eta)}{\cosh(\eta) - \cos(2p_j)} \\
&- 2\cosh(\eta) \sum_{\alpha=1}^{M} \left(2 - \frac{\sinh^2(\eta)}{\sin^2(\lambda_\alpha) + \sinh^2(\eta)}\right) \ .
\end{aligned}
\tag{4.133}
$$

In the thermodynamic limit (*i.e.*, $L, N, M \to \infty$ with the ratios N/L and M/L remaining fixed) we introduce densities for rapidities, $\rho(p)$ and

$\sigma'(\lambda)$, and their holes, $\rho_h(p)$ and $\sigma'_h(\lambda)$. Bethe ansatz equations satisfied by densities are

$$\begin{aligned}\Theta'[p,\eta/2] &= \int d\lambda \Theta'[p-\lambda,\eta/2]\,\sigma'(\lambda) + 2\pi[\rho(p)+\rho_h(p)]\ ,\\ \Theta'[\lambda,\eta] &= \int dp \Theta'[\lambda-p,\eta/2]\,\rho(p) \\ &\quad + \int d\lambda' \Theta'[\lambda-\lambda',\eta]\,\sigma'(\lambda') + 2\pi[\sigma'(\lambda)+\sigma'_h(\lambda)]\ ,\end{aligned} \tag{4.134}$$

where the prime at $\Theta[x,y]$ denotes derivative with respect to the first argument (x). Dressed energies, $\varepsilon(p)$ for unbound electron states and $\psi(\lambda)$ for singlet pairs, satisfy following integral equations in the ground state

$$\begin{aligned}\Theta'[p,\eta/2] - \mu - \frac{H}{2} &= \frac{1}{2\pi}\int d\Lambda \Theta'[p-\lambda,\eta/2]\,\psi(\lambda) + \varepsilon(p)\ ,\\ \Theta'[\lambda,\eta] - 2\mu &= \frac{1}{2\pi}\int dv \Theta'[\lambda-p,\eta/2]\,\varepsilon(p) \\ &\quad + \frac{1}{2\pi}\int d\lambda' \Theta'[\lambda-\lambda',\eta]\,\psi(\lambda') + \psi(\lambda)\ .\end{aligned} \tag{4.135}$$

All states with negative (positive) dressed energy are populated (empty). The bands $\varepsilon(p)$ and $\psi(\lambda)$ can form Dirac seas with the filling beginning at the edges of the interval $[-\pi,\pi]$, where dressed energies have their minimum. In the thermodynamic limit the internal ground state energy of the system is

$$\begin{aligned}e_0 &= -2\int \rho(p)\left[\frac{1-\cos(2p)\cosh(\eta)}{\cosh(\eta)-\cos(2p)}\right] dp \\ &\quad - 2\cosh(\eta)\int \sigma'(\lambda)\left[2 - \frac{\sinh^2(\eta)}{\sin^2(\lambda)+\sinh^2(\eta)}\right] d\lambda\ .\end{aligned} \tag{4.136}$$

The number of electrons and the z-projection of the magnetization per site are given by

$$(N/L) = 2\int d\lambda \sigma'(\lambda) + \int dp \rho(p)\ ,\ m^z = (1/2)\int dp \rho(p)\ , \tag{4.137}$$

respectively.

The energy of unbound electron states are gapped for an external magnetic field less than a critical value, H_c, given by

$$H_c = -2\mu + 2\Theta'[\pi,\eta/2] - \frac{1}{\pi}\int d\lambda \Theta'[\pi-\lambda,\eta/2]\,\psi(\lambda)\ . \tag{4.138}$$

In other words, H_c is one half of the minimal external magnetic field necessary to depair a singlet bound state. The presence of a spin gap and, hence, this quantum phase transition in the ground state of an anisotropic supersymmetric t-J chain at H_c distinguishes it from the isotropic situation (at which $\eta = 0$, and, thus, $H_c = 0$). The presence of a spin gap is similar to a Hubbard chain with an attraction between electrons. However, the drastic difference exists between the present model and an attractive Hubbard chain. Namely, the attractive Hubbard chain respects SU(2) magnetic symmetry, while the present model reveals only U(1) spin symmetry.

If the value of an external magnetic field is larger than H_s, given by

$$H_s = -2\mu + 2\Theta' \left[\pi, \eta/2\right] \ , \tag{4.139}$$

the magnetization is maximal, *i.e.*, saturated. At this saturation field the system undergoes a second order quantum phase transition into the ferromagnetic spin-polarized state, in which there are no pairs because the dressed energy of bound electrons is gapped. This behaviour is also similar to a type-II superconductor in a magnetic field: for $H \leq H_c$ there are only Cooper-pairs, while for $H_c \leq H \leq H_s$ pairs and unbound electrons coexist, which is reminiscent of the Meissner effect. Note, however, that in a one-dimensional electron model there is no true superconducting order with off-diagonal long range orderings, but the correlation functions of singlet pairs and/or unbound electrons fall off with power-laws for long times and/or distances. For $H \geq H_s$, it is straightforward to obtain the ground state energy. In the intermediate phase, $H_c \leq H \leq H_s$, however, the ground state energy depends on the filling of both Dirac seas.

We first study the case $H < H_c$, where the ground state Dirac sea only of singlet pairs ($2M = N$) is present. In this case Bethe ansatz equations reduce to only one set of equations,

$$\Theta' \left[\lambda, \eta\right] = 2\pi[\sigma'(\lambda) + \sigma_h'(\Lambda)] + \left[\int_{-\pi}^{-Q} + \int_{Q}^{\pi}\right] d\lambda' \Theta' \left[\lambda - \lambda', \eta\right] \sigma'(\lambda') \ . \tag{4.140}$$

Here $\pm Q$ are Fermi points, because in the ground state only states with $\lambda \in [-\pi, -Q] \cup [Q, \pi]$ are filled. The wave functions of pairs are symmetric (pairs of electrons form bosons), but these bosons are hard-core ones, satisfying an *anyon-like exclusion statistics*, as a consequence of interactions among pairs. Because they are hard-core bosons, Cooper-pairs form a Dirac sea. The parameter Q is related to the chemical potential, μ, *via* $\psi(\pm Q) = 0$.

The energy and the total number of electrons are now given by Eq. (4.136) with only the σ density term integrated over occupied states.

Next we consider the situation $H_c \leq H \leq H_s$, where both, unbound electrons and singlet pairs, have gapless low-lying excitations, *i.e.*, form Dirac seas. Due to the van Hove singularity of the empty band of unpaired electron states, the magnetization is proportional to $\sqrt{H - H_c}$ for fields H slightly larger than H_c. This feature is characteristic of a Pokrovsky–Talapov level-crossing transition, which is the analog of a second order phase transitions in one-dimension. With increasing magnetic field the population of the Dirac sea of singlet pairs gradually decreases until H_s is reached, which is the field at which the band is empty. For fields larger than the saturation field H_s the magnetization is equal to $m^z = (N/2L)$.

At finite but low temperatures the magnetic susceptibility is exponentially small for $H < H_c$ and $H > H_s$. At $H = H_c$ or H_s the magnetic susceptibility and the Sommerfeld coefficient of the specific heat display $\sqrt{T}$ features corresponding to van Hove singularities of empty bands. For $H_c < H < H_s$, on the other hand, the magnetic susceptibility is finite as $T \to 0$. The specific heat is proportional to temperature everywhere away from van Hove singularities.

To summarize, in this chapter we presented the derivations of exact solutions of stationary Schrödinger equations for several models of highly correlated electrons: one-dimensional Hubbard repulsive and attractive chains and supersymmetric t-J chains (with and without anisotropy of interactions) with periodic boundary conditions in the framework of the co-ordinate nested Bethe ansätze. Thermodynamic Bethe ansatz equations are derived and solved analytically in several important cases. The ground state behaviour and low-energy behaviour of thermodynamic characteristics of these models are analyzed.

The importance of interactions in electron systems was pointed out in [Anderson (1959)]. The Hubbard model was introduced in [Hubbard (1963); Gutzwiller (1963); Kanamori (1963)]. The co-ordinate Bethe ansatz solution for a repulsive Hubbard chain is given in [Lieb and Wu (1968)]. A nested Bethe ansatz scheme for correlated electron systems was pioneered in [Gaudin (1967); Yang (1967)]. The ground state magnetic susceptibility of a repulsive Hubbard chain was given in [Shiba (1970)]. For thermodynamics of a one-dimensional repulsive Hubbard model in the framework of the string hypothesis, consult [Takahashi (1999)], see also [Jüttner, Klümper and Suzuki (1998)] for thermodynamic characteristics of a repulsive

Hubbard chain. The exact solution of an attractive Hubbard chain can be found in [Bahder and Woynarovich (1986); Lee and Schlottmann (1988); Lee and Schlottmann (1989)], see also [Sacramento (1994); Sacramento (1995)] for the numerical solution of thermodynamic Bethe ansatz equations in this case. The completeness of the Bethe ansatz solution for a Hubbard chain was proved in [Eßler, Korepin and Schoutens (1992)]. The one-dimensional supersymmetric t-J model was solved in [Lai (1974); Sutherland (1975); Schlottmann (1987)] (notice that the SU(3)-symmetric t-J model was solved even earlier, in [Uimin (1970)]). The completeness of the Bethe ansatz solution for a supersymmetric t-J chain was proved in [Foerster and Karowski (1993a)]. The reader can find the exact solution of an anisotropic t-J chain in [Bariev (1994); Bariev, Klümper, Schadschneider and Zittartz (1993); Foerster and Karowski (1993b)]. The description of the crossover one-dimensional phase transition can be found in [Pokrovsky and Talapov (1979)]. The exclusion statistics was introduced in [Haldane (1991)].

Chapter 5

Algebraic Bethe Ansatz

In this chapter we shall describe the *algebraic Bethe ansatz*. This version of the Bethe ansatz will be very useful for following studies of inhomogeneous quantum chains, the main topic of this book. Generally speaking, it is very important for the search of models which permit Bethe ansatz solutions.

5.1 The Algebraic Bethe Ansatz for a Spin-$\frac{1}{2}$ Chain

As we pointed out in the previous chapter, when we studied the nested Bethe ansatz for a Hubbard chain, the problem for a quantum spin-$\frac{1}{2}$ chain can be solved in a different way than using the co-ordinate Bethe ansatz from Chapter 3. Namely, we take into account that the condition on a Bethe ansatz two-spin nested wave function for a Hubbard chain (which is, in fact, the wave function of an inhomogeneous one-dimensional Heisenberg spin-$\frac{1}{2}$ model) can be written as

$$\begin{aligned}|\dots, k_{P_m}, \dots, k_{P_n} \dots\rangle \\ = Y_{mn}(\sin k_{P_m}, \sin k_{P_n})|\dots, k_{P_n}, \dots, k_{P_m}, \dots\rangle \ . \qquad (5.1)\end{aligned}$$

Here we used the wave function of a Hubbard chain

$$|k_{P_1}, \dots, k_{P_N}\rangle = \sum_{\sigma_1, \cdots, \sigma_N} A_{\sigma_1, \cdots, \sigma_N}(k_{P_1}, \dots, k_{P_N})|x_1, \dots, x_M\rangle \ , \qquad (5.2)$$

where $|x_1, \dots, x_M\rangle$ denotes the wave function with M down spins (of N electrons) at positions $x_1 < \cdots < x_M$ and the two-particle scattering matrix Y of a Hubbard chain.

In general for a SU(2)-symmetric spin-$\frac{1}{2}$ chain of length L one can introduce spectral parameters λ (which, *e.g.*, can be related to momenta of

Bethe ansatz wave functions), so that

$$Y_{mn}(\lambda) = \frac{\lambda \hat{I}_{mn} + ic\hat{P}_{mn}}{\lambda + ic} , \tag{5.3}$$

with $\hat{I}_{mn}$ and $\hat{P}_{mn}$ being the identity and permutation operators acting in the Hilbert space of the m-th and n-th spin $(n, m = 1, \ldots, L)$ $V_n \otimes V_n$, where V_n is isomorphic to C^2, and c is some coupling constant. Here the identity and permutation operators in that Hilbert space are $\hat{I}^{\beta\beta'}_{\alpha\alpha'} = \delta_{\alpha\alpha'}\delta_{\beta\beta'}$ and $\hat{P}^{\beta\beta'}_{\alpha\alpha'} = \delta_{\alpha\beta'}\delta_{\alpha'\beta}$. For example, for the Heisenberg spin-$\frac{1}{2}$ model the permutation operator is equal to $\hat{P} = (\hat{I} + 4\vec{S}_m\vec{S}_n)/2$. The permutation operator can be written as the 4×4 matrix in $C^2 \otimes C^2$

$$P^{\beta\beta'}_{\alpha\alpha'} = \begin{pmatrix} 1 & 0 & 0 & 0 \\ 0 & 0 & 1 & 0 \\ 0 & 1 & 0 & 0 \\ 0 & 0 & 0 & 1 \end{pmatrix} . \tag{5.4}$$

In the previous chapter we also showed that the Yang–Baxter relation for two-particle scattering matrices takes place

$$Y^{\sigma_1\sigma_1'}_{\sigma_2\sigma_2'}(\lambda)Y^{\sigma_1'\sigma_1''}_{\sigma_3\sigma_3'}(\lambda+\mu)Y^{\sigma_2'\sigma_2''}_{\sigma_3'\sigma_3''}(\mu) = Y^{\sigma_2\sigma_2'}_{\sigma_3\sigma_3'}(\mu)Y^{\sigma_1\sigma_1'}_{\sigma_3'\sigma_3''}(\lambda+\mu)Y^{\sigma_1'\sigma_1''}_{\sigma_2'\sigma_2''}(\lambda) , \tag{5.5}$$

where summation over repeated indices is understood. The use of the co-ordinate Bethe ansatz implies that all two-particle scattering matrices satisfy Yang–Baxter relations. [These relations are called sometimes "triangular", or "star-triangular" ones.]

Generally speaking, we can introduce some R-matrix (the central object of the algebraic Bethe ansatz, which is often also called the *quantum inverse scattering method*). This method was developed mostly by the Leningrad (St. Petersburg) group headed by L. D. Faddeev. For example, for spin-$\frac{1}{2}$ system it can be the 4×4 matrix acting on the tensor product space $V_0 \otimes V_0$, where V_0 is isomorphic to C^2,

$$R(\lambda) = \begin{pmatrix} a(\lambda) & 0 & 0 & 0 \\ 0 & b(\lambda) & c(\lambda) & 0 \\ 0 & c(\lambda) & b(\lambda) & 0 \\ 0 & 0 & 0 & a(\lambda) \end{pmatrix} , \tag{5.6}$$

where $a(\lambda)$, $b(\lambda)$ and $c(\lambda)$ are c-number functions. Actually, this structure is implied by the structure of the two-particle scattering matrix for spin-$\frac{1}{2}$ system

$$S(\lambda) = \begin{pmatrix} a(\lambda) & 0 & 0 & 0 \\ 0 & c(\lambda) & b(\lambda) & 0 \\ 0 & b(\lambda) & c(\lambda) & 0 \\ 0 & 0 & 0 & a(\lambda) \end{pmatrix} , \tag{5.7}$$

which is the generalization of the two-particle scattering matrix of the Heisenberg chain Y acting in $V_0 \otimes V_0$. It turns out that $R(\lambda) = \hat{P}S(\lambda)$. Suppose these two-particle scattering matrices satisfy the Yang–Baxter equation

$$S^{\sigma_1\sigma_1'}_{\sigma_2\sigma_2'}(\lambda) S^{\sigma_1'\sigma_1''}_{\sigma_3\sigma_3'}(\lambda+\mu) S^{\sigma_2'\sigma_2''}_{\sigma_3'\sigma_3''}(\mu) = S^{\sigma_2\sigma_2'}_{\sigma_3\sigma_3'}(\mu) S^{\sigma_1\sigma_1'}_{\sigma_3'\sigma_3''}(\lambda+\mu) S^{\sigma_1'\sigma_1''}_{\sigma_2'\sigma_2''}(\lambda) , \tag{5.8}$$

cf. Fig. 5.1 (from that figure, the reader understands why the Yang–Baxter relations are often referred to as "triangular" equations).

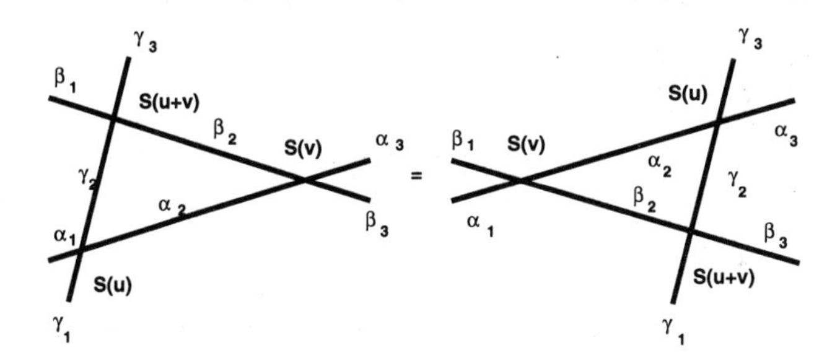

Fig. 5.1 Illustration of Yang–Baxter relations for two-particle scattering matrices.

We can demand from R-matrices to satisfy the Yang–Baxter equation which can be written in the symbolic form for the R-matrices acting in the space $V_0 \otimes V_0 \otimes V_0$ as

$$R_{23}(\lambda)R_{12}(\lambda+\mu)R_{23}(\mu) = R_{12}(\mu)R_{23}(\lambda+\mu)R_{12}(\lambda) , \tag{5.9}$$

where the subscripts indicate in which spaces the R-matrix acts nontrivially. For the concrete example, considered here, these equations imply the following relations between the functions $a(x)$, $b(x)$ and $c(x)$:

$$\frac{c(x)}{b(x)} = \frac{c(y)}{b(y)} + \frac{c(x-y)}{b(x-y)} , \tag{5.10}$$

for which a solution in the most simple form for a linear function of the spectral parameter for the quantity $c(x)/b(x)$ is

$$\frac{c(x)}{b(x)} = \frac{x}{ic} , \tag{5.11}$$

where c is the coupling constant. Taking into account that $S(\lambda = 0) = \hat{P}$ and, hence, $R(\lambda = 0) = \hat{I}$ (in the Hilbert space $V_0 \otimes V_0$) we can choose, *e.g.*, $a(\lambda) = 1$, $b(\lambda) = ic/(\lambda + ic)$ and $c(\lambda) = \lambda/(\lambda + ic)$, but such a choice is not necessary: the only condition for the R-matrices with the structure equation (5.6) to satisfy the Yang–Baxter equation is $a(\lambda) : b(\lambda) : c(\lambda) = (\lambda + ic) : ic : \lambda$. Naturally, our choice of the solution of the Yang–Baxter equation for R-matrices is not unique. In general, what is necessary for the algebraic Bethe ansatz is to have R-matrices which satisfy the Yang–Baxter relation.

We can now define an L-operator acting on the tensor product between the "matrix-space" V_0 (it is often called the *auxiliary subspace*) and the *quantum space* V_n, which is identified with the Hilbert space over the n-th site of our lattice. For example, for our case of spins $\frac{1}{2}$, we define

$$\begin{aligned} L_n(\lambda) &= \frac{\lambda \hat{I}_{0,n} + ic\hat{P}_{0,n}}{\lambda + ic} \\ &= (\lambda + ic)^{-1} \begin{pmatrix} \lambda I_n + i[(I_n/2) + S_n^z]c & 2iS_n^- c \\ 2iS_n^+ c & \lambda I_n + i[(I_n/2) - S_n^z]c \end{pmatrix} , \end{aligned} \tag{5.12}$$

where $\hat{I}_{0,n}$ and $\hat{P}_{0,n}$ are the identity and permutation operators in the space $V_0 \otimes V_n$, respectively, and I_n and $S_n^{z,+,-}$ are the identity and spin operators acting in the quantum Hilbert space V_n. We can see that the following Yang–Baxter relations for L-operators (they are often called *intertwining relations*) hold

$$R(\lambda - \mu)(L_n(\lambda) \otimes L_n(\mu)) = (L_n(\mu) \otimes L_n(\lambda))R(\lambda - \mu) , \tag{5.13}$$

where the tensor product is between quantum spaces, *i.e.*, these Yang–Baxter relations are nontrivial over the space $V_0 \otimes V_0 \otimes V_n$. In fact, intertwining relations determine the structure of L-operators, if one already knows R-matrices. Actually, in the Hilbert space of all L spins of the Heisenberg spin-$\frac{1}{2}$ chain the L-operator is the matrix $(L_j)_{\sigma_1,\ldots,\sigma_L,\tau}^{\sigma_1',\ldots,\sigma_L',\tau'}(\lambda)$ where the indices $\sigma_1, \ldots, \sigma_L$ and $\sigma_1', \ldots, \sigma_L'$ denote the state of spins at sites $1, \ldots, L$ before and after scattering (in the quantum space), while τ and τ' denote the states before and after scattering in the auxiliary subspace, *i.e.*, a L-operator acts in the space $V_0 \otimes V_1 \otimes \cdots \otimes V_L$. For a more general case indices of the quantum space define the state of all quantum particles in the system. However, the L-operator acts nontrivially only in the quantum subspace V_j:

$$(L_j)_{\sigma_1,\dots,\sigma_L,\tau}^{\sigma'_1,\dots,\sigma'_L,\tau'}(\lambda) = I_1 \otimes I_2 \otimes \cdots \otimes S_j(\lambda) \otimes \cdots \otimes I_L$$
$$= \delta_{\sigma_1,\sigma'_1}\delta_{\tau,\gamma_1} \cdots S_{\sigma_j\sigma'_j}^{\gamma_{j-1}\gamma_j}(\lambda) \cdots \delta_{\sigma_L,\sigma'_L}\delta_{\gamma_{L-1},\tau'} , \quad (5.14)$$

i.e., it is diagonal over the indices $\sigma_n\sigma'_n$ $(n = 1, \dots, L)$ except of $n = j$.

Then we can introduce the monodromy matrix acting in the space $V_0 \otimes V_1 \otimes \cdots \otimes V_L$ as

$$T_{\sigma_1,\dots,\sigma_L,\tau}^{\sigma'_1,\dots,\sigma'_L,\tau'}(\lambda) = L_1(\lambda) \cdots L_L(\lambda)$$
$$= S_{\sigma_1,\sigma'_1}^{\tau,\gamma_1}(\lambda) \cdots S_{\sigma_j\sigma'_j}^{\gamma_{j-1}\gamma_j}(\lambda) \cdots S_{\sigma_L,\sigma'_L}^{\gamma_{L-1},\tau'}(\lambda) , \quad (5.15)$$

cf. Fig. 5.2.

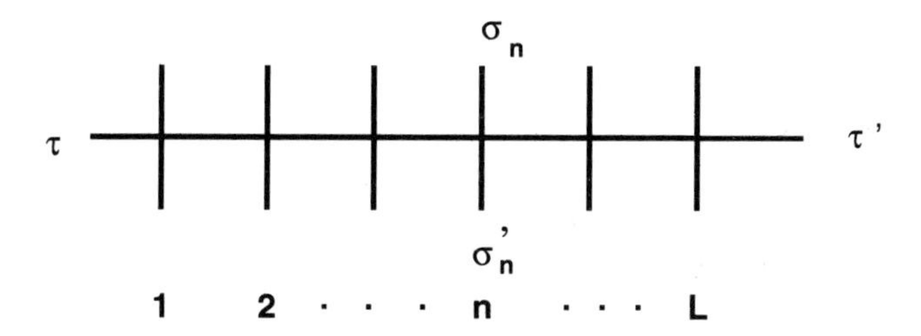

Fig. 5.2 Illustration of the monodromy operator of an integrable model.

Due to the definition and intertwining relations for L-operators the monodromy matrices also satisfy Yang–Baxter (intertwining) relations

$$R(\lambda - \mu)(T(\lambda) \otimes T(\mu)) = (T(\mu) \otimes T(\lambda))R(\lambda - \mu) , \quad (5.16)$$

see Fig. 5.3.

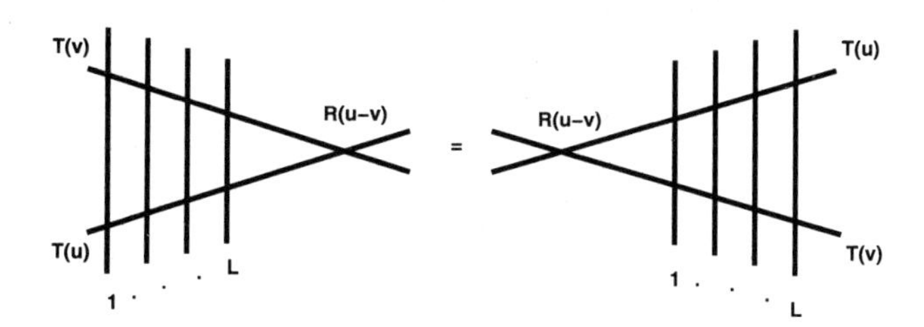

Fig. 5.3 Illustration of the intertwining relations for monodromy operators.

By tracing the monodromy matrix over the auxiliary space V_0, we get the transfer matrix

$$\hat{\tau}_{\sigma_1,\ldots,\sigma_L}^{\sigma'_1,\ldots,\sigma'_L}(\lambda) = \mathrm{Tr}L_1(\lambda)\cdots L_L(\lambda)$$
$$= S_{\sigma_1,\sigma'_1}^{\tau,\gamma_1}(\lambda)\cdots S_{\sigma_j\sigma'_j}^{\gamma_{j-1}\gamma_j}(\lambda)\cdots S_{\sigma_L,\sigma'_L}^{\gamma_{L-1},\tau}(\lambda) , \quad (5.17)$$

for the graphical illustration see Fig. 5.4.

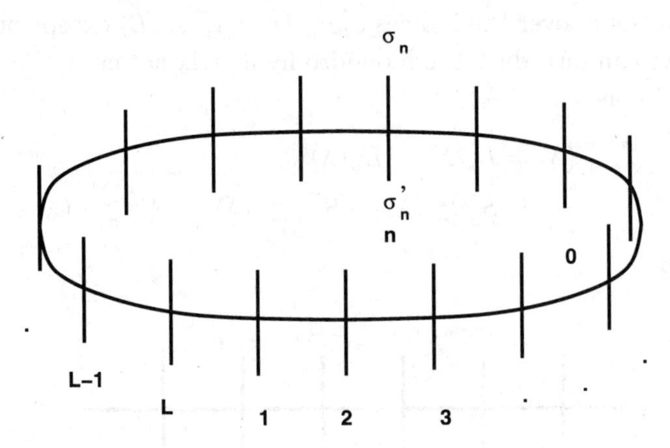

Fig. 5.4 Illustration of the transfer matrix of an integrable model.

Equations (5.16) and (5.17) imply that

$$[\hat{\tau}(\lambda), \hat{\tau}(\mu)] = 0 , \quad (5.18)$$

i.e., transfer matrices with different spectral parameters commute. This property is the fundamental property, which implies the exact integrability of a system. The exact integrability means that there exists infinitely many (for a system with infinitely many degrees of freedom) integrals of motion which commute mutually, and, hence, have the common set of eigenfunctions. Then one can construct any function of the transfer matrix, and due to the property equation (5.18), all such functions will commute mutually and with the transfer matrix.

Generally speaking, one can introduce any function of $\hat{\tau}(\lambda)$ as integrals of motion. In practice, the following series is used for the determination of integrals of motion (which was given by M. Lüscher, the Hamiltonian from this series was first introduced by B. Sutherland)

$$\hat{Q}_n(\lambda) = A_n \frac{\partial^{n-1} \ln \hat{\tau}(\lambda)}{\partial \lambda^{n-1}}|_{\lambda=0} , \quad (5.19)$$

where finally, after taking derivatives, the spectral parameter is taken to be equal to its value, at which the R-matrix is unity (or two-particle scattering matrix is the permutation operator), and A_n are constants. This series is chosen because of the *locality property*: the integral of motion $\hat{Q}_n$ acts nontrivially only on n sites of a chain.

Let us consider some integrals of motion of a Heisenberg spin-$\frac{1}{2}$ chain. For $n = 1$, we have

$$\hat{Q}_1 = A_1 \ln \hat{\tau}(0) = A_1 \ln \hat{P}_{01} \cdots \hat{P}_{0L} = \text{const}\hat{P} \ , \tag{5.20}$$

(where the subscript 0 denotes the auxiliary subspace), which is (up to a constant) the total momentum operator (where $\hat{\tau}(0)$ is the cyclic shift operator). The total momentum operator has the property

$$\exp(-i\hat{P}) S_j^{x,y,z} \exp(i\hat{P}) = S_{j+1}^{x,y,z} \ , \tag{5.21}$$

which follows from the definitions. For $n = 2$, we need to consider an operator which has the form

$$\begin{aligned}
&\hat{\tau}^{-1}(\lambda)\frac{\partial \hat{\tau}(\lambda)}{\partial \lambda}|_{\lambda=0} \\
&= \text{const} \sum_{j=1}^{L} \delta_{\sigma_1,\sigma_1'} \cdots \delta_{\sigma_{j-1},\sigma_{j-1}'} \frac{\partial S_{\sigma_{j+1}\sigma_j'}^{\sigma_j \sigma_{j+1}'}(\lambda)}{\partial \lambda}|_{\lambda=0} \delta_{\sigma_{j+2},\sigma_{j+2}'} \cdots \delta_{\sigma_L \sigma_L'} \\
&= -\text{const}\frac{i}{2c}\left[4\sum_{j=1}^{L-1} \vec{S}_j \vec{S}_{j+1} - L\hat{I}\right] \ ,
\end{aligned} \tag{5.22}$$

where I is the identity operator and we used the property

$$\frac{\partial S_{\sigma_{l+1}\sigma_l'}^{\sigma_l \sigma_{l+1}'}(\lambda)}{\partial \lambda}|_{\lambda=0} = -\frac{2i}{c}\left[\vec{S}_{\sigma_{l+1}\sigma_{l+1}'} \vec{S}_{\sigma_l \sigma_l'} - \frac{1}{4}\delta_{\sigma_{l+1},\sigma_{l+1}'} \delta_{\sigma_l \sigma_l'}\right] . \tag{5.23}$$

We see that $\hat{Q}_2$ coincides (up to constants) with the Hamiltonian of the spin-$\frac{1}{2}$ Heisenberg model.

The reader can check that for a Heisenberg–Ising spin-$\frac{1}{2}$ chain one needs to take the following parametrization of the trigonometric solution of the Yang–Baxter equation

$$a(\lambda) : c(\lambda) : b(\lambda) = \sin[\lambda + (\eta/2)] : \sin[\lambda - (\eta/2)] : \sin \eta \ , \tag{5.24}$$

which, after some straightforward calculations along the described above lines produces $\hat{Q}_2$ proportional to the Heisenberg–Ising Hamiltonian with the parameter of the magnetic anisotropy $(J_z/J) = \cos \eta$.

To summarize, the strategy of the algebraic Bethe ansatz is as follows. One starts with the solution of the Yang–Baxter equation for R-matrices. Using that solution one constructs L-operators and monodromy operators of a Bethe ansatz-integrable system. The trace of the monodromy operator over the auxiliary subspace is the transfer matrix. As the consequence of the Yang–Baxter relations for L-operators and monodromies (intertwining relations), transfer matrices with different spectral parameters commute, which constitutes the exact integrability of the model. Finally, the integrals of motion (including the operator of the energy, the Hamiltonian) can be constructed from the expression for the transfer matrix.

This is why, now the task of the algebraic Bethe ansatz is to find the eigenfunctions and eigenstates of the transfer matrix. For the spin-$\frac{1}{2}$ chain it was done in the previous chapter for the nested Bethe ansatz of the Hubbard chain. Here the reader can use those results taking them for $N = L$ and $\lambda^0_{1,\dots,N} = 0$, *i.e.*, for the homogeneous chain. Nevertheless, it is better to repeat here the main steps, because this procedure is the main issue of the algebraic Bethe ansatz.

Let us denote the elements of the 2×2 monodromy matrix in the auxiliary subspace V_0 as

$$T^{\tau}_{\tau'}(\lambda) = \begin{pmatrix} \hat{A} & \hat{B} \\ \hat{C} & \hat{D} \end{pmatrix} , \tag{5.25}$$

where the operators $\hat{A}$, $\hat{B}$, $\hat{C}$ and $\hat{D}$ in the matrix representation have indices $\sigma'_1, \dots, \sigma'_L, \sigma_1, \dots, \sigma_L$. It follows from the definition that

$$\hat{\tau}(\lambda) = \hat{A} + \hat{D} . \tag{5.26}$$

The operators $\hat{A}$, $\hat{B}$, $\hat{C}$ and $\hat{D}$ obey the following commutation relations, which stem from the intertwining relations for monodromies,

$$\begin{aligned} [\hat{A}(x), \hat{A}(y)] &= [\hat{D}(x), \hat{D}(y)] = [\hat{A}(x), \hat{D}(y)] = 0 , \\ c(x-y)\hat{A}(x)\hat{B}(y) &= \hat{B}(y)\hat{A}(x) - b(x-y)\hat{B}(x)\hat{A}(y) , \\ c(y-x)\hat{D}(x)\hat{B}(y) &= \hat{B}(y)\hat{D}(x) - b(y-x)\hat{B}(x)\hat{D}(y) , \\ [\hat{B}(x), \hat{B}(y)] &= [\hat{C}(x), \hat{C}(y)] = 0 . \end{aligned} \tag{5.27}$$

Let us denote the mathematical vacuum as $|0\rangle$. The action of the L-operator $L_{0j}(\lambda)$ on this vacuum state is

$$L_{0j}(\lambda)|0\rangle = (\lambda + ic)^{-1} \begin{pmatrix} \lambda + ic & 2icS^-_j \\ 0 & \lambda \end{pmatrix} |0\rangle . \tag{5.28}$$

The lower left element of this matrix is zero. From the definition of the monodromy matrix the operators $\hat{A}$, $\hat{B}$, $\hat{C}$ and $\hat{D}$ can be obtained by successive multiplications of such L-operators for each site of the lattice of L sites. We have $\hat{C}|0\rangle = 0$, while action of the operators $\hat{A}$ and $\hat{D}$ on the mathematical vacuum state is diagonal:

$$\hat{A}(\lambda)|0\rangle = a(\lambda)|0\rangle = |0\rangle \ , \ \hat{D}(\lambda)|0\rangle = c^L(\lambda)|0\rangle \ . \tag{5.29}$$

The operator $\hat{B}$ plays the role of a "spin-lowering" operator, hence we consider the state with M down spins as a result of action of M operators $\hat{B}$:

$$|\lambda_1, \ldots, \lambda_M\rangle = \prod_{\beta=1}^{M} \hat{B}(\lambda_\beta)|0\rangle \ . \tag{5.30}$$

Let us act with the operator $\hat{\tau} = \hat{A} + \hat{D}$ on the state equation (5.30) using the commutation relations (5.27). We get

$$\begin{aligned}\hat{\tau}(\lambda)|\lambda_1, \ldots, \lambda_M\rangle &= \Lambda(\lambda, \lambda_1, \ldots, \lambda_M)|\lambda_1, \ldots, \lambda_M\rangle \\ &\quad + \sum_{\gamma=1}^{M} \Lambda_\gamma(\lambda, \lambda_1, \ldots, \lambda_M) \prod_{\substack{\beta=1, \\ \beta\neq\gamma}}^{M} \hat{B}(\lambda_\beta)B(\lambda)|0\rangle \ ,\end{aligned} \tag{5.31}$$

where

$$\Lambda(\lambda, \lambda_1, \ldots, \lambda_M) = \prod_{\beta=1}^{M} c^{-1}(\lambda_\beta - \lambda) + c^L(\lambda) \prod_{\beta=1}^{M} c^{-1}(\lambda - \lambda_\beta) \tag{5.32}$$

and

$$\begin{aligned}&\Lambda_\gamma(\lambda, \lambda_1, \ldots, \lambda_M) \\ &\quad = \frac{b(\lambda - \lambda_\gamma)}{c(\lambda - \lambda_\gamma)} \left(\prod_{\substack{\beta=1 \\ \beta\neq\gamma}}^{M} c^{-1}(\lambda_\beta - \lambda_\gamma) - c^L(\lambda_\gamma) \prod_{\substack{\beta=1 \\ \beta\neq\gamma}}^{M} c^{-1}(\lambda_\gamma - \lambda_\beta) \right) \ .\end{aligned} \tag{5.33}$$

The state $|\lambda_1, \ldots, \lambda_M\rangle$ is the eigenstate of the transfer matrix if $\Lambda_\gamma = 0$. It is true if

$$c^{-L}(\lambda_\gamma) = \prod_{\substack{\beta=1 \\ \beta\neq\gamma}}^{M} \frac{c(\lambda_\beta - \lambda_\gamma)}{c(\lambda_\gamma - \lambda_\beta)} \ , \tag{5.34}$$

which holds for any $\gamma = 1, \ldots, M$. The energy is the logarithmic derivative of the eigenvalue of the transfer matrix taken at the value of the spectral parameter $\lambda = 0$

$$E = A_2 \frac{\partial \ln \Lambda(\lambda, \lambda_1, \ldots, \lambda_M)}{\partial \lambda}|_{\lambda=0}$$
$$= \frac{2i}{c} A_2 \left[\frac{L}{4} - \sum_{j=1}^{M} \left(\frac{1}{\lambda_j} - \frac{1}{(\lambda_j + ic)} \right) \right] . \tag{5.35}$$

Taking the values $\lambda_j \to \lambda_j - i(c/2)$, and using $c = 1$ and $A_2 = -i(J/2)$, we obtain the Bethe ansatz equations and the expression for the energy of a Heisenberg chain, which coincide with the ones obtained in Chapter 3 in the framework of the co-ordinate Bethe ansatz:

$$\left(\frac{\lambda_j + (i/2)}{\lambda_j - (i/2)} \right)^L = \prod_{\substack{l=1,\\ l \neq j}}^{M} \frac{\lambda_j - \lambda_l + i}{\lambda_j - \lambda_l - i} , \tag{5.36}$$

where $j = 1, \ldots, M$, and

$$E = \frac{LJ}{4} - 2J \sum_{j=1}^{M} (4\lambda_j^2 + 1)^{-1} . \tag{5.37}$$

5.2 SU(2)-Symmetric Spin-S Chain

The usefulness of the algebraic Bethe ansatz reveals itself, naturally, for other models, different from the models which possess co-ordinate Bethe ansatz solutions. A good example of such a model is the SU(2)-symmetric spin-S chain, for which the Bethe ansatz solution was provided by H. M. Babujian and L. A. Takhtajan (Takhtadzhan). Here we shall follow the strategy of the algebraic Bethe ansatz, described in the previous section. We start with the Yang–Baxter equation for R-matrices, Eq. (5.9). Suppose we obtain the solution of this equation in the form Eq. (5.6). Now let us look for the solution of the intertwining equation for L-operator, Eq.(5.13). In the previous section we considered this equation in $V_0 \otimes V_n$, where the dimension of V_0 coincided with the dimension of the quantum Hilbert space for a spin $\frac{1}{2}$. Let us write the solution of this equation as:

$${}_{\sigma}L_{0n}(\lambda) = \frac{(2\lambda + ic)}{2ic} I_0 \otimes I_n + \frac{1}{2} \vec{\sigma}_0 \otimes \vec{\sigma}_n , \tag{5.38}$$

where we used the Pauli matrices for the definition of spins $\frac{1}{2}$ for convenience, see below. The notations are similar to the previous section, *i.e.*, we use the subscript 0 to define the auxiliary subspace and subscript n to define the quantum subspace of the n-th spin, where the L-operator acts nontrivially. It turns out that such a form does not contradict the intertwining relations for L-operators, because the solution of the latter is determined up to a multiplier, see the previous section. Using this L-operator, according to the rules of the previous section, we can obtain the Hamiltonian of the spin-$\frac{1}{2}$ chain. However it is easy to check that the L-operator

$$ {}_{\sigma S}L_{0n}(\lambda) = \frac{(2\lambda + ic)}{2ic} I_0 \otimes I_n + \vec{\sigma}_0 \otimes \vec{S}_n \ , \tag{5.39} $$

where we introduced the operator $\vec{S}_n$ acting in the subspace V_n with the dimension $2S+1$, is also the solution of the intertwining relations. This solution, as well as Eq. (5.38), respects the SU(2) symmetry. It is also possible to consider the L-operator of the form

$$ {}_SL_{0n}(\lambda) = -\sum_{j=0}^{2S} \prod_{l=1}^{j} \frac{\lambda + ilc}{\lambda - ilc} P_j \ , \tag{5.40} $$

where P_j is the projection operator acting in the space which is a tensor product of two spin S spaces (*i.e.*, we consider here not only the quantum space V_n, but also the auxiliary space V_0 being of the dimension $2S+1$). Naturally, here we need to look for the solution of the Yang–Baxter equation for R-matrices in the same auxiliary subspace of the dimension $(2S+1) \times (2S+1)$. The projection operator fixes the state with the total spin j, *i.e.*, if $|m\rangle$ is a state with the total spin m, then $P_j|m\rangle = \delta_{m,j}|m\rangle$. This operator can be written as

$$ P_j = \prod_{\substack{m=0,\\ m\neq j}}^{2S} \frac{\vec{S}_0 \otimes \vec{S}_j - x_m}{x_j - x_m} \ , \tag{5.41} $$

where

$$ x_m = \frac{1}{2}m(m+1) - S(S+1) \ . \tag{5.42} $$

The illustration of the Yang–Baxter relations for the operators ${}_{\sigma S}L(u)$ and ${}_SL(u+v)$ see Fig. 5.5.

By using the construction of the previous section the reader can show after some lengthy but straightforward calculations that the second integral

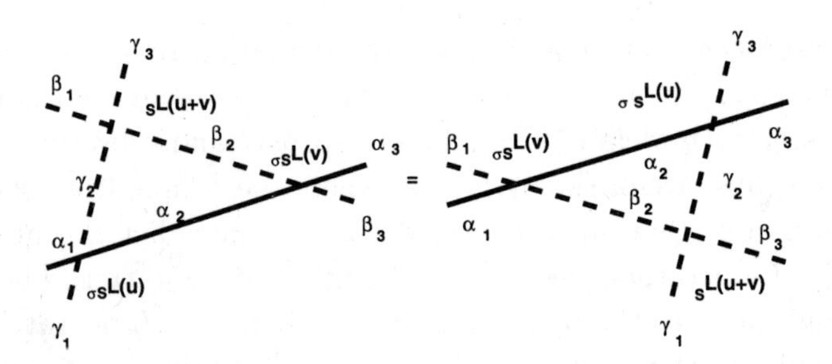

Fig. 5.5 Illustration of Yang–Baxter relations for L-operators related to a SU(2)-symmetric spin-S Hamiltonian.

of motion of the transfer matrix, which is constructed using the L-operators as in Eq. (5.40), has the form

$$\mathcal{H}_{SU(2)} = A_2 \sum_{j=1}^{L-1} A_{2S}(\vec{S}_j \vec{S}_{j+1}) \ , \tag{5.43}$$

where A_2 is a constant and

$$A_{2S}(x) = -\frac{i}{c} \sum_{m=1}^{2S} \left(\sum_{f=1}^{m} \frac{1}{f} \right) \prod_{\substack{l=0,\\ l\neq m}}^{2S} \frac{x - x_l}{x_m - x_l} \ . \tag{5.44}$$

One can use the conditions $c = i$ and $A_2 = -(J/4)$ to obtain the standard Heisenberg Hamiltonian (up to a constant shift) for $S = \frac{1}{2}$. For example, for $S = 1$ the Hamiltonian is

$$\mathcal{H}_{SU(2)} = \frac{J}{4} \sum_{j=1}^{L-1} [\vec{S}_j \vec{S}_{j+1} - (\vec{S}_j \vec{S}_{j+1})^2] \ . \tag{5.45}$$

Please notice that this SU(2)-symmetric spin-1 exactly solvable Hamiltonian differs from the SU(3)-symmetric spin-1 exactly solvable one, introduced in the previous chapter, by the value of the coefficient in front of the biquadratic exchange term.

Actually, the Yang–Baxter equation which is necessary to be solved by the L-operator Eq. (5.40) is

$${}_{\sigma S}L_{12}(\lambda)_S L_{13}(\lambda+\mu)_{\sigma S} L_{23}(\mu) = {}_{\sigma S}L_{23}(\mu)_S L_{13}(\lambda+\mu)_{\sigma S} L_{12}(\lambda) \ , \tag{5.46}$$

with obvious notations. This implies that for intertwining relations for L-operators Eq. (5.40) one can use the R-matrices acting in the auxiliary subspace $V_0 \otimes V_0$ with the dimension $2 \times (2S+1)$.

Now we can construct two monodromy matrices:

$$T_S(\lambda) = {}_SL_1(\lambda)\cdots {}_SL_L(\lambda) , \tag{5.47}$$

which acts in the auxiliary subspace with the dimension $(2S+1)\times(2S+1)$, and

$$T_\sigma(\lambda) = {}_{\sigma S}L_1(\lambda)\cdots {}_{\sigma S}L_L(\lambda) , \tag{5.48}$$

which acts in the auxiliary subspace with the dimension $2\times(2S+1)$, and two transfer matrices $\hat{\tau}_S(\lambda)$ and $\hat{\tau}_{\sigma S}(\lambda)$, as the traces of monodromies Eqs. (5.47) and (5.48), respectively. As follows from Eq. (5.46) and the intertwining relation for the operator ${}_\sigma L(\lambda)$ we have the Yang–Baxter (intertwining) relations for these monodromies

$$\begin{aligned}
{}_SL(\lambda-\mu))(T_S(\lambda)\otimes T_S(\mu)) &= (T_S(\mu)\otimes T_S(\lambda))_SL(\lambda-\mu)) ,\\
{}_{\sigma S}L(\lambda-\mu))(T_\sigma(\lambda)\otimes T_S(\mu)) &= (T_S(\mu)\otimes T_\sigma(\lambda))_{\sigma S}L(\lambda-\mu)) ,\\
{}_\sigma L(\lambda-\mu))(T_\sigma(\lambda)\otimes T_\sigma(\mu)) &= (T_\sigma(\mu)\otimes T_\sigma(\lambda))_\sigma L(\lambda-\mu)) ,
\end{aligned} \tag{5.49}$$

in which the L-operators Eqs. (5.40), (5.39) and (5.38) play the roles of R-matrices (in fact one can rewrite these equations using the R-matrices, obtained by the multiplication of L-operators by the permutation operators of the subspace with the dimensions $(2S+1)\times(2S+1)$, $2\times(2S+1)$ and 2×2, respectively). Then, multiplying from the left Eq. (5.49) by ${}_SL^{-1}(\lambda)$, ${}_\sigma L^{-1}(\lambda)$ and ${}_{\sigma S}L^{-1}(\lambda)$, respectively, and taking the traces over the auxiliary subspaces, we get

$$[\hat{\tau}_S(\lambda),\hat{\tau}_S(\mu)] = 0\ ,\ [\hat{\tau}_\sigma(\lambda),\hat{\tau}_S(\mu)] = 0\ ,\ [\hat{\tau}_\sigma(\lambda),\hat{\tau}_\sigma(\mu)] = 0\ . \tag{5.50}$$

At the points, where ${}_SL^{-1}(\lambda)$, ${}_\sigma L^{-1}(\lambda)$ and ${}_{\sigma S}L^{-1}(\lambda)$ do not exist, the commutativity follows from the analytical continuation principle. The first of these equations means that transfer matrices $\hat{\tau}_S(\lambda)$ with different spectral parameters commute, which justifies the construction of the Hamiltonian. On the other hand, the second equation (*i.e.*, the mutual commutation of $\hat{\tau}_\sigma(\lambda)$ and $\hat{\tau}_S(\mu)$) implies that these transfer matrices have the same set of eigenfunctions. The consideration of the eigenfunctions for $\hat{\tau}_\sigma(\lambda)$ is simpler than for $\hat{\tau}_S(\mu)$, and we concentrate on this case in what follows.

Let us again denote the elements of the 2×2 monodromy matrix $T_\sigma(\lambda)$ as

$$T_\sigma(\lambda) = \begin{pmatrix} \hat{A} & \hat{B} \\ \hat{C} & \hat{D} \end{pmatrix} . \tag{5.51}$$

Hence, the transfer matrix $\hat{\tau}_\sigma(\lambda)$ is

$$\hat{\tau}_\sigma(\lambda) = \hat{A} + \hat{D} \ . \tag{5.52}$$

The operators $\hat{A}$, $\hat{B}$, $\hat{C}$ and $\hat{D}$ obey the commutation relations, which stem from the intertwining relations for the monodromies $T_\sigma(\lambda)$

$$\begin{aligned}
[\hat{A}(x), \hat{A}(y)] &= [\hat{D}(x), \hat{D}(y)] = [\hat{A}(x), \hat{D}(y)] = 0 \ , \\
\hat{B}(x)\hat{A}(y) &= b(y-x)\hat{B}(y)\hat{A}(x) + c(y-x)\hat{A}(y)\hat{B}(x) \ , \\
\hat{B}(y)\hat{D}(x) &= b(y-x)\hat{B}(x)\hat{D}(y) + c(y-x)\hat{D}(x)\hat{B}(y) \ , \\
[\hat{B}(x), \hat{B}(y)] &= [\hat{C}(x), \hat{C}(y)] = 0 \ ,
\end{aligned} \tag{5.53}$$

where $b(\lambda) = ic/(\lambda + ic)$ and $c(\lambda) = \lambda/(\lambda + ic)$ are related to the elements of the L-operator Eq. (5.38) (or the R-matrix). Let us again denote the mathematical vacuum (the state with all spins up) as $|0\rangle$. Then the action of the L-operator ${}_{\sigma S}L_{0j}(\lambda)$ on this vacuum state can be written as

$${}_{\sigma S}L_{0j}(\lambda)|0\rangle = \frac{1}{2ic}\begin{pmatrix} 2\lambda + ic(2S+1) & i4cS_j^- \\ 0 & 2\lambda - ic(2S-1) \end{pmatrix}|0\rangle \ . \tag{5.54}$$

The lower left element of this matrix is zero. Then the reader can see that

$$\hat{C}|0\rangle = 0 \ , \tag{5.55}$$

and the action of the operators $\hat{A}$ and $\hat{D}$ on the mathematical vacuum state is diagonal:

$$\begin{aligned}
\hat{A}(\lambda)|0\rangle &= \left(\frac{2\lambda + ic(2S+1)}{2ic}\right)^L |0\rangle \ , \\
\hat{D}(\lambda)|0\rangle &= \left(\frac{2\lambda - ic(2S-1)}{2ic}\right)^L |0\rangle \ .
\end{aligned} \tag{5.56}$$

The operator $\hat{B}$ again plays the role of a "spin-lowering" operator. Hence, one again can consider the state as an eigenstate of the transfer matrix $\hat{\tau}_\sigma(\lambda)$

$$|\lambda_1, \ldots, \lambda_M\rangle = \prod_{\beta=1}^{M} \hat{B}(\lambda_\beta)|0\rangle \ . \tag{5.57}$$

Let us act with the operator $\hat{\tau} = \hat{A} + \hat{D}$ on the state Eq. (5.57) using the commutation relations Eq. (5.53). We get

$$\hat{\tau}_\sigma(\lambda)|\lambda_1, \ldots, \lambda_M\rangle = \Lambda_\sigma(\lambda, \lambda_1, \ldots, \lambda_M)|\lambda_1, \ldots, \lambda_M\rangle \,, \tag{5.58}$$

where

$$\Lambda_\sigma(\lambda, \lambda_1, \ldots, \lambda_M) = \left(\frac{2\lambda + ic(2S+1)}{2ic}\right)^L \prod_{\beta=1}^{M} \frac{\lambda_\beta - \lambda + ic}{\lambda_\beta - \lambda} + \left(\frac{2\lambda - ic(2S-1)}{2ic}\right)^L \prod_{\beta=1}^{M} \frac{\lambda - \lambda_\beta + ic}{\lambda - \lambda_\beta} \,. \tag{5.59}$$

The state $|\lambda_1, \ldots, \lambda_M\rangle$ is the eigenstate of the transfer matrix if

$$\left(\frac{\lambda_\gamma + ic(2S+1)/2}{\lambda_\gamma - ic(2S-1)/2}\right)^L = \prod_{\substack{\beta=1 \\ \beta\neq\gamma}}^{M} \frac{\lambda_\beta - \lambda_\gamma - ic}{\lambda_\beta - \lambda_\gamma + ic} \,, \tag{5.60}$$

which holds for any $\gamma = 1, \ldots, M$. Taking the values $\lambda_\gamma \to \lambda_j - i(c/2)$, and using $c = 1$ we finally get the Bethe ansatz equations for the SU(2)-symmetric spin-S chain

$$\left(\frac{\lambda_j + iS}{\lambda_j - iS}\right)^L = \prod_{\substack{l=1, \\ l\neq j}}^{M} \frac{\lambda_j - \lambda_l + i}{\lambda_j - \lambda_l - i} \,, \tag{5.61}$$

where $j = 1, \ldots, M$.

The energy is the logarithmic derivative of the eigenvalue of the transfer matrix $\hat{\tau}_S(\lambda)$ taken at the value of the spectral parameter $\lambda = 0$. That transfer matrix can be written as

$$\hat{\tau}_S(\lambda) = \sum_{m=-S}^{S} (T_s)_{mm}(\lambda) \,, \tag{5.62}$$

i.e., the sum of diagonal matrix elements in the auxiliary subspace with the dimension $(2S+1) \times (2S+1)$. The intertwining relations for this matrix imply the following relations

$$\begin{aligned}(T_S)_{mm}(x)\hat{B}(y) &= c_m(x-y)\hat{B}(y)(T_S)_{mm}(x) \\ &\quad + c_{m1}(x-y)(T_S)_{mm-1}(x)\hat{A}(y) \\ &\quad + c_{m2}(x-y)(T_S)_{m+1m}(x)\hat{D}(y) \\ &\quad + c_{m3}(x-y)(T_S)_{m+1m-1}(x)\hat{C}(y) \,,\end{aligned} \tag{5.63}$$

where

$$\begin{aligned}
c_m(x) &= \frac{[x+ic(2S+1)/2][x-ic(2S-1)/2]}{[x+ic(2m+1)/2][x+ic(2m-1)/2]} , \\
c_{m1}(x) &= -\frac{\sqrt{(S+m)(S-m+1)}}{x+ic(2m-1)/2} , \\
c_{m2}(x) &= \frac{\sqrt{(S-m)(S+m+1)}}{x+ic(2m+1)/2} , \\
c_{m3}(x) &= \frac{\sqrt{(S^2-m^2)[(S+1)^2-m^2]}}{[x+ic(2m+1)/2][x+ic(2m-1)/2]} .
\end{aligned} \tag{5.64}$$

Now it is necessary to find the matrix elements of the operator ${}_SL(\lambda)$. Re-denoting $|0\rangle = |S,S\rangle$ (which respects the fact of the highest eigenvalue of the operators $\hat{S}$ and S^z acting on this state) we obtain from the definition of the projection operator P_j

$$\begin{aligned}
\langle S,m'|{}_SL(\lambda)|S,m\rangle|S,S\rangle &= -\sum_{j=0}^{2S}\prod_{l=1}^{j}\frac{\lambda+icl}{\lambda-icl}(j,m+S|m,S) \\
&\times (m',m+S-m'|j,m+S)|m+S-m',m+S-m'\rangle ,
\end{aligned} \tag{5.65}$$

where $(j,l|m,S) = \langle j,l|m,S\rangle$ are, in fact, the short-hand notations for the Clebsch–Gordan coefficients. Then the diagonal matrix elements are

$$\begin{aligned}
\langle S,S|\langle S,m|{}_SL(\lambda)|S,m\rangle|S,S\rangle &\equiv ({}_SL(\lambda))^{S,m}_{S,m} \\
&= -\sum_{j=0}^{2S}\prod_{l=1}^{j}\frac{\lambda+icl}{\lambda-icl}(j,m+S|m,S)^2\langle m,S|S,S\rangle \\
&= \prod_{l=m+1}^{S}\frac{\lambda+ic(l-S)}{\lambda+ic(l+S)}({}_SL(\lambda))^{S,S}_{S,S} ,
\end{aligned} \tag{5.66}$$

where

$$({}_SL(\lambda))^{S,S}_{S,S} = -\prod_{l=1}^{2S}\frac{\lambda+icl}{\lambda-icl} . \tag{5.67}$$

Then the action of the diagonal elements of ${}_SL(\lambda)$ on the mathematical vacuum is

$$({}_SL(\lambda))^{S,m}_{S,m} = -\prod_{l=m+1}^{S}\frac{\lambda+ic(l-S)}{\lambda+ic(l+S)}\prod_{l'=1}^{2S}\frac{\lambda+icl'}{\lambda-icl'} . \tag{5.68}$$

Using this formula and taking into account the Bethe ansatz equations Eq. (5.60) we finally obtain

$$\hat{\tau}_S(\lambda)|\lambda_1,\ldots,\lambda_M\rangle = \Lambda_S(\lambda,\lambda_1,\ldots,\lambda_M)|\lambda_1,\ldots,\lambda_M\rangle \ , \tag{5.69}$$

where

$$\Lambda_S(\lambda,\lambda_1,\ldots,\lambda_M) = \sum_{m=-S}^{S}\left(\prod_{l=m+1}^{S}\frac{\lambda+ic(l-S)}{\lambda+ic(l+S)}\prod_{l'=1}^{2S}\frac{\lambda+icl'}{\lambda-icl'}\right)^L\prod_{\gamma=1}^{M}c_m(\lambda-\lambda_\gamma) \ . \tag{5.70}$$

Unwanted terms become zero due to Bethe ansatz equations. Then, taking the logarithmic derivative we obtain the energy of a SU(2)-symmetric spin-S chain (notice that only the term with $m = S$ contributes) as

$$\begin{aligned} E &= \text{const} + A_2\frac{\partial\ln\Lambda_S(\lambda,\lambda_1,\ldots,\lambda_M)}{\partial\lambda}\bigg|_{\lambda=0} \\ &= \text{const} - \frac{2i}{c}A_2\sum_{j=1}^{M}\left(\frac{1}{\lambda_j+ic(2S-1)/2} - \frac{1}{\lambda_j+ic(2S+1)/2}\right) \ . \end{aligned} \tag{5.71}$$

Taking the values $\lambda_j \to \lambda_j - i(c/2)$, and using $c = 1$ and const $= -i(J/2)$ we obtain the expression for the energy of the SU(2)-symmetric spin-S chain

$$E = E_0 - SJ\sum_{j=1}^{M}(\lambda_j^2+S^2)^{-1} \ . \tag{5.72}$$

The total magnetization of this chain is equal to $M^z = SL - M$. We would also like to present some interesting property of the monodromy matrix:

$$\left[\left(\sum_{j=1}^{L}S_j^{x,y,z}+\frac{1}{2}\sigma_0^{x,y,z}\right), T_\sigma(\lambda)\right] = 0 \ . \tag{5.73}$$

In the framework of the string hypothesis we look for the solution of Eq. (5.61) in the form of strings of length $2m+1$ (we denote by μ_m the number of strings of length m). Introducing strings as $\lambda_j = \lambda_{j,m} + i[(m+1)/2-\nu]$ with $\nu = 1,\ldots,m$ and summing Eq. (5.61) for a distinguished string of length m for the parameters λ_j occurring in it, we obtain

$$\theta_{m,2S}(\lambda_j^m) = \frac{2\pi}{L}I_{j,m} + \frac{1}{L}\sum_{n=1}^{\infty}\sum_{l=1}^{\mu_n}\Theta_{mn}(\lambda_j^m-\lambda_l^n) \ , \tag{5.74}$$

where $\theta_n(x) = 2\tan^{-1}(x/n)$,

$$\theta_{m,n}(x) = \sum_{l=1}^{\min(m,n)} \theta_{m+1+n-2l}(x) \ , \tag{5.75}$$

and

$$\begin{aligned}\Theta_{mn}(x) &= (1-\delta_{m,n})\theta_{|m-n|}(x) + 2\theta_{|m-n|+2}(x) + \cdots \\ &+ 2\theta_{m+n-2}(x) + \theta_{m+n}(x) \ ,\end{aligned} \tag{5.76}$$

and $I_{j,m}$ are integers or non-integers, which appear because the logarithm is the multi-valued function. Then we look for solutions to Bethe ansatz equations in the thermodynamic limit $L \to \infty$, introducing densities for particles and holes, corresponding to strings of length m: $\rho_m(x)$ and $\rho_{mh}(x)$, respectively. Then, differentiating Eq. (5.74) with respect to real parts of λ_j and introducing continuous distributions of those real parts we obtain

$$\rho_{mh}(\lambda) + \sum_{n=1}^{\infty} A_{m,n} * \rho_n(\lambda) = \sum_{l=1}^{\min(m,2S)} a_{m+2S+1-2l} \ , \tag{5.77}$$

where $*$ denotes the convolution,

$$A_{m,n}(x) = a_{|m-n|}(x) + 2\sum_{l=1}^{\min(n,m)-1} a_{m+n-2l}(x) + a_{m+n}(x) \tag{5.78}$$

and $a_m(x) = 2m/[\pi(4x^2+m^2)]$. Then the internal energy and magnetization per site can be written as

$$\begin{aligned} e &= e_0 - \frac{1}{2}\sum_{m=1}^{\infty}\int_{-\infty}^{\infty} d\lambda \theta'_{m,2S}(\lambda)\rho_m(\lambda) \ , \\ m^z &= S - \sum_{m=1}^{\infty} m \int_{-\infty}^{\infty} d\lambda \rho_m(\lambda) \ , \end{aligned} \tag{5.79}$$

where $e_0 \equiv (E_0/L)$.

Then, the set of equations for dressed energies $\varepsilon_n(\lambda) = T\ln[\rho_{nh}(\lambda)/\rho_n(\lambda)] = \eta_n(\lambda)$ is

$$Hm - J\theta'_{m,2S}(\lambda) = T\ln[1+\eta_m(\lambda)] - T\sum_n A_{n,m} * \ln[1+\eta_n^{-1}(\lambda)] \ , \tag{5.80}$$

which set of equations completes the set Eq. (5.77). These equations can be rewritten in the following useful form

$$\varepsilon_m(\lambda) = T(2\cosh[\pi(\lambda-\lambda')])^{-1} * \ln[1+\eta_{m-1}(\lambda')][1+\eta_{m+1}(\lambda')] \\ - \frac{\pi J\delta_{m,2S}}{\cosh[\pi(\lambda)]} \, , \quad \lim_{m\to\infty} \frac{\varepsilon_m(\lambda)}{m} = H \, . \tag{5.81}$$

Inserting the thermal equilibrium density functions into the expression for the Helmholtz free energy we obtain

$$f = e_0 + \frac{J}{2}(\psi(1/2) - \psi[(1/2)+S]) \\ -T\int_{-\infty}^{\infty} d\lambda \frac{\ln(1+\exp[\varepsilon_{2S}(\lambda)/T])}{2\cosh(\pi\lambda)} \, , \tag{5.82}$$

where ψ are digamma functions.

In the limit of high temperatures $T \gg |J|$, but keeping the ratio H/T finite, we can consider $\varepsilon_m^{(0)}(\lambda) \approx mH$. In this case driving terms do not depend on λ and we can solve Eqs. (5.80) and (5.77) exactly. This high temperature solution describes the behaviour of a free spin S in an external magnetic field H. The magnetic susceptibility is the smooth function of the temperature at high T, as expected.

In the limit of $T \to 0$ from Eq. (5.80) we see that $\varepsilon_m(\lambda) > 0$ for $m \neq 2S$ for the most interesting antiferromagnetic situation $J > 0$. In the ferromagnetic situation, $J < 0$, $\varepsilon_m(\lambda) > 0$ for all lengths of strings $m = 1, 2, 3, \ldots$, and the ground state energy is just e_0. Then, introducing the positive and negative parts of $\varepsilon_{2S}(\lambda)$ we obtain

$$\varepsilon_{2S}(\lambda) + A_{2S,2S} * \varepsilon_{2S}^-(\lambda) = 2SH - J\theta'_{2S,2S}(\lambda) \, . \tag{5.83}$$

This equation determines the Dirac sea for quasiparticles with dressed energies $\varepsilon_{2S}(\lambda)$. The ground state pertains to the situation in which all states with negative energies are filled and all states with positive energies are empty. The Fermi points (related to the limits of integration) are determined from the conditions $\varepsilon_{2S}(\pm B) = 0$. The equations for densities stem from the equations for dressed energies. In the ground state, $T = 0$, we have

$$\pi\rho_{2Sh}(\lambda) = \theta'_{2S,2S}(\lambda) - \pi\int_{-B}^{B} d\lambda' A_{2S,2S}(\lambda-\lambda')\rho_{2S}(\lambda') \, . \tag{5.84}$$

The ground state internal energy can be written as

$$e_{T=0} = e_0 + \int_{-B}^{B} d\lambda [2SH - J\theta'_{2S,2S}(\lambda)]\rho_{2S}(\lambda) \tag{5.85}$$

and the ground state magnetization is equal to

$$m^z = S - \int_{-B}^{B} d\lambda \rho_{2S}(\lambda) \ . \tag{5.86}$$

The value of the magnetic field H determines these limits of integration (*i.e.*, Fermi points). For the antiferromagnetic case large values of the external magnetic field $|H| > H_s$ the system is in the ferromagnetic state and $B = 0$. In these regions of values of H the ground state energy is equal to e_0, the magnetization has its nominal values S, and the magnetic susceptibility is zero. On the other hand, in zero magnetic field, for the antiferromagnetic situation at $H = 0$ we have $B = \infty$, *i.e.*, rapidities fill the total interval in the thermodynamic limit. In the ferromagnetic situation, $J < 0$, the point of the quantum phase transition is $H = 0$: any infinitesimal magnetic field removes the degeneracy of the ferromagnetic spin-S chain and the magnetization of the latter becomes nominal.

The behaviour of the internal energy for small values of the magnetic field can be found analytically, by using the Wiener–Hopf method. First, we rewrite the integral equation for dressed energies in the following way in the Fourier space

$$\varepsilon_m^+(\omega) = 2\pi m H\delta(\omega) - \frac{\pi J A_{m,2S}(\omega)}{2\cosh(\omega/2)} - A_{2S,m}(\omega)\varepsilon_m^-(\omega) \ , \tag{5.87}$$

where

$$A_{n,m}(\omega) = \coth(\omega/2)[\exp[-|\omega||n-m|/2] - [\exp[-|\omega|(n+m)/2]] \ . \tag{5.88}$$

Notice that the solution for $H = 0$ immediately follows: $\varepsilon_{2S}(\lambda) = -\pi J/2\cosh(\pi\lambda)$ and $\varepsilon_m(\lambda) = 0$ for any $m > 1$. Let us multiply Eq. (5.87) by $A_{2S,2S}(\omega)^{-1}$, where

$$A_{n,m}(\omega)^{-1} = \delta_{n,m} - (\delta_{n,m+1} + \delta_{n,m-1})/2\cosh(\omega/2) \ . \tag{5.89}$$

Then we take the inverse Fourier transform which yields

$$\varepsilon_{2S}(\lambda) = \frac{H}{2} - \frac{\pi J}{2\cosh(\pi\lambda)} + \left(\int_{-\infty}^{-B} + \int_{B}^{\infty}\right) d\lambda' J(\lambda - \lambda')\varepsilon_{2S}(\lambda) \ , \tag{5.90}$$

where the kernel is the Fourier transform of $1-A_{2S,2S}(\omega)^{-1}$. Now we define $y(\lambda)=\varepsilon_{2S}(\lambda+B)$, so that the Fermi point for dressed energies corresponds to $y(0)$. Then we rewrite Eq. (5.90) as follows (here we use the identity $\varepsilon_{2S}(\lambda)=\varepsilon_{2S}(-\lambda)$)

$$\begin{aligned} y(\lambda) = \frac{H}{2} - \frac{\pi J}{2\cosh[\pi(\lambda+B)]} + \int_0^\infty d\lambda' J(\lambda-\lambda')y(\lambda') \\ + \int_0^\infty d\lambda J(\lambda+\lambda'+2B)y(\lambda') \, . \end{aligned} \tag{5.91}$$

If $H \ll J$, than B is very large and $J(\lambda+\lambda'+2B)\sim B^{-1}$. Hence, the last term in Eq. (5.91) is order B^{-1} smaller than the previous ones. We can, then, solve Eq. (5.91) iteratively $y(\lambda)=y_1(\lambda)+y_2(\lambda)+\ldots$, where

$$\begin{aligned} y_1(\lambda) &= \frac{H}{2} - \frac{\pi J}{2\cosh[\pi(\lambda+B)]} + \int_0^\infty d\lambda' J(\lambda-\lambda')y_1(\lambda') \\ y_2(\lambda) &= \int_0^\infty d\lambda' J(\lambda-\lambda')y_2(\lambda') + \int_0^\infty d\lambda J(\lambda+\lambda'+2B)y_1(\lambda') \end{aligned} \tag{5.92}$$

etc. One divides y into positive y^+ $(\lambda>0)$ and negative y^- $(\lambda<0)$ parts. The Fourier transform of the equation for y_1 is

$$\frac{y_1^+(\omega)}{A_{2S,2S}(\omega)} + y_1^-(\omega) = \pi H\delta(\omega) - \frac{\pi J\exp(i\omega B)}{2\cosh(\omega/2)} \, . \tag{5.93}$$

To apply the Wiener–Hopf method we rewrite the kernel $(A_{2S,2S})^{-1}$ as a product of two functions, one, $G_{2S}^+(\omega)$, being analytic in the upper half-plane, and the other one, $G_{2S}^-(\omega)$, being analytic in the lower half-plane, where

$$\begin{aligned} G_{2S}^+(\omega) &= G_{2S}^-(-\omega) \\ &= \frac{1}{2\sqrt{\pi S}}\left(\frac{S(-i\omega+0)}{\pi e}\right)^{iS\omega/\pi} \frac{\Gamma[(1/2)-i(\omega/2\pi)]\Gamma[1-i(\omega S/\pi)]}{\Gamma[1-i(\omega/2\pi)]} \, . \end{aligned} \tag{5.94}$$

Observe that $G_{2S}^\pm(\infty)$ is a constant. The Wiener–Hopf method uses the fact that from the analyticity of the functions $y_1^\pm(\omega)$ and $G_{2S}^\pm(\omega)$ it follows that

$$y_1^+(\omega) = -\frac{q^+(\omega)}{G_{2S}^+(\omega)} \, , \; y_1^-(\omega) = \frac{q^-(\omega)}{G_{2S}^-(\omega)} \, , \tag{5.95}$$

where

$$q^{\pm}(\omega) = -\frac{iH\sqrt{S}}{(\omega \pm i0)} - \frac{iJ}{2}\int_{-\infty}^{\infty}\frac{d\omega'}{2\pi} \times \frac{\Gamma[(1/2)+i(\omega'/2\pi)]\Gamma[(1/2)-i(\omega'/2\pi)]e^{-i\omega' B}}{G_{2S}^{-}(\omega')(\omega'-\omega\mp i0)} . \quad (5.96)$$

We are interested in the results for large positive B, hence, the contour of integration can be closed through the lower half-plane. Then, the value of the integral can be given as the sum of the residua of $\Gamma[(1/2)-i(\omega'/2\pi)]$. The leading term, *i.e.*, the pole closest to the real axes, yields the result (the next term is of the order of $\exp(-2\pi B)$ smaller)

$$y_1^{+}(\omega) = \frac{i\sqrt{S}}{G_{2S}^{+}(\omega)}\left(\frac{H}{\omega+i0} - \frac{\pi^2 J S^S \exp(-\pi B)}{e^S\Gamma(1+S)(\omega+i\pi)}\right) . \quad (5.97)$$

The Fourier transform of the equation for y_2 yields

$$y_2^{+}(\omega)G_{2S}^{+}(\omega) + \frac{y_2^{-}(\omega)}{G_{2S}^{-}(\omega)} = \frac{[1-G_{2S}^{+}(\omega)G_{2S}^{-}(\omega)]\exp(-i2\pi B\omega)y_1^{+}(-\omega)}{G_{2S}^{-}(\omega)} . \quad (5.98)$$

The analyticity of $y_{1,2}^{\pm}(\omega)$ and $G_{2S}^{\pm}(\omega)$ implies

$$y_2^{+}(\omega)G_{2S}^{+}(\omega) = -i\int_{-\infty}^{\infty}\frac{d\omega'}{2\pi}\frac{[1-G_{2S}^{+}(\omega')G_{2S}^{-}(\omega')]\exp(-i2\pi B\omega')y_1^{+}(-\omega')}{G_{2S}^{-}(\omega')(\omega'-\omega-i0)} . \quad (5.99)$$

Again, we use the fact that $0 < 1 \ll B$ and close the contour through the lower half-plane. In this half-plane only $G_{2S}^{+}(\omega)$ has singularities. The leading singularity is the cut along the imaginary axis.

The parameter B is the function of the applied magnetic field H. It is determined from the condition $y(\lambda = 0) = 0$, which is equivalent to the condition $\lim_{\omega\to\infty}\omega y^{+}(\omega) = 0$. Using the results for $y_{1,2}^{+}(\omega)$ we get

$$H\left[1 + \frac{S}{\pi B} + \frac{S^2\ln(S/\pi B)}{2(\pi B)^2} + \cdots\right] = \frac{\pi^2 S^S J\exp(-\pi B)}{e^S\Gamma(1+S)} . \quad (5.100)$$

Then the Helmholtz free energy per site can be written as

$$f(T=0,H\ll J)=e_0+\frac{J}{2}(\psi(1/2)-\psi[(1/2)+S]) \\ -J\int_{-\infty}^{\infty}\frac{d\omega}{2\pi}\frac{e^{i\omega B}y^{+}(\omega)}{2\cosh(\omega/2)}\,. \tag{5.101}$$

The contour has to be closed through the upper half-plane. Then the value of the integral is given by the sum of the residua of the poles of $1/\cosh(\omega/2)$: any pole $\omega=i(2n+1)\pi$ yields the term $\sim (H/J)^{2n+2}$. The leading contribution arises from the closest to the real axis pole, $\omega=i\pi$, and it gives $-\exp(-\pi B)y^{+}(i\pi)\sim (H/J)^2$. Then the final answer for the Helmholtz free energy of the Heisenberg antiferromagnetic spin chain in a weak magnetic field is

$$f(T=0,H\ll J)=e_0+\frac{J}{2}(\psi(1/2)-\psi[(1/2)+S]) \\ -\frac{SH^2}{\pi^2 J}\left(1+\frac{S}{\ln(AH/J)}-\frac{S^2\ln|\ln(AH/J)|}{\ln^2(AH/J)}+\cdots\right)\,, \tag{5.102}$$

where $A=e^{S}\Gamma(1+S)/\pi^2 S^S$ is a constant.

5.3 The Algebraic Bethe Ansatz for Correlated Electron Models

As a good example of the power of the algebraic Bethe ansatz for correlated electron models we study its solution to the supersymmetric t-J chain. Here we also introduce to the reader the *graded* version of the Bethe ansatz.

Let us start with the graded linear space $V^{(n|m)}=V^{(n)}\oplus V^{(m)}$, where n and m denote the dimensions of the parts of this space and $\oplus$ denotes the direct sum. Let $\{e_1,\ldots,e_{n+m}\}$ be a basis of $V^{(n|m)}$, such that $\{e_1,\ldots,e_n\}$ is a basis of $V^{(n)}$ and $\{e_{n+1},\ldots,e_{n+m}\}$ is a basis of $V^{(m)}$. The *Grassmann parities* of the basis vectors (they are often called the grading) can be given by $\epsilon_1=\cdots=\epsilon_n=0$ and $\epsilon_{n+1}=\cdots=\epsilon_{n+m}=1$. Then any linear operator on $V^{(n|m)}$ can be represented in a block form as

$$M=\begin{pmatrix} A & B \\ C & D\end{pmatrix}\,,\ \varepsilon\begin{pmatrix} A & 0 \\ 0 & D\end{pmatrix}=0\,,\ \varepsilon\begin{pmatrix} 0 & B \\ C & 0\end{pmatrix}=1\,, \tag{5.103}$$

and the *supertrace* (the graded trace) of this matrix is defined as

$$\mathrm{str}M = \mathrm{tr}A - \mathrm{tr}D \ , \tag{5.104}$$

where the traces on the right hand side are the usual operator traces in $V^{(n)}$ and $V^{(m)}$, respectively. The graded tensor product $V^{(n|m)} \otimes V^{(n|m)}$ in terms of its basis vectors $\{e_a \otimes e_b\}$ (where $a, b = 1, \dots, m+n$) can be defined as

$$v \otimes w = (e_a v_a) \otimes (e_b w_b) = (e_a \otimes e_b) v_a w_b (-1)^{\epsilon_{v_a} \epsilon_b} \ , \tag{5.105}$$

i.e., the additional factor $(-1)^{\epsilon_{v_a}\epsilon_b}$ occurs comparing to the standard tensor product. This factor originates from passing v_a past e_b. The action of the right linear operator $F \otimes G$ on the vector $v \otimes w$ in $V^{(n|m)} \otimes V^{(n|m)}$ has the form $(F \otimes G)(v \otimes w) = F(v) \otimes G(w)$ with its matrix elements

$$(F \otimes G)^{ab}_{cd} = F_{ab} G_{cd} (-1)^{\epsilon_c(\epsilon_a + \epsilon_b)} \ . \tag{5.106}$$

Then the identity operator in $V^{(n|m)} \otimes V^{(n|m)}$ is $\hat{I}^{ab}_{cd} = \delta_{a,b}\delta_{c,d}$ and the permutation operator is $\hat{P}^{ab}_{cd} = \delta_{a,d}\delta_{c,b}(-1)^{\epsilon_b\epsilon_d}$.

Using the above definitions of graded operators, it is easy to see that the operator $R(\lambda) = b(\lambda)\hat{I} + c(\lambda)\hat{P}$, with $c(\lambda) = \lambda/(\lambda + ic)$ and $b(\lambda) = ic/(\lambda + ic)$ satisfies the Yang–Baxter equations for R-matrices Eq. (5.9) acting in $V^{(n|m)} \otimes V^{(n|m)}$.

Why is this mathematical construction relevant for physics? It is helpful when one considers a system of n species of bosons and m species of fermions. In such a case $V_j^{(n|m)}$ denotes the quantum Hilbert space of configurations at every site of the lattice (if we consider a lattice situation). For example, for the supersymmetric t-J model with $V = -J/4$ we have one boson (an empty state) and two fermions (electrons with spins directed upward and downward) at each site. Then the quantum space for each site for such a model can be considered as $V_j^{(1|2)}$.

In the previous chapter we promised to explain why the t-J model (here we shall mostly concentrate on the $V = -J/4$ case) is called supersymmetric.

Let us consider nine operators at each site j (here we keep the notations of the previous chapter). The first of these operators is the unity operator I_j. The second one is related to the operator of the number of electrons in site j, $n_j = n_{j\uparrow} + n_{j\downarrow}$, as $N_j = 1 - (1/2)n_j$. Three other operators, $S_j^z = (1/2)(n_{j\uparrow} - n_{j\downarrow})$, $S_j^+ = a^\dagger_{j\uparrow}a_{j\downarrow}$, and $S_j^- = a^\dagger_{j\downarrow}a_{j\uparrow}$, form the SU(2) algebra. Finally, there are four more operators: $Q_{j\uparrow} = (1 - n_{j\downarrow})a^\dagger_{j\uparrow}$, $Q_{j\downarrow} = (1 - n_{j\uparrow})a^\dagger_{j\downarrow}$, $Q^\dagger_{j\uparrow} = (1 - n_{j\downarrow})a_{j\uparrow}$ and $Q^\dagger_{j\downarrow} = (1 - n_{j\uparrow})a_{j\downarrow}$. All these nine operators (let us call them, *e.g.*, J_j^α, where $\alpha = 1, \ldots, 9$) are the generators of the algebra gl(1|2) which can be written in the form

$$J_j^\alpha J_j^\beta - (-1)^{\epsilon_\alpha \epsilon_\beta} J_j^\beta J_j^\alpha = f_{\alpha\beta}^\gamma J_j^\gamma \ , \tag{5.107}$$

where $f_{\alpha\beta}^\gamma$ are the structure constants of gl(1|2), and $\epsilon_\alpha = 0$ for the first five generators (*i.e.*, they are bosonic operators), and $\epsilon_\alpha = 1$ for the last four (fermionic) operators. The fundamental matrix representation of the generators is in the basis, in which the fermionic states are $e_{j1} = (1\ 0\ 0)_j^T$ for the electron with spin down, $e_{j2} = (0\ 1\ 0)_j^T$ for the electron with spin up, and the bosonic state is $e_{j3} = (0\ 0\ 1)_j^T$ (empty state) is:

$$\begin{aligned}
S_j^- &= \begin{pmatrix} 0 & 0 & 0 \\ 1 & 0 & 0 \\ 0 & 0 & 0 \end{pmatrix} , \qquad S_j^+ = \begin{pmatrix} 0 & 1 & 0 \\ 0 & 0 & 0 \\ 0 & 0 & 0 \end{pmatrix} , \\
S_j^z &= \frac{1}{2} \begin{pmatrix} -1 & 0 & 0 \\ 0 & 1 & 0 \\ 0 & 0 & 0 \end{pmatrix} , \quad N_j = \frac{1}{2} \begin{pmatrix} 1 & 0 & 0 \\ 0 & 1 & 0 \\ 0 & 0 & 2 \end{pmatrix} , \\
Q_{j\uparrow}^\dagger &= \begin{pmatrix} 0 & 0 & 0 \\ 0 & 0 & 0 \\ 0 & 1 & 0 \end{pmatrix} , \qquad Q_{j\uparrow} = \begin{pmatrix} 0 & 0 & 0 \\ 0 & 0 & 1 \\ 0 & 0 & 0 \end{pmatrix} , \\
Q_{j\downarrow}^\dagger &= \begin{pmatrix} 0 & 0 & 0 \\ 0 & 0 & 0 \\ 1 & 0 & 0 \end{pmatrix} , \qquad Q_{j\downarrow} = \begin{pmatrix} 0 & 0 & 1 \\ 0 & 0 & 0 \\ 0 & 0 & 0 \end{pmatrix} .
\end{aligned} \tag{5.108}$$

We can introduce the invariant nondegenerate bilinear form $K_{\alpha\beta}$, given as the supertrace over two generators

$$K_{\alpha\beta} = (K^{\alpha\beta})^{-1} = \mathrm{str}\ J_j^\alpha J_j^\beta \ . \tag{5.109}$$

It is easy to show that using these operators we can write the Hamiltonian of the supersymmetric t-J chain at $V = -J/4$ as

$$\begin{aligned}
\mathcal{H}_{tJ} = &- \sum_{j=1}^{L-1} \sum_\sigma (Q_{j\sigma} Q_{j+1\sigma}^\dagger + Q_{j+1\sigma} Q_{j\sigma}^\dagger) + \frac{J}{2} \sum_{j=1}^{L-1} (S_j^+ S_{j+1}^- + S_j^- S_{j+1}^+ \\
&+ 2 S_j^z S_{j+1}^z - 2 N_j N_{j+1} + I_j I_{j+1}) - 2J \sum_{j=1}^{L} (n_j - 1) \ ,
\end{aligned} \tag{5.110}$$

which at the supersymmetric point $J = 2$ can be rewritten as

$$\mathcal{H}_{tJ} = -\sum_{j=1}^{L}\left(\sum_{\alpha,\beta=1}^{9} K^{\alpha\beta} J_j^{\alpha} J_{j+1}^{\beta} - 2n_j + 1\right)$$
$$= -\sum_{j=1}^{L}[\Pi_{j,j+1} - 2n_j + 1] \ , \tag{5.111}$$

where the graded operator $\Pi_{j,j+1}$ permutes the three possible configurations (empty state and states with electrons with spins up or down) between sites j and $j+1$, picking up a minus sign if both of the permuted configurations are fermionic. This Hamiltonian is obviously supersymmetric, because it is the quadratic form of the generators of the gl(1|2) algebra with the coefficients being the invariant nondegenerate bilinear form of those generators. Notice that the sums of all nine generators over all sites of the lattice commute with the Hamiltonian of the supersymmetric t-J model, $[\mathcal{H}_{tJ}, \sum_{j=1}^{L} J_j^{\alpha}] = 0$ for $\alpha = 1, \ldots, 9$.

One can show that the case $V = 3J/4$ ($J = \pm 2t = \pm 2$) pertains to the grading, in which all nine Grassmann parities of generators are bosonic. This is clear, because they form SU(3), but not gl(1|2) algebra.

Let us now consider the algebraic Bethe ansatz for the gl($n|m$)-symmetric correlated electron chain. For the gl(1|2)-symmetric chain the Hilbert space at each site is isomorphic to C^3 and is spanned by the above mentioned three basis vectors. In the *FFB grading, i.e.*, in which e_{j1} and e_{j2} are fermionic (the Grassmann parities are $\varepsilon_{1,2} = 1$) and e_{j3} is bosonic (the Grassmann parity is $\varepsilon_3 = 0$) we can start from the mathematical vacuum state $|0\rangle = \prod_{j=1}^{L} e_{j3}$. This choice of the grading implies that R-matrix for the gl(1|2)-symmetric chain has the form

$$R(\lambda) = b(\lambda)I + c(\lambda)$$
$$\times \begin{pmatrix} -1 & 0 & 0 & 0 & 0 & 0 & 0 & 0 & 0 \\ 0 & 0 & 0 & -1 & 0 & 0 & 0 & 0 & 0 \\ 0 & 0 & 0 & 0 & 0 & 0 & 1 & 0 & 0 \\ 0 & -1 & 0 & 0 & 0 & 0 & 0 & 0 & 0 \\ 0 & 0 & 0 & 0 & -1 & 0 & 0 & 0 & 0 \\ 0 & 0 & 0 & 0 & 0 & 0 & 0 & 1 & 0 \\ 0 & 0 & 1 & 0 & 0 & 0 & 0 & 0 & 0 \\ 0 & 0 & 0 & 0 & 0 & 1 & 0 & 0 & 0 \\ 0 & 0 & 0 & 0 & 0 & 0 & 0 & 0 & 1 \end{pmatrix} . \tag{5.112}$$

Let us consider the monodromy matrix $T(\lambda)$ for the gl$(n|m)$ symmetric model as the $(n+m) \times (n+m)$ matrix in the auxiliary subspace

$$T(\lambda) = \begin{pmatrix} \hat{A} & \hat{B} \\ \hat{C} & \hat{D} \end{pmatrix} , \tag{5.113}$$

where $\hat{A}$ is $n \times n$ matrix, $\hat{D}$ is $m \times m$ matrix, $\hat{B}$ is $n \times m$ matrix and $\hat{C}$ is $m \times n$ matrix. For the gl(1|2)-symmetric chain we have

$$\hat{A}(\lambda) = \begin{pmatrix} \hat{A}_{11} & \hat{A}_{12} \\ \hat{A}_{21} & \hat{A}_{22} \end{pmatrix} , \tag{5.114}$$

and

$$\hat{B} = (\hat{B}_1 \; \hat{B}_2)^T \; , \; \hat{C} = (\hat{C}_1 \; \hat{C}_2) \; . \tag{5.115}$$

Hence, the transfer matrix $\hat{\tau}(\lambda)$ is

$$\hat{\tau}(\lambda) = \mathrm{str}\, T(\lambda) = -\mathrm{tr}\, \hat{A} + \mathrm{tr}\, \hat{D} \tag{5.116}$$

where for the gl(1|2)-symmetric chain we have

$$\hat{\tau}(\lambda) = -\hat{A}_{11} - \hat{A}_{22} + \hat{D} \; . \tag{5.117}$$

Let us define the action of the monodromy matrix on the mathematical vacuum so that the action of diagonal matrix elements $T_{\alpha\alpha}$ produces c-numbers, *i.e.*, $T_{\alpha\alpha}(\lambda)|0\rangle = a_\alpha(\lambda)|0\rangle$ and the mathematical vacuum is the eigenstate for these diagonal components, and the action of all upper elements $T_{\alpha\beta}$ with $\alpha < \beta$ is zero, *i.e.*, $T_{\alpha\beta}(\lambda)|0\rangle = 0$ for $\alpha < \beta$. Then the monodromy matrix has the triangular form. Such a monodromy matrix satisfies the intertwining relation for monodromy matrices Eq. (5.16) with our R-matrix in the graded space. Then it is not difficult to show that transfer matrices with different spectral parameters commute, which constitutes the exact integrability of the problem.

For the gl(1|2)-symmetric correlated electron chain $\hat{C}_{1,2}$ operators play the role of "creation operators" (please do not confuse: now we define the operators $\hat{C}$ as "creation operators", unlike the operators $\hat{B}$ of the previous sections, it is dictated by our choice of the mathematical vacuum). Let us construct the states

$$|\lambda_1^0, \ldots, \lambda_N^0|F\rangle = F_{a_1 \ldots a_N} \prod_{j=1}^{N} C_{a_j}(\lambda_j^0)|0\rangle \; . \tag{5.118}$$

Using the intertwining relations for the monodromy matrices we obtain the following nontrivial commutation relations for the operators $\hat{A}$, $\hat{B}$, $\hat{C}$ and $\hat{D}$

$$\begin{aligned}(-1)^{\epsilon_a\epsilon_f} r^{dc}_{fb}(y-x)\hat{C}_f(x)\hat{A}_{ad}(y) &= -b(y-x)\hat{C}_b(y)\hat{A}_{ac}(x) \\ &\quad + c(y-x)\hat{A}_{ab}(y)\hat{C}_c(x)\ , \\ \hat{C}_a(y)\hat{D}(x) &= b(y-x)\hat{C}_a(x)\hat{D}(y) \\ &\quad + c(y-x)\hat{D}(x)\hat{C}_a(y)\ , \\ \hat{C}_a(x)\hat{C}_b(y) &= r^{db}_{ca}(x-y)\hat{C}_c(y)\hat{C}_d(x)\ ,\end{aligned} \tag{5.119}$$

where $a, b, c, d, f = 1, 2$ and

$$r^{ab}_{cd}(x) = b(x)\delta_{a,b}\delta_{c,d} - c(x)\delta_{a,d}\delta_{c,b} = b(x)I^{(2)} + c(x)P^{(2)}\ . \tag{5.120}$$

Here the operators $I^{(2)}$ and $P^{(2)}$ play the role of the 4×4 identity and permutation operators corresponding to the grading $\epsilon_{1,2} = 1$. One can check that the operator $r(\lambda)$ satisfies the Yang–Baxter equation (5.9) (with the replacements $R \to r$; such a Yang–Baxter equation is called a *graded* one) and can be identified as the R-matrix of a fundamental spin model describing two species of electrons (with spins up and down).

Let us act with the transfer matrix on the state Eq. (5.118). We have

$$\begin{aligned}D(\lambda)|\lambda^0_1,\ldots,\lambda^0_N|F\rangle &= a_3(\lambda)\prod_{j=1}^{N} c^{-1}(\lambda^0_j-\lambda)|\lambda^0_1,\ldots,\lambda^0_N|F\rangle \\ &\quad + \sum_{k=1}^{N}(\tilde{T}_k)^{b_1\ldots b_N}_{a_1\ldots a_N} C_{b_k}(\lambda)\prod_{\substack{j=1,\\ j\neq k}}^{N} C_{b_j}(\lambda^0_j))F_{a_1\ldots a_N}|0\rangle\ ,\end{aligned} \tag{5.121}$$

and

$$\begin{aligned}&(-1)^N[\hat{A}_{11}(\lambda)+\hat{A}_{22}(\lambda)]|\lambda^0_1,\ldots,\lambda^0_N|F\rangle \\ &= \prod_{j=1}^{N} c^{-1}(\lambda-\lambda^0_j)(\mathrm{tr}[\hat{A}(\lambda)\hat{\tau}^{(2)}(\lambda)])^{b_1\ldots b_N}_{a_1\ldots a_N}\prod_{j=1}^{N} C_{b_j}(\lambda^0_j)F_{a_1\ldots a_N}|0\rangle \\ &\quad - (T_k)^{b_1\ldots b_N}_{a_1\ldots a_N}\sum_{k=1}^{N} C_{b_k}(\lambda)\prod_{\substack{j=1,\\ j\neq k}}^{N} C_{b_j}(\lambda^0_j)F_{a_1\ldots a_N}|0\rangle\ .\end{aligned} \tag{5.122}$$

Here the matrix

$$(\hat{\tau}^{(2)})^{b_1\ldots b_N}_{a_1\ldots a_N}(\lambda) = \mathrm{str}T^{(2)}(\lambda) \tag{5.123}$$

plays the role of the graded transfer matrix of a fundamental spin model describing two species of electrons (with spins up and down), with the monodromy of the fundamental inhomogeneous spin model

$$T^{(2)}(\lambda) = L_N^{(2)}(\lambda - \lambda_N^0) \otimes \cdots \otimes L_1^{(2)}(\lambda - \lambda_1^0) , \tag{5.124}$$

where

$$L_j^{(2)}(\lambda) = b(\lambda)P_j^{(2)} + c(\lambda)I_j^{(2)} . \tag{5.125}$$

Again, $L_j^{(2)}(\lambda)$ can be interpreted as the L-operator of the fundamental spin model on the inhomogeneous lattice of the length N. The eigenvalue condition

$$\hat{\tau}(\lambda)|\lambda_1^0, \ldots, \lambda_N^0|F\rangle = \Lambda(\lambda, \lambda_1^0, \ldots, \lambda_N^0)|\lambda_1^0, \ldots, \lambda_N^0|F\rangle \tag{5.126}$$

implies that $F_{a_1\ldots a_N}$ to be an eigenvector of the nested transfer matrix $\hat{\tau}^{(2)}(\lambda)$, and that the cancellation of the unwanted terms

$$[(\tilde{T}_k)^{b_1\ldots b_N}_{a_1\ldots a_N} - (T_k)^{b_1\ldots b_N}_{a_1\ldots a_N}]F_{a_1\ldots a_N} = 0 . \tag{5.127}$$

The unwanted terms in Eqs. (5.121) and (5.122) are

$$\begin{aligned}(\tilde{T}_k F)_{b_1\ldots b_N} &= -S^{b_1\ldots b_k}_{a_1\ldots a_k}(\lambda_k^0)F_{a_1\ldots a_k b_{k+1}\ldots b_N} \\ &\quad \times a_3(\lambda_k^0)\frac{b(\lambda_k^0 - \lambda)}{c(\lambda_k^0 - \lambda)}\prod_{\substack{j=1 \\ j\neq k}}^N c^{-1}(\lambda_j^0 - \lambda_k^0) ,\end{aligned} \tag{5.128}$$

and

$$\begin{aligned}(T_k F)_{b_1\ldots b_N} &= \mathrm{tr}[\hat{A}(\lambda_k^0)G(\lambda_k^0)]^{b_1\ldots b_k}_{a_1\ldots a_k}F_{b_1\ldots b_{k-1}a_k\ldots a_N} \\ &\quad \times (-1)^{k+1}\frac{b(\lambda - \lambda_k^0)}{c(\lambda - \lambda_k^0)}\prod_{\substack{j=1 \\ j\neq k}}^N c^{-1}(\lambda_k^0 - \lambda_j^0) ,\end{aligned} \tag{5.129}$$

where

$$S(x) = r^{b_{k-1}a_k}_{c_{k-1}a_{k-1}}(\lambda_{k-1}^0 - x)r^{b_{k-2}c_{k-1}}_{c_{k-2}a_{k-2}}(\lambda_{k-2}^0 - x)\cdots r^{b_1 c_2}_{b_k a_1}(\lambda_1^0 - x) \tag{5.130}$$

and

$$G(x) = (L_N^{(2)})_{b_N a_N}^{b_k d_{N-2}}(x - \lambda_N^0)(L_{N-1}^{(2)})_{b_{N-1} a_{N-1}}^{d_{N-2} d_{N-3}}(x - \lambda_{N-1}^0) \cdots \times (L_{k+1}^{(2)})_{b_{k+1} a_{k+1}}^{d_k a_k}(x - \lambda_{k+1}^0) . \tag{5.131}$$

It is easy to derive the following intertwining relation for the monodromy matrices of the fundamental spin model

$$r(x-y)(T^{(2)}(x) \otimes T^{(2)}(y)) = (T^{(2)}(y) \otimes T^{(2)}(x))r(x-y) . \tag{5.132}$$

Let us now define

$$T^{(2)}(\lambda) = \begin{pmatrix} \hat{A}^{(2)} & \hat{B}^{(2)} \\ \hat{C}^{(2)} & \hat{D}^{(2)} \end{pmatrix} . \tag{5.133}$$

Then the transfer matrix of the fundamental spin model can be written as $\hat{\tau}^{(2)}(\lambda) = -\hat{A}^{(2)}(\lambda) - \hat{D}^{(2)}(\lambda)$. The intertwining relation implies the following commutation relations

$$\begin{gathered} \frac{1}{c(x-y)}\hat{C}^{(2)}(x)\hat{A}^{(2)}(y) + \frac{b(y-x)}{c(y-x)}\hat{C}^{(2)}(y)\hat{A}^{(2)}(x) = \hat{A}^{(2)}(y)\hat{C}^{(2)}(x) , \\ \frac{1}{c(y-x)}\hat{C}^{(2)}(x)\hat{D}^{(2)}(y) + \frac{b(x-y)}{c(x-y)}\hat{C}^{(2)}(y)\hat{D}^{(2)}(x) = \hat{D}^{(2)}(y)\hat{C}^{(2)}(x) , \\ [\hat{C}^{(2)}(x), \hat{C}^{(2)}(y)] = 0 . \end{gathered} \tag{5.134}$$

Let us take as the mathematical vacuum for this nested fundamental spin problem the state $|0\rangle^{(2)} = \prod_{j=1}^N |0\rangle_j^{(2)}$. This state has the property $\hat{B}^{(2)}(\lambda)|0\rangle^{(2)} = 0$. Then it follows that

$$\begin{aligned} \hat{A}^{(2)}(\lambda)|0\rangle^{(2)} &= (-1)^N \prod_{j=1}^N c(\lambda - \lambda_j^0)|0\rangle^{(2)} , \\ \hat{D}^{(2)}(\lambda)|0\rangle^{(2)} &= (-1)^N \prod_{j=1}^N \frac{c(\lambda - \lambda_j^0)}{c(\lambda_j^0 - \lambda)}|0\rangle^{(2)} . \end{aligned} \tag{5.135}$$

Notice that $[\hat{A}_{ab}(\lambda), T_{cd}^{(2)}(\lambda)] = 0$, for $a, b, c, d = 1, 2$ (*i.e.*, in the auxiliary subspace), which can be proved using the definition of the monodromy matrix of the fundamental spin model. It is important, because we are also interested in the action of operators

$$[\hat{B}^{(2)}(\lambda)A_{21}(\lambda)+\hat{D}^{(2)}(\lambda)A_{22}(\lambda)]|0\rangle^{(2)}=(-1)^N a_2(\lambda)\prod_{j=1}^{N}\frac{c(\lambda-\lambda_j^0)}{c(\lambda_j^0-\lambda)}|0\rangle^{(2)}\ ,$$

$$[\hat{A}^{(2)}(\lambda)A_{11}(\lambda)+\hat{C}^{(2)}(\lambda)A_{12}(\lambda)]|0\rangle^{(2)}=(-1)^N a_1(\lambda)\prod_{j=1}^{N}c(\lambda-\lambda_j^0)|0\rangle^{(2)}\ ,$$

$$[\hat{A}^{(2)}(\lambda)A_{21}(\lambda)+\hat{C}^{(2)}(\lambda)A_{22}(\lambda)]|0\rangle^{(2)}=0\ . \tag{5.136}$$

Also, similar to the previous cases we can consider $\hat{C}^{(2)}$ operators as "creation operators" for the fundamental spin model and study the action of the transfer matrix $\hat{\tau}^{(2)}(\lambda)$ on the state $\prod_{\gamma=1}^{M}\hat{C}^{(2)}(\lambda_\gamma)|0\rangle^{(2)}$. This state is the eigenstate of the matrix $\mathrm{tr}[\hat{A}(\lambda)\hat{\tau}^{(2)}(\lambda)]$, which appears in Eq. (5.122), with the eigenvalue

$$\bar{\Lambda}^{(2)}(\lambda,\lambda_1^0,\lambda_N^0)=(-1)^{N-1}\Bigg(a_2(\lambda)\prod_{j=1}^{N}\frac{c(\lambda-\lambda_j^0)}{c(\lambda_j^0-\lambda)}\prod_{\gamma=1}^{M}c^{-1}(\lambda-\lambda_\gamma)$$
$$+a_1(\lambda)\prod_{j=1}^{N}c(\lambda-\lambda_j^0)\prod_{\gamma=1}^{M}c^{-1}(\lambda_\gamma-\lambda)\Bigg)\ , \tag{5.137}$$

if the following conditions for the rapidities λ_γ are satisfied

$$\frac{a_2(\lambda_\gamma)}{a_1(\lambda_\gamma)}\prod_{j=1}^{N}c^{-1}(\lambda_j^0-\lambda_\gamma)=\prod_{\substack{\beta=1\\ \beta\neq\gamma}}^{M}\frac{c(\lambda_\gamma-\lambda_\beta)}{c(\lambda_\beta-\lambda_\gamma)}\ ,\ \gamma=1,\ldots,M\ . \tag{5.138}$$

The cancellation of the first set of the unwanted terms yields (those equations can be obtained after some lengthy but straightforward calculations, which we drop here)

$$a_3(\lambda_k)\prod_{\substack{j=1\\ J\neq k}}^{N}\frac{c(\lambda_k^0-\lambda_j^0)}{c(\lambda_j^0-\lambda_k^0)}F_{b_1\ldots b_N}=(-1)^N[\mathrm{tr}A(\lambda_k)\hat{\tau}^{(2)}(\lambda_k)]_{a_1\ldots a_N}^{b_1\ldots b_N}F_{a_1\ldots a_N}\ , \tag{5.139}$$

where $k=1,\ldots,N$. Inserting Eq. (5.137) here we obtain

$$\frac{a_3(\lambda_j^0)}{a_2(\lambda_j^0)}=\prod_{\gamma=1}^{M}c^{-1}(\lambda_j^0-\lambda_\gamma)\ ,\ j=1,\ldots,N\ . \tag{5.140}$$

Then the eigenvalue of the transfer matrix $\hat{\tau}(\lambda)$ is equal to

$$\Lambda(\lambda) = a_3(\lambda)\prod_{j=1}^{N} c^{-1}(\lambda_j^0 - \lambda) - a_2(\lambda)\prod_{j=1}^{N} c^{-1}(\lambda_j^0 - \lambda)\prod_{\gamma=1}^{M} c^{-1}(\lambda - \lambda_\gamma) - a_1(\lambda)\prod_{\gamma=1}^{M} c^{-1}(\lambda_\gamma - \lambda)\,. \quad (5.141)$$

Taking the logarithmic derivative of the eigenvalue $\Lambda(\lambda)$ at $\lambda = 0$ we get

$$E = \sum_{j=1}^{N}\left[A\frac{ic}{(\lambda_j^0 + ic)\lambda_j^0} - 2\right] + Aa_3^{-1}(0)\frac{da_3(\lambda)}{d\lambda}\bigg|_{\lambda=0}, \quad (5.142)$$

where A is a constant.

It turns out that these derivations never used the concrete form of L-operators of the problem, but rather used the triangular property of the monodromy matrix: the action of diagonal matrix elements $T_{\alpha\alpha}$ produces c-numbers, $T_{\alpha\alpha}(\lambda)|0\rangle = a_\alpha(\lambda)|0\rangle$, and the action of all upper elements $T_{\alpha\beta}$ with $\alpha < \beta$ is zero, *i.e.*, $T_{\alpha\beta}(\lambda)|0\rangle = 0$ for $\alpha < \beta$.

Let us consider the L-operator of the supersymmetric t-J chain for $V = -J/4$ and $J = 2$ as

$$L_j(\lambda) = c(\lambda)I_j^{(1|2)} - b(\lambda) \times \begin{pmatrix} (N_j + S_j^z)(I_j - N_j + S_j^z) & -S_j^+ & -Q_{j\uparrow} \\ S_j^- & (N_j - S_j^z)(I_j - N_j - S_j^z) & -Q_{j\downarrow} \\ -Q_{j\uparrow}^\dagger & -Q_{j\downarrow}^\dagger & -(N_j - S_j^z)(N_j + S_j^z) \end{pmatrix}. \quad (5.143)$$

It is easy to check that such an L-operator satisfies the graded intertwining relations (Yang–Baxter relations) for L-operators with the graded R-matrix, Eq. (5.112), from which we started. The corresponding monodromy matrix $T(\lambda) = L_L(\lambda)\otimes\cdots\otimes L_1(\lambda)$ also satisfies the graded intertwining relations. The reader can check that the action of the L-operator on our mathematical vacuum $|0\rangle$ is

$$L_j(\lambda)|0\rangle = \begin{pmatrix} c(\lambda) & 0 & 0 \\ 0 & c(\lambda) & 0 \\ b(\lambda)Q_{j\uparrow}^\dagger & b(\lambda)Q_{j\downarrow}^\dagger & 1 \end{pmatrix}, \quad (5.144)$$

i.e., it has the triangular form, which implies

$$T(\lambda)|0\rangle = \begin{pmatrix} c^L(\lambda) & 0 & 0 \\ 0 & c^L(\lambda) & 0 \\ C_1(\lambda) & C_2(\lambda) & 1 \end{pmatrix} . \tag{5.145}$$

Eq. (5.145) means the triangular action of the monodromy matrix on the mathematical vacuum, *i.e.*, this choice of the L-operator can be used for the above described scheme with $a_1(\lambda) = a_2(\lambda) = c^L(\lambda)$ and $a_3(\lambda) = 1$. The Hamiltonian of the supersymmetric t-J chain for $V = -J/4$ and $J = 2$ can be obtained (up to constants) as

$$\mathcal{H}_{tJ} = -icA\frac{\partial}{\partial\lambda}\ln[\mathrm{str}\,\hat{\tau}(\lambda)]|_{\lambda=0} . \tag{5.146}$$

One can check that the Hamiltonian, constructed this way, coincides with the Hamiltonian of the supersymmetric t-J chain, and Eqs. (5.138), (5.140), and (5.142) for shifted rapidities $\lambda_j^0 \to \lambda_j^0 - ic/2$ and $c = 1$ coincide (up to constants) with the Bethe ansatz equations and the expression for the energy of the supersymmetric t-J chain presented in the previous chapter.

It is interesting to notice that the Bethe ansatz equations and the expression for the energy of the Hubbard chain can be obtained (up to constants) from Eqs. (5.138), (5.140), and (5.142) with the choice $\lambda_j^0 = \sin k_j$ ($j = 1, \ldots, N$), $\lambda_\gamma \to \lambda_\gamma - ic/2$ ($\gamma = 1, \ldots, M$) $c = U/2$, together with $a_1(\lambda - ic/2) = a_2(\lambda - ic/2)$, and $a_3(\sin k) = a_1(\sin k)\exp(ikL)$. This reflects the fact that the Hubbard chain also respects SU(2) (here equivalent to gl(2)) spin symmetry. However, we have to point out that the algebraic Bethe ansatz analysis for the Hubbard chain is, generally speaking, more complicated, and we shall not present it here, referring the reader to the original paper.

To summarize, in this chapter we presented the algebraic version of the Bethe ansatz: the quantum inverse scattering method. The main feature of the Bethe ansatz solvable models, *i.e.*, factorization of multi-particle scatterings into a two-particle one is discussed in the formalism of Yang–Baxter relations. In the framework of the algebraic Bethe ansatz we re-derived Bethe ansatz equations for a Heisenberg spin-$\frac{1}{2}$ chain and for a supersymmetric t-J chain (with graded Bethe ansätze), and derived the exact solution for higher-S SU(2)-symmetric spin chains (Takhtajan–Babujian model). Using the algebra of operators we showed the supersymmetry of the considered t-J chain.

The algebraic Bethe ansatz (or the quantum inverse scattering method) was formulated for lattice integrable models, *e.g.*, in [Takhtadzhan and Faddeev (1979)]. The Yang–Baxter relations were introduced by M. Gaudin, C. N. Yang and R. J. Baxter [Gaudin (1967); Yang (1967); Baxter (1982)]. The reader can find the description of the algebraic Bethe ansatz in [Korepin, Bogoliubov and Izergin (1993)], see also [Izyumov and Skryabin (1990)]. The construction of the integrals of motion for Bethe ansatz-solvable models was proposed in [Lüsher (1978)]. The Bethe ansatz solution of the SU(2)-symmetric spin-S chain was obtained in [Takhtajan (1982); Babujian (1983)]. The graded Bethe ansatz solution of the supersymmetric t-J chain can be found in [Eßler and Korepin (1992); Göhmann (2001)]. For the algebra of generators of the gl(1|2)-symmetric models, please, consult [Scheunert, Nahm and Rittenberg (1977)]. The reader can find the algebraic Bethe ansatz solution of the Hubbard chain in [Ramos and Martins (1997)].

Chapter 6

Correlated Quantum Chains with Open Boundary Conditions

In this chapter we shall present the results of the Bethe ansatz studies for quantum spin and correlated electron chains with open boundary conditions and shall discuss the effects of local potentials or fields applied to open edges of quantum chains. The important generalization of the algebraic Bethe ansatz for open chains will also be presented here.

6.1 Open Boundaries. XY and Ising Chains

So far we considered spin and correlated electron chains with periodic boundary conditions, *i.e.*, we considered chains in the ring geometry. However, the reader can ask the question: what will happen if a chain is not closed in a ring?

At the level of the Hamiltonian it means that we have to equate the terms like $S_L^{x,y,z}S_1^{x,y,z}$, $a_{L\sigma}^{\dagger}a_{1\sigma}$, $a_{1\sigma}^{\dagger}a_{L\sigma}$ and $n_{L\sigma}n_{1\sigma}$ to zero in previous formulae. To have more general results, we can add some *boundary magnetic fields* to the Hamiltonian of a spin chain with open boundary conditions as

$$\mathcal{H}_{bs} = -h_1 S_1^z - h_L S_L^z \ . \tag{6.1}$$

For the case of a correlated electron chain we can add not only the term Eq. (6.1), but also the term

$$\mathcal{H}_{bc} = -\mu_1 n_1 - \mu_L n_L \ , \tag{6.2}$$

which determines the action of *boundary potentials* (for example, related to point contact potentials applied only to edges of a chain). We shall call the case with $h_1 = h_L = \mu_1 = \mu_L = 0$ the situation with *free boundary conditions*, or *free edges* of an open quantum chain.

Let us start with the simplest spin-$\frac{1}{2}$ chains. Consider, first, the isotropic XY chain, $J_z = 0$. According to the procedure of Chapter 2, we can use the Jordan–Wigner transformation from spin operators to spinless Fermi operators. The Hamiltonian of the XY chain after such a transformation becomes a quadratic form of the Fermi operators. The difference between the periodic and open boundary conditions appears to be not very dramatic: one has to remove the terms $-(J/2)[\nu_{L+1}a_L^\dagger a_1 - \nu_{L+1}La_1^\dagger a_L]$ from Eq. (2.12), but to add instead the terms $-(h_1/2)(1-2a_1^\dagger a_1) - (h_L/2)(1-2a_L^\dagger a_L)$. Actually, it implies even simpler consideration than what was used in Chapter 2. Namely, the Hamiltonian of the isotropic spin-$\frac{1}{2}$ XY open chain can be straightforwardly diagonalized with the help of the Fourier transform. The dispersion law for these spinless fermions after the diagonalization becomes $\epsilon_k = H - 2J\cos k$, *i.e.*, the same as for the periodic case. Here H describes, as in previous chapters, the homogeneous magnetic field acting on all spins of a chain, *i.e.*, the field $H + h_{1,L}$ acts on edge spins. However, for the open boundary situation the values k (which play the role of quasimomenta) are determined from the quantization conditions

$$e^{ik(2L+2)}\left(\frac{J \pm 2h_1 e^{-ik}}{J \pm 2h_1 e^{ik}}\right)\left(\frac{J \pm 2h_L e^{-ik}}{J \pm 2h_L e^{ik}}\right) = 1 , \qquad (6.3)$$

where we can choose either plus, or minus signs. The reader can see that even for free edges $h_1 = h_L = 0$ quantization conditions are different from the case of a periodic chain. However, in this case the only real ks are solutions to Eq. (6.3). Actually, these solutions (with real ks) define standing waves, unlike plane waves for the periodic boundary conditions. On the other hand, nonzero boundary fields produce features, which are not present in the periodic chain. Namely, one can see that complex ks can be solutions to Eq. (6.3) for $h_1 \neq 0$, or $h_L \neq 0$. These complex solutions are called *boundary bound states.* Their wave functions decay with distances from edges, unlike the case with real ks. Notice, that even for complex solutions for ks, energy eigenvalues are ever real. We shall consider features of those solutions in what follows more precisely.

Let us study the effect of open boundaries for an isotropic XY chain in the situation with the large length of the chain. In this case we can write the expression for the internal ground state energy as

$$E = eL + f + o(L^{-1}) \qquad (6.4)$$

where e coincides with the ground state energy for a periodic chain of the length L, and f is the energy of open edges themselves. For example, for the homogeneous magnetic field equal to zero, $H = 0$, we obtain after some straightforward calculations

$$f = -\frac{h_1 + h_L}{2} - \frac{J}{\pi} \sum_{j=1,L} \int_0^\infty dx \frac{\sinh[2 - 2S_j(h_j)]x}{\sinh 2x \cosh x} , \tag{6.5}$$

where

$$S_j(h_j) = \frac{2}{\pi} \tan^{-1}\left(\frac{J + 2h_j}{|J - 2h_j|}\right) , \tag{6.6}$$

and $\tan^{-1}$ takes the principal branch for $h_j \leq 2J$, and takes the branch with the principal value minus π for $h_j > 2J$. The reader can see that the difference, as expected, is of the order of 1 (compared to the main contributions of order of L, which are equal for open and periodic chains; we shall show that the latter is true for any exactly solvable quantum chain). It is clear, because the difference between the periodic and open chains is connected with only one link, *i.e.*, it is the finite size effect. Even for $h_j = 0$ the ground state energy of the open XY chain differs from the periodic one, closed into a ring. The difference is equal to $J[1 - (2/\pi)]$. This is, in fact, the energy of infinitely large potential walls at the edges of an open chain, which reflect waves and produce standing waves instead of plane waves. The reader can also see that for $H = 0$ the magnetic moments of edge spins are

$$m^z_{1,L} = \frac{1}{2} - \frac{8}{\pi^2} \frac{J^2}{J^2 + 4h^2_{1,L}} \int_0^\infty dx \frac{x \cosh x[2 - 2S_{1,L}(h_{1,L})]}{\sinh 2x \cosh x} , \tag{6.7}$$

which implies $m^z_{1,L}|_{h_{1,L}=0} = 0$. On the other hand, local magnetic susceptibilities of edge spins are

$$\chi_{1,L}|_{h_{1,L}=0} = \frac{8}{3\pi J} , \tag{6.8}$$

i.e., they are finite.

Now let us study the difference in the behaviours of the Ising model (*e.g.*, with $J_x = J$, $J_y = J_z = 0$) in the transverse magnetic field for periodic and open boundary conditions.

Consider the behaviour at the quantum critical point $H = J/2$. Here the dispersion law is $\epsilon_k = J\cos(k/2)$, where the quantization conditions are

$$e^{ik(2L+2)}\left(\frac{J^2-[(J+2h_1)^2-J^2]e^{-ik}}{J^2-[(J+2h_1)^2-J^2]e^{ik}}\right) \times\left(\frac{J^2-[(J+2h_L)^2-J^2]e^{-ik}}{J^2-[(J+2h_L)^2-J^2]e^{ik}}\right)=1\,. \tag{6.9}$$

Again, we see that for free edges $h_1 = h_L = 0$ only possible solutions for this equation are real ks, but for $h_1 \neq 0$, or $h_L \neq 0$ complex solutions, *i.e.*, boundary bound states, can appear.

Again, the main contribution to the ground state (internal) energy, e, coincides with the one for the Ising chain in the transverse critical field with periodic boundary conditions. The difference (in the quantum critical point $H = J/2$ for $h_1 = h_L = h$) is

$$f = -\frac{J}{4}\left(1-\frac{2}{\pi}\right)+\frac{J}{\pi}F([1-(2h/J)]^2-1)\,, \tag{6.10}$$

where

$$F(x) = -1+\frac{1-x^2}{2\sqrt{x}|1-x|}\tan^{-1}\left(\frac{2\sqrt{x}}{|1-x|}\right)\,, \tag{6.11}$$

and $\tan^{-1}$ takes the principal branch for $x \leq 1$ and takes the branch with the principal value minus π for $x > 1$.

Out of the quantum critical point the boundary bound states appear even for free edges $h_1 = h_L = 0$. Here the dispersion law for free fermions after the diagonalization is $\epsilon_k = \sqrt{(J/2)^2 - HJ\cos k + H^2}$, but quantization conditions are

$$\frac{\sin k(L+1)}{\sin kL} = \frac{J}{2H}\,. \tag{6.12}$$

The reader can see that for $H < -J/2$ (or $H > J/2$)) one of L roots of this equation becomes complex: $k_0 = \pi + iv$, where

$$\frac{\sinh v(L+1)}{\sinh vL} = -\frac{J}{2H}\,. \tag{6.13}$$

The excitation with k_0 carries the energy

$$\epsilon_{k_0} = \frac{J}{2}(-1)^L(2H/J)^L\sum_{p=0}^{\infty}a_p(-2H/J)^{2p}\,. \tag{6.14}$$

Why is this one state so important? For $H < -(J/2)$ this equation reveals the asymptotic degeneracy of the ground state leading to the onset of an ordering in the system when $L \to \infty$ (the gap of order of $(2H/J)^L$ tends to zero when $L \to \infty$ more rapidly than $1/L$). Suppose L is finite. The ground state of the Ising model at $H = 0$ is doubly degenerate with $\langle 0|S_j^x|0\rangle = \pm\frac{1}{2}$. When a transverse magnetic field is switched on, this degeneracy is removed, and, thus, one has $\langle 0|S_j^x|0\rangle = 0$, because the new ground state is symmetric and remains unchanged when S_j^x is changed to $-S_j^x$, while the first excited state is antisymmetric and changes its sign. If L tends to infinity, *i.e.*, we are in the thermodynamic limit, for $H < -(J/2)$ (or $H > J/2$) the ground state becomes degenerate with the boundary bound state and $\langle 0|S_j^x|0\rangle \neq 0$. On the other hand, if $H < |J/2|$ the ground state remains non-degenerate and no order appears. At any nonzero temperature, naturally, the long-range order vanishes and there is no phase transition (*i.e.*, in this system only quantum critical point exists).

6.2 Open Boundaries: Co-ordinate Bethe Ansatz for the Heisenberg–Ising Chain

Let us now study the behaviour of the Heisenberg–Ising spin-$\frac{1}{2}$ chain with open boundary conditions using the co-ordinate Bethe ansatz. The z-projection of the total spin commutes with the Hamiltonian $\mathcal{H}_{HI} + \mathcal{H}_Z + \mathcal{H}_{bs}$. This is why, we classify all states of the Hamiltonian by the eigenvalue of the operator $\sum_{j=1}^{L} S_j^z$, as for periodic boundary conditions. It is convenient to choose the basis functions of the form

$$|x_1, \ldots, x_M\rangle \equiv e_1^+ \otimes \cdots \otimes e_{x_1}^- \otimes \cdots \otimes e_j^+ \otimes \cdots e_{x_M}^- \otimes \cdots \otimes e_L^+ \ , \tag{6.15}$$

where x_j determine M coordinates of sites with spins down (all other spins are directed up). We suppose that $1 \leq x_1 < x_2 < \cdots < x_M \leq L$. Then the wave function can be written as

$$\Psi_M = \sum_{x_1 < x_2 < \cdots < x_M} a(x_1, \ldots, x_M)|x_1, \ldots, x_M\rangle \ , \tag{6.16}$$

where $a(x_1, \ldots, x_M)$ is the wave function in the co-ordinate representation. First, suppose $M = 1$. In this case we have for $x \neq 1, L$

$$\begin{aligned} Ea(x) = &-\frac{H(L-2)}{2}a(x) + \frac{J_z L}{4}a(x) - \frac{5J_z}{4}a(x) \\ &- \frac{h_1 + h_L}{2}a(x) + \frac{J}{2}a(x-1) + \frac{J}{2}a(x+1) \ . \end{aligned} \tag{6.17}$$

At the boundaries we have slightly different equations

$$\begin{aligned} Ea(1) &= -\frac{H(L-2)}{2} + \frac{J_z(L-3)}{4}a(1) - J_z a(1) \\ &\quad - \frac{-h_1 + h_L}{2}a(1) + \frac{J}{2}a(2) , \\ Ea(L) &= -\frac{H(L-2)}{2} + \frac{J_z(L-3)}{4}a(L) - J_z a(L) \\ &\quad - \frac{h_1 - h_L}{2}a(L) + \frac{J}{2}a(L-1) . \end{aligned} \tag{6.18}$$

We want the solution

$$\begin{aligned} a(x) &= A(k)\exp(ikx) - A(-k)\exp(-ikx) , \\ E &= -\frac{H(L-2) + h_1 + h_L}{2} + \frac{J_z L}{4} - \frac{5J_z}{4} + J\cos k \end{aligned} \tag{6.19}$$

to be valid for $x = 1$ and $x = L$. This happens if the following equations hold

$$Ja(0) = (J_z + 2h_1)a(1) , \qquad Ja(L+1) = (J_z + 2h_L)a(L) . \tag{6.20}$$

This can be re-formulated as

$$A(k)\alpha(-k) - A(-k)\alpha(k) = 0 , \qquad A(k)\beta(k) - A(-k)\beta(-k) = 0 , \tag{6.21}$$

where

$$\begin{aligned} \alpha(k) &= -\frac{1}{2}[J + (J_z - 2h_1)\exp(-ik)] , \\ \beta(k) &= -\frac{1}{2}[J + (J_z - 2h_L)\exp(ik)]\exp[ik(L+1)] . \end{aligned} \tag{6.22}$$

This yields

$$\alpha(k)\beta(k) = \alpha(-k)\beta(-k) , \tag{6.23}$$

or this quantization condition can be written as

$$e^{2ik(L-1)}\frac{(Je^{ik} - 2h_1 - J_z)(Je^{ik} - 2h_L - J_z)}{(Je^{-ik} - 2h_1 - J_z)(Je^{-ik} - 2h_L - J_z)} = 1 . \tag{6.24}$$

The solution for $A(k)$ is $A(k) = \beta(-k)$, but it should be noted that this coefficient is determined up to a factor that is invariant under the change $k \leftrightarrow -k$.

For the case $M = 2$ we have to distinguish two situations. If $x_2 \neq x_1+1$ one has the equation

$$\begin{aligned} Ea(x_1, x_2) = &-\frac{H(L-4)}{2}a(x_1,x_2) + \frac{J_zL}{4}a(x_1,x_2) \\ &- \frac{9J_z}{4}a(x_1,x_2) + \frac{J}{2}[a(x_1-1,x_2) + a(x_1+1,x_2) \\ &+ a(x_1,x_2-1) + a(x_1,x_2+1)] \ . \end{aligned} \tag{6.25}$$

On the other hand, when $x_2 = x_1 + 1$ (*i.e.*, down spins are situated at the nearest neighbour sites) one obtains

$$\begin{aligned} Ea(x,x+1) = &-\frac{H(L-4)}{2}a(x,x+1) + \frac{J_zL}{4}a(x,x+1) - \frac{5J_z}{4}a(x,x+1) \\ &+ \frac{J}{2}[a(x-1,x+1) + a(x,x+2)] \ . \end{aligned} \tag{6.26}$$

These equations coincide with Eqs. (6.25), if the following condition is satisfied

$$Ja(x_1,x_1) + Ja(x_1+1,x_1+1) = 2J_za(x_1,x_1+1) \ . \tag{6.27}$$

As for $M = 1$, these equations in the case of open boundary conditions are added by the following equations

$$\begin{aligned} Ja(0,x_2) &= (J_z + 2h_1)a(1,x_2) \ , \\ Ja(x_1,L+1) &= (J_z + 2h_L)a(x_1,L) \ . \end{aligned} \tag{6.28}$$

We can look for the solution of these equations in the form, similar to the previous case

$$a(x_1,x_2) = \sum_P \epsilon_P A(k_1,k_2)e^{i(k_1x_1+k_2x_2)} \ , \tag{6.29}$$

where the sum extends over the permutations and the negations of k_1 and k_2, and ϵ_P is the sign factor, that changes sign on negation or pair interchange. The eigenvalue for the energy is

$$E = -\frac{H(L-4)+h_1+h_L}{2} + \frac{J_zL}{4} - \frac{9J_z}{4} + J(\cos k_1 + \cos k_2) \ . \tag{6.30}$$

Then the conditions on the coefficients $A(k_1, k_2)$, which follow from Eqs. (6.24) and (6.28) are

$$\begin{aligned} A(k_1,k_2)s(k_1,k_2) &= A(k_2,k_1)s(k_2,k_1) \ , \\ A(k_1,k_2)\alpha(-k_1) &= A(-k_1,k_2)\alpha(k_1) \ , \\ A(k_1,k_2)\beta(k_2) &= A(k_1,-k_2)\beta(-k_2) \ , \end{aligned} \tag{6.31}$$

together with nine other equations that can bc obtained from the above formulae by applying permutations and negations. Here the coefficient is

$$s(k_1,k_2) = -\frac{1}{2}\left(J - 2J_z e^{ik_2} + Je^{i(k_1+k_2)}\right) . \tag{6.32}$$

It implies

$$\frac{\alpha(k_1)\beta(k_1)}{\alpha(-k_1)\beta(-k_1)} = \frac{B(-k_1,k_2)}{B(k_1,k_2)} \ , \tag{6.33}$$

where

$$B(k,k') = s(k,k')s(k',-k) \ . \tag{6.34}$$

Finally, the quantization condition can be written as

$$e^{2ik_1(L-1)}\frac{(Je^{ik_1} - 2h_1 - J_z)(Je^{ik_1} - 2h_L - J_z)}{(Je^{-ik_1} - 2h_1 - J_z)(Je^{-ik_1} - 2h_L - J_z)} = \frac{B(-k_1,k_2)}{B(k_1,k_2)} \ . \tag{6.35}$$

Due to the symmetries $k_{1,2} \leftrightarrow -k_{1,2}$ the eight more functional relations result in only one additional equation (quantization condition)

$$e^{2ik_2(L-1)}\frac{(Je^{ik_2} - 2h_1 - J_z)(Je^{ik_2} - 2h_L - J_z)}{(Je^{-ik_2} - 2h_1 - J_z)(Je^{-ik_2} - 2h_L - J_z)} = \frac{B(-k_2,k_1)}{B(k_2,k_1)} \ . \tag{6.36}$$

The coefficient $A(k_1, k_2)$ has the form:

$$A(k_1,k_2) = \beta(-k_1)\beta(-k_2)B(-k_1,k_2)e^{-ik_2} \ . \tag{6.37}$$

This implies the general form of the eigenfunction in the co-ordinate representation as a superposition of waves

$$a(x_1,\ldots,x_M) = \sum_P \epsilon_P A(k_1,\ldots,k_M) \exp\left(i\sum_{j=1}^{M} k_j x_j\right) \ , \tag{6.38}$$

where the sum extends over all permutations and negations of of $k_1, \ldots, k_M$ and ϵ_P changes sign at each such "mutation". The coefficients are given as

$$A(k_1, \ldots, k_M) = \prod_{j=1}^{M} \beta(-k_j) \prod_{1 \le j < l \le M} B(-k_j, k_l) \exp(-ik_l) \ . \tag{6.39}$$

The quantization conditions for the quasimomenta k_j can be written as

$$\frac{\alpha(k_j)\beta(k_j)}{\alpha(-k_j)\beta(-k_j)} = \prod_{\substack{l=1 \\ l \neq j}} \frac{B(-k_j, k_l)}{B(k_j, k_l)} , \tag{6.40}$$

where $j = 1, \ldots, M$. This can be re-written in the following way:

$$e^{2ik_j(L-1)} \frac{(Je^{ik_j} - 2h_1 - J_z)(Je^{ik_j} - 2h_L - J_z)}{(Je^{-ik_j} - 2h_1 - J_z)(Je^{-ik_j} - 2h_L - J_z)} = \prod_{\substack{l=1, \\ l \neq j}}^{M} \exp(-i[\theta(k_j, k_l) + \theta(k_j, -k_l)]) , \tag{6.41}$$

where $j = 1, \ldots, M$ and

$$\theta(k_1, k_2) = \tan^{-1}\left(\frac{J_z \sin \frac{k_1 - k_2}{2}}{J \cos \frac{k_1 + k_2}{2} - J_z \cos \frac{k_1 - k_2}{2}} \right) , \tag{6.42}$$

which coincides with the definition of this function for periodic boundary conditions, *cf.* Chapter 3. Equation (6.41) is the Bethe ansatz equation for the Heisenberg–Ising spin-$\frac{1}{2}$ chain with open boundary conditions. The eigenvalue of the Hamiltonian $\mathcal{H}_{HI} + \mathcal{H}_Z + \mathcal{H}_{bs}$, which corresponds to the eigenfunction Eq. (6.38) is

$$E = -\frac{H(L - 2M) + h_1 + h_l}{2} + \frac{(L-1)J_z}{4} - \sum_{j=1}^{M} (J_z - J \cos k_j) \ . \tag{6.43}$$

Taking the logarithm, *e.g.*, for $h_{1,L} \to \pm\infty$, we get

$$(L-1)k_j = \pi I_j - \frac{1}{2} \sum_{\substack{l=1, \\ l \neq j}}^{M} [\theta(k_j, k_l) + \theta(k_j, -k_l)] \tag{6.44}$$

with positive integers I_j. It is easy to check that equation (6.41) for $J_z = 0$ agree with the quantization conditions and the definition of energies for the isotropic spin-$\frac{1}{2}$ XY chain with open boundary conditions.

These Bethe ansatz equations can be again re-written in the following way. For the case $J = J_z$ (*i.e.*, the isotropic Heisenberg chain) one can introduce the set of rapidities $\{\lambda_j\}_{j=1}^M$ instead of $\{k_j\}_{j=1}^M$ as in Chapter 3 (*e.g.*, $\lambda_j = \frac{1}{2}\cot\frac{k_j}{2}$ *etc.*). Then the Bethe ansatz equations can be written in the differential form

$$\left(\frac{\lambda_j + (i/2)}{\lambda_j - (i/2)}\right)^{2L} \frac{\lambda_j + iS_1}{\lambda_j - iS_1}\frac{\lambda_j + iS_L}{\lambda_j - iS_L} = \prod_{\substack{l=1,\\ l\neq j}}^{M} \frac{\lambda_j - \lambda_l + i}{\lambda_j - \lambda_l - i}\frac{\lambda_j + \lambda_l + i}{\lambda_j + \lambda_l - i} . \tag{6.45}$$

The energy is

$$E = -\frac{H(L-2M) - h_1 - h_2}{2} + \frac{(L-1)J}{4} - 2J\sum_{j=1}^{M}(4\lambda_j^2 + 1)^{-1} \tag{6.46}$$

where H is the value of the homogeneous magnetic field and

$$2S_{1,L} = \frac{J}{h_{1,L}} - 1 . \tag{6.47}$$

For the anisotropic chain with $\cos\eta = J_z/J$ the set of rapidities is introduced *via* $k_j = -i\sin[\lambda_j + (\eta/2)]/\sin[\lambda_j - (\eta/2)]$. Bethe ansatz equations and the energy of an open chain become

$$\left(\frac{\sin[\lambda_j + (\eta/2)]}{\sin[\lambda_j - (\eta/2)]}\right)^{2L} \frac{\sin[\lambda_j + iS_1']}{\sin[\lambda_j - iS_1']}\frac{\sin[\lambda_j + iS_L']}{\sin[\lambda_j - iS_L']} = \prod_{\substack{l=1,\\ l\neq j}}^{M} \frac{\sin[\lambda_j - \lambda_l + \eta]}{\sin[\lambda_j - \lambda_l - \eta]}\frac{\sin[\lambda_j + \lambda_l + \eta]}{\sin[\lambda_j + \lambda_l - \eta]} \tag{6.48}$$

where

$$S_{1,L}' = \frac{1}{2}\ln\left(\frac{\sin[\ln\sqrt{\cos\eta + (2h_{1,2}/J)} - (\eta/2)]}{\sin[\ln\sqrt{\cos\eta + (2h_{1,2}/J)} + (\eta/2)]}\right) , \tag{6.49}$$

and the energy is

$$E = -\frac{HL + h_1 + h_2}{2} + \frac{(L-1)J_z}{4} - \sum_{j=1}^{M}\left(J_z - H - \frac{J\sin[\lambda_j + (\eta/2)]}{2\sin[\lambda_j - (\eta/2)]} - \frac{J\sin[\lambda_j - (\eta/2)]}{2\sin[\lambda_j + (\eta/2)]}\right) . \tag{6.50}$$

The quantities $S'_{1,L}$ are defined in such a way that at $h_{1,L} = 0$ they are $2S'_{1,L} = \pi - \eta$, and for $h_{1,L} \to \infty$ one has $S'_{1,L} = 2\pi - \eta$.

We see that Bethe ansatz equations for an open chain differ from the ones for a closed geometry, Eqs. (3.20)–(3.23), by:

- There are not only differences but also sums of rapidities on the right hand sides of Bethe ansatz equations for an open case;
- The effective length of a chain is doubled, *i.e.*, L is replaced by $2L$ for a system with open boundaries;
- On the left hand sides of Bethe ansatz equations for an open case there are multipliers connected with nonzero boundary fields.

Now, let us study how these differences affect thermodynamic characteristics in the limit of large L and M (with M/L fixed), concentrating on the case of the antiferromagnetic Heisenberg spin-$\frac{1}{2}$ chain (*i.e.*, $J_z = J > 0$).

In the framework of the string hypothesis we look for the solution of Eq. (6.45) in the form of strings. Introducing strings as $\lambda_j = \lambda_{j,m} + i[(m+1)/2 - \nu]$ with $\nu = 1, \dots, m$, we get

$$\theta_{m,1}(\lambda_j^m) + \frac{1}{2L}\left(\theta_{m,2S_1}(\lambda_j^m) + \theta_{m,2S_2}(\lambda_j^m)\right)$$
$$= \frac{\pi}{L} I_{j,m} + \frac{1}{2L} \sum_{\substack{n=1 \\ n\neq m}}^{\infty} \sum_{\substack{l=1 \\ l\neq j}}^{M_n} [\Theta_{mn}(\lambda_j^m - \lambda_l^n) + \Theta_{mn}(\lambda_j^m + \lambda_l^n)] \ , \quad (6.51)$$

where $\theta_n(x) = 2\tan^{-1}(x/n)$,

$$\theta_{m,n}(x) = \sum_{l=1}^{\min([m],[n])} \theta_{m+1+n-2l}(x) \ , \quad (6.52)$$

$[x]$ denotes the integer part of x,

$$\Theta_{mn}(x) = (1 - \delta_{m,n})\theta_{|m-n|}(x) + 2\theta_{|m-n|+2}(x) + \cdots$$
$$+ 2\theta_{m+n-2}(x) + \theta_{m+n}(x) \ , \quad (6.53)$$

and integers $1 \le I_{j,m} \le (L + M_m - 2\sum_{n=1}^{\infty} \min(m,n)M_n)$ appear because the logarithm is the multi-valued function. Again, we introduce two sets of these quantum numbers. The first set, $I_{j,m}$ which characterizes strings present in the given configuration, parametrizes quasiparticles, and the second set, $I_{j,m}^{(h)}$, which characterizes unoccupied vacancies of the given configuration of strings, parametrizes quasiholes.

Then we look for solutions to thermodynamic Bethe ansatz equations for large L, keeping corrections of order of L^{-1}, too. In the framework of the string hypothesis it is justified, because we dropped only terms of order of $\exp(-L)$, when we used string solutions, *cf.* Chapter 3. In this limit we introduce distribution functions (densities) for particles and holes, corresponding to strings of length m: $\rho_m(x)$ and $\rho_{mh}(x)$, respectively. It yields

$$\rho_{mh}(\lambda) + \frac{1}{2}\sum_{n=1}^{\infty}[A_{mn}(\lambda-\lambda') + A_{mn}(\lambda+\lambda')] * [\rho_n(\lambda') - p(\lambda')\delta_{m,1}]$$
$$= \frac{1}{2L}\sum_{n=1}^{\infty}[A_{m,n}(\lambda-\lambda') + A_{m,n}(\lambda+\lambda')] * p(\lambda')(\delta_{m,[2S_1]}+\delta_{m,[2S_2]}) \ , \quad (6.54)$$

where $p(\lambda) = 1/4\cosh(\pi\lambda/2)$, $*$ denotes the convolution,

$$A_{m,n}(x) = a_{|m-n|}(x) + 2\sum_{l=1}^{\min(n,m)-1} a_{m+n-2l}(x) + a_{m+n}(x) \ , \quad (6.55)$$

and $a_m(x) = 2m/[\pi(4x^2+m^2)]$. The internal energy E and the total magnetic moment M^z are given as

$$E = E_0 - \frac{1}{2}\sum_{m=1}^{\infty}\int_0^{\infty} d\lambda\theta'_{m,1}(\lambda)\rho_m(\lambda) \ ,$$
$$M^z = \frac{L}{2} - L\sum_{m=1}^{\infty} m\int_0^{\infty} d\lambda\rho_m(\lambda) \ , \quad (6.56)$$

where $E_0 = -[2(HL+h_1+h_2) - (L-1)J_z]/4$.

The set of thermodynamic equations for dressed energies $\varepsilon_n(\lambda) = T\ln[\rho_{nh}(\lambda)/\rho_n(\lambda)] = \eta_n(\lambda)$ is

$$Hm - J\theta'_{m,1}(\lambda) = T\ln[1+\eta_m(\lambda)]$$
$$-\frac{T}{2}\sum_n[A_{n,m}(\lambda-\lambda') + A_{n,m}(\lambda+\lambda')] * \ln[1+\eta_n^{-1}(\lambda')] \ , \quad (6.57)$$

which completes the set Eq. (6.54). We, actually, see that the set of equations for dressed energies for the chains with open and periodic boundary conditions coincide, up to the change $A_{n,m}(\lambda-\lambda') \to (1/2)[A_{n,m}(\lambda-\lambda') + A_{n,m}(\lambda+\lambda')]$.

The reader can observe that thermodynamic Bethe ansatz equations for densities are linear integral equations. There are two kinds of driving terms:

the ones of order of 1, and the ones of order of L^{-1}. This is why, one can divide densities as $\rho_n(\lambda) = \rho_n^{(0)}(\lambda) + L^{-1}\rho_n^{(1)}(\lambda)$ (and the same for densities of holes). Then one can separate Bethe ansatz equations for densities into two sets: one of the scale 1 for the main (of order of L) contribution to the energy, magnetization, *etc.*, *i.e.*, for $\rho_n^{(0)}(\lambda)$ only, and the other one of the scale L^{-1} for the finite contribution (of order of 1) to the energy, magnetic moment, *etc.*, *i.e.*, for $\rho_n^{(1)}(\lambda)$ only. The former describes thermodynamics of the bulk, while the latter reveals the contribution from edge spins. It turns out that the set of equations for dressed energies does not have terms of order of L^{-1} explicitly.

We already showed that the most interesting behaviour of one-dimensional quantum systems is in the ground state and at low temperatures. The reader knows that for a spin-$\frac{1}{2}$ Heisenberg chain only spinons (*i.e.*, strings of length 1) have negative energies (the Dirac sea). The latter is defined as the solution of the equation

$$\varepsilon_1(\lambda) + \frac{1}{2}[A_{1,1}(\lambda - \lambda') + A_{1,1}(\lambda + \lambda')] * \varepsilon_1^-(\lambda') = H - J\theta'_{1,1}(\lambda) \qquad (6.58)$$

with the same notations as in the previous chapters. The Fermi point (related to the limit of integration) is determined from the condition $\varepsilon_1(B) = 0$. The equations for densities in the ground state are

$$\begin{aligned}\rho_1(\lambda) + \rho_{1h}(\lambda) &= a_1(\lambda) + \frac{1}{2L}\left(a_{2S_1}(\lambda) + a_{2S_2}(\lambda)\right) \\ &\quad - \frac{1}{2}\int_0^B d\lambda'[a_2(\lambda - \lambda') + a_2(\lambda + \lambda')]\rho_1(\lambda') \, . \end{aligned} \qquad (6.59)$$

The ground state internal energy can be written as

$$E_{T=0} = E_0 + \int_0^B d\lambda[H - J\theta'_{1,1}(\lambda)]\rho_1(\lambda) \qquad (6.60)$$

and the ground state magnetization is equal to

$$M^z = \frac{L}{2} - L\int_0^B d\lambda\rho_1(\lambda) \, . \qquad (6.61)$$

Additional terms in Eq. (6.59) comparing Eq. (3.54) describe boundary fields. There $S_{1,L}$ play the role of effective "boundary spins". These "boundary spins" depend on the values of the boundary fields $h_{1,L}$. For $h_{1,L} = 0$ these boundary spins are infinite, leading to the effective twists of π at each edge. At $h_{1,L} = \pm J$ these effective "boundary spins" change their signs. This situation is related to the effective addition or removal of one site to or from the chain, respectively, with finite zero-field magnetic susceptibility. It leads to onsets of complex roots of Bethe ansatz equations (6.45) in the ground state: for $-\frac{1}{2} < S_{1,L} < 0$ there appear bound states parametrized by complex rapidities $\lambda_j = (i/2)[1 - (J/h_{1,L})]$, localized at edges. Finally, for $h_{1,L} \to \pm\infty$ we have $S_{1,L} \to -\frac{1}{2}$, effectively removing one site, respectively, from the system.

The value of a homogeneous magnetic field H determines the limit of integration, *i.e.*, the Fermi point. For the antiferromagnetic case large values of the external magnetic field $|H| > H_s = 2J$ the system is in the ferromagnetic state and $B = 0$. In these regions of values of H the ground state energy is equal to E_0, the magnetic moments of all spins have their nominal values $\frac{1}{2}$, and the magnetic susceptibility is zero. On the other hand, in zero magnetic field, $H = 0$, for the antiferromagnetic situation we get $B = \infty$.

Since $\epsilon_1(\lambda)$ and $\rho_1(\lambda)$ are even functions, one can re-write the equation for dressed energies as

$$\varepsilon_1(\lambda) + A_{1,1}(\lambda - \lambda') * \varepsilon_1^-(\lambda') = H - J\theta'_{1,1}(\lambda) \ . \tag{6.62}$$

The main contribution to the equations of densities, which describes the behaviour of the bulk, can be written as

$$\rho_1^{(0)}(\lambda) + \rho_{1h}^{(0)}(\lambda) = a_1(\lambda) - \int_B^B d\lambda' a_2(\lambda - \lambda')\rho_1(\lambda') \ . \tag{6.63}$$

It is easy to check that the answers for the main contribution for the open chain coincide with those for the periodic chain, as expected.

Let us then concentrate on the finite size contributions, for which we have the equation for dressed densities:

$$\rho_1^{(1)}(\lambda) + \rho_{1h}^{(1)}(\lambda) = \frac{1}{2}\left[a_2(\lambda) + a_1(\lambda) + a_{2S_1}(\lambda) + a_{2S_2}(\lambda)\right] - \int_B^B d\lambda' a_2(\lambda - \lambda')\rho_1^{(1)}(\lambda') \ , \tag{6.64}$$

where we introduced the term $a_1(\lambda) + a_2(\lambda)$ to avoid double counting due to the symmetrization of functions (with $\lambda = 0$) and to take into account the term with $\lambda_\alpha = \lambda_\beta$ in the right hand side of Eq. (6.45). In fact, the limits of integration are already determined by the main contribution, see Chapter 3. Combining all contributions we obtain for the vanishing homogeneous magnetic field $H = 0$, where $B = \infty$ and for positive $S_{1,L}$:

$$E = E_0 - \frac{2L+1}{2} J \ln 2 + \frac{\pi J}{4} - \frac{J}{4}\Big[\psi(3/4) - \psi(1/4) + \psi[(2S_1+3)/4]$$

$$- \psi[(2S_1+1)/4] + \psi[(2S_L+3)/4] - \psi[(2S_1+3)/4]\Big] , \qquad (6.65)$$

where $\psi(x)$ are digamma functions. The reader can see that there is a difference of order of 1 between this expression and the one for a periodic chain. This is the ground state energy of free edges of the chain (the surface energy) and the energy of applied boundary fields. The nature of the former contribution is as follows: edges are affected by infinitely large potentials, which do not permit waves to propagate through them. This ground state corresponds to the total magnetic moment $-\frac{1}{2}$ for nonzero $h_{1,L}$ and odd L, and to zero for $h_{1,L} = 0$ and even L.

For small values of the homogeneous magnetic field H we can apply the Wiener–Hopf technique, described in Chapter 3. The magnetic moment of open boundaries consists of two contributions: from free edges themselves and from boundary fields. The magnetic moment of free edges of the open chain is

$$m^z_{edges} = \frac{1}{2}\left(\frac{1}{2|\ln\sqrt{e}H/\sqrt{\pi^3}J|} - \frac{\ln\frac{1}{2}|\ln\sqrt{e}H/\sqrt{\pi^3}J|}{4(\ln\sqrt{e}H/\sqrt{\pi^3}J)^2} + \cdots\right) . \qquad (6.66)$$

This contribution is different from the linear in H contribution for the magnetization per site of bulk spins (which is the same as for periodic spin-$\frac{1}{2}$ chain). Naturally, the magnetic susceptibility of free edges is also different from the finite value of the one for bulk spins:

$$\chi_{edges} = -\frac{1}{4H\ln^2(\sqrt{e}H/\sqrt{\pi^3}J)} . \qquad (6.67)$$

Magnetic moments, caused by finite boundary magnetic fields $h_{1,L}$ are given by

$$m^z_h = \frac{1}{4}\left(-1 + \frac{2S_{1,L}-1}{|\ln\sqrt{e}H/\sqrt{\pi^3}J|} + \cdots\right) , \qquad (6.68)$$

for $2S_{1,L} \ll |\ln\sqrt{e}H/\sqrt{\pi^3}J|$, and

$$m^z_h = -\frac{1}{\pi^2(2S_{1,L}-1)}|\ln\sqrt{e}H/\sqrt{\pi^3}J| + \cdots , \qquad (6.69)$$

for $2S_{1,L} \gg |\ln \sqrt{e}H/\sqrt{\pi^3}J|$. As discussed above, the contribution $-\frac{1}{4}$ for each boundary in a vanishing bulk field H is due to the fact that we consider a chain of odd size in the presence of boundary fields. Magnetic moments of boundary spins and their magnetic susceptibilities can be extracted for $H = 0$ by differentiating Eq. (6.65) with respect to $h_{1,L}$. Boundary magnetic moments vanish for $H = h_{1,L} = 0$, but local boundary magnetic susceptibilities are finite. It is necessary to note that for large boundary fields the result coincides with Eq. (6.69), but for small boundary fields the expectation value of an edge spin is only half of the expression in Eq. (6.69). The reason for such a difference is the non-commutativity of the limits $H \to 0$ and $h_{1,L} \to 0$.

It turns out that if the homogeneous magnetic field H and the boundary ones $h_{1,L}$ are connected to each other, *e.g.*, *via* $h_{1,L} = (1 - \mu_{1,L})H$, where $\mu_{1,L}$ are effective magnetons of edge spins of an open chain, we can use the above result for H small enough and/or $\mu_{1,L} \sim 1$. For larger field H the perturbative solution outlined above fails and integral equations should be studied numerically.

It is important to point out that for negative $S_{1,L}$, where boundary bound states appear, the energy and the magnetic field dependence of magnetic moments are the same as above, despite appearing of local levels.

At low temperatures the contribution of free edges for $h_{1,L} = 0$ can produce for the most interesting case $\eta = 0$ the divergent local magnetic susceptibilities $\chi \sim (8T|\ln x|)^{-1}[1 - (\ln|\ln x|/2|\ln x)] + \ldots$, where $x = a\pi J\sqrt{\pi/e}/T$, a is a constant, the divergent Sommerfeld coefficient of the low-temperature specific heat $\gamma \sim 3\pi^2/32T\ln^4 x$; the boundary entropy is $\mathcal{S} \sim \pi^2/32|\ln x|^3$.

6.3 Open Boundaries. The Algebraic Bethe Ansatz

In the previous section we studied the behaviour of an open chain using the co-ordinate Bethe ansatz. It is interesting now to show how the algebraic version of the Bethe's ansatz (the quantum inverse scattering method) is modified due to the presence of open edges of a quantum chain and boundary potentials. This technique was developed mostly by I. V. Cherednik and E. K. Sklyanin.

As for the periodic chain we start with the R-matrix acting in the space $V_1 \otimes V_2 \otimes V_3$, which satisfies the Yang–Baxter equation

$$R_{23}(\lambda)R_{12}(\lambda + \mu)R_{23}(\mu) = R_{12}(\mu)R_{23}(\lambda + \mu)R_{12}(\lambda) \ , \qquad (6.70)$$

where the subscripts indicate in which spaces the R-matrix acts nontrivially. One can also require from the R-matrix the unitarity $R_{12}(x)R_{12}(-x) = \rho(x)$, crossing-unitarity $R_{12}(x)R_{12}(-x-ic) = \tilde{\rho}(x)$ (where ρ and $\tilde{\rho}$ are scalar c-functions), and we want the R-matrix to be symmetric $P_{12}R_{12}(x)P_{12} = R_{12}(x)$ and $R_{12}^{t_1}(x) = R_{12}^{t_2}(x)$, where P_{12} is the permutation operator $P(x \otimes y) = y \otimes x$, and $t_{1,2}$ denotes the transposition in the space $V_{1,2}$. One can consider the monodromy matrix $T(x)$, which satisfies the intertwining relations

$$R(\lambda - \mu)(T(\lambda) \otimes T(\mu)) = (T(\mu) \otimes T(\lambda))R(\lambda - \mu) . \tag{6.71}$$

Now let us introduce two new algebras $\mathcal{K}^{\pm}$ *via* so called *reflection equations* introduced by Cherednik and generalized by Sklyanin:

$$\begin{aligned} &R_{12}(x-y)\mathcal{K}_1^-(x)R_{12}(x+y)\mathcal{K}_2^-(y) = \mathcal{K}_2^-(y)R_{12}(x+y)\mathcal{K}_1^-(x)R_{12}(x-y) , \\ &R_{12}(y-x)(\mathcal{K}_1^+)^{t_1}(x)R_{12}(-x-y-2ic)(\mathcal{K}_2^+)^{t_2}(y) \\ &\quad = (\mathcal{K}_2^+)^{t_2}(y)R_{12}(-x-y-2ic)(\mathcal{K}_1^+)^{t_1}(x)R_{12}(y-x) , \end{aligned} \tag{6.72}$$

respectively. The illustration of a reflection equation is presented in Fig. 6.1.

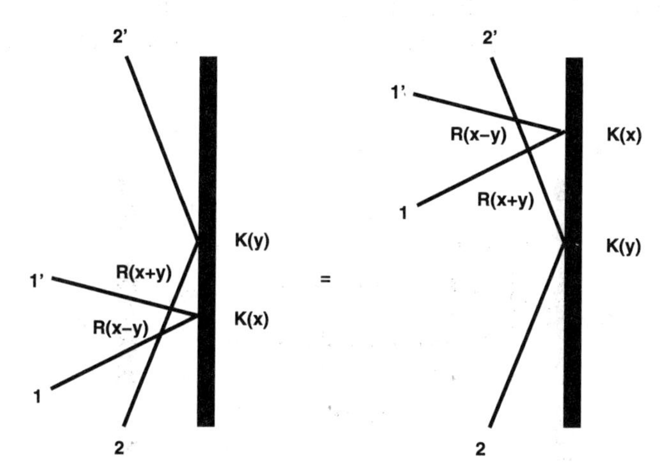

Fig. 6.1 Illustration of a reflection equation for an integrable model with open boundary conditions.

Then it is easy to show that the quantities $\hat{\tau}_o(\lambda) = \operatorname{tr} \mathcal{K}^+(\lambda)\mathcal{K}^-(\lambda)$, can be considered as transfer matrices for open chains and commute with different spectral parameters: $[\hat{\tau}_o(\lambda), \hat{\tau}_o(\mu)] = 0$. The latter property constitutes the exact integrability of the problem, as for the situation with periodic boundary conditions. Actually it is possible to consider the matrices

$$\begin{aligned}\mathcal{K}^-(\lambda) &= T^-(\lambda)\tilde{\mathcal{K}}^-(\lambda)(T^-)^{-1}(-\lambda) \ , \\ (\mathcal{K}^+)^t(\lambda) &= (T^+)^t(\lambda)(\tilde{\mathcal{K}}^-)^t(\lambda)(T^-)^a(-\lambda) \ ,\end{aligned} \tag{6.73}$$

where $T^a(x) = (T^{-1})^t(x)$, $\mathcal{K}(x) = \mathcal{K}^{-1}(-x)$ and

$$\begin{aligned}&T^-(\lambda) = L_N(\lambda)\cdots L_1(\lambda) \ , \qquad T^+(\lambda) = L_L(\lambda)\cdots L_{N+1}(\lambda) \ , \\ &\tilde{\mathcal{K}}^\pm(\lambda) = K^\pm(\lambda) \ ,\end{aligned} \tag{6.74}$$

where $L_n(\lambda)$ are L-operators (which satisfy intertwining relations with R-matrix), and $K^\pm(\lambda)$ are representations of $\mathcal{K}^\pm$ in C^1, *i.e.*, c-number matrices. One can prove that

$$\hat{\tau}(\lambda) = \mathrm{tr}\, K^+(\lambda)T(\lambda)K^-(\lambda)T^{-1}(-\lambda) \ , \tag{6.75}$$

where $T(\lambda) = T^+(\lambda)T^-(\lambda) = L_L(\lambda)\cdots L_1(\lambda)$ is the monodromy matrix, and it is independent of the factorization of $T(\lambda)$ into $T^+(\lambda)$ and $T^-(\lambda)$. The illustration of a transfer matrix of an open integrable chain is presented in Fig. 6.2.

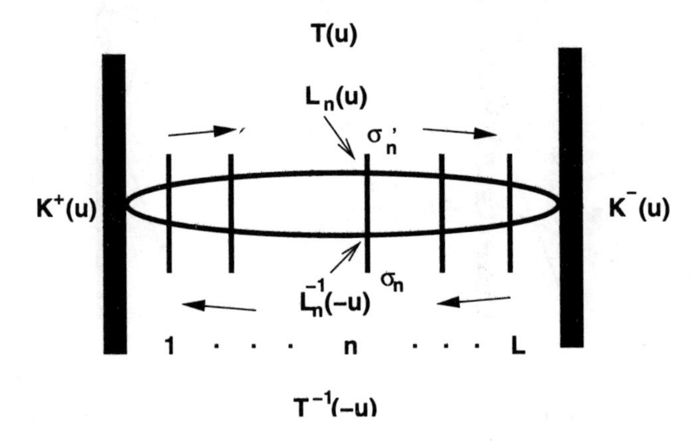

Fig. 6.2 Illustration of a transfer matrix of an integrable model with open boundary conditions.

Taking $K^-(0) = I$ and using the properties of L-operators and R-matrices we can construct the Hamiltonian of an open chain as the logarithmic derivative of the transfer matrix of an open chain

$$\mathcal{H}_o = \text{const}\left(\sum_{j=1}^{L-1} \frac{d}{d\lambda} L_{j,j+1}(\lambda)|_{\lambda=0} + \frac{1}{2}\frac{d}{d\lambda} K_1^-(\lambda)|_{\lambda=0} \right.$$
$$\left. + \frac{\mathrm{tr}_0\, K_0^+(0)\frac{d}{d\lambda}L_{N,0}(\lambda)|_{\lambda=0}}{\mathrm{tr}\, K^+(0)}\right) , \tag{6.76}$$

where the index 0 denotes the auxiliary subspace and we used the notations from Chapter 5.

Let us now apply this scheme to the open Heisenberg spin-$\frac{1}{2}$ chain, considered in the previous section with the help of the co-ordinate Bethe ansatz.

We shall use the expressions for the R-matrix and L-operator from the previous chapter, Eqs. (5.6) and (5.12), respectively. Reflection equations then yield

$$K^-(\lambda) = K(\lambda, \xi_-) \ , \qquad K^+(\lambda) = K(\lambda + ic, \xi_+) \ ,$$
$$K(\lambda, \xi) = \begin{pmatrix} \xi + \lambda & 0 \\ 0 & \xi - \lambda \end{pmatrix} . \tag{6.77}$$

Then the application of Eq. (6.76) produces (up to constants)

$$\mathcal{H}_o = \sum_{j=1}^{L-1} \vec{S}_j \vec{S}_{j+1} + ic\left(\frac{1}{\xi_-} S_1^z + \frac{1}{\xi_+} S_1^L\right) , \tag{6.78}$$

which coincides with the Hamiltonian of the Heisenberg spin-$\frac{1}{2}$ open chain with $h_{1,L} \to -ic\xi_{\mp}^{-1}$.

It is possible to write down

$$T^a(x) = \frac{\sigma^y T(x - ic)\sigma^y}{\delta\{T(x - i(c/2))\}} , \tag{6.79}$$

where

$$\delta\{T(x)\} = \mathrm{tr}_{12}\, P_{12}^- T_1(x - i(c/2)) T_2(x + i(c/2)) \tag{6.80}$$

is the *quantum determinant* of $T(x)$. Here $P_{12}^- = \frac{1}{2}(1 - P_{12})$ is the antisymmetrizing operator. Then we can re-write

$$\mathcal{K}^-(\lambda) = T^-(\lambda) K^-(\lambda) \frac{\sigma^y (T^-)^t(-\lambda - ic)\sigma^y}{\delta\{T(-\lambda - i(c/2))\}} . \tag{6.81}$$

Then, let us introduce the matrices

$$U^-(x) = T^-(x)K^-(x - i(c/2), \xi_-)\sigma^y (T^-)^t(-x - ic)\sigma^y \tag{6.82}$$

and

$$(U^+)^t(x) = (T^+)^t(x)(K^+)^t(x + i(c/2), \xi_+)\sigma^y T^+(-x)\sigma^y \ , \tag{6.83}$$

which satisfy the reflection equations

$$\begin{aligned} &L_{12}(x-y)U_1^-(x)L_{12}(x+y-ic)U_2^-(y) \\ &\quad = U_2^-(y)L_{12}(x+y-ic)U_1^-(x)L_{12}(x-y) \ , \\ &L_{12}(y-x)(U_1^+)^{t_1}(x)L_{12}(-x-y-ic)(U_2^+)^{t_2}(y) \\ &\quad = (U_2^+)^{t_2}(y)L_{12}(-x-y-ic)(U_1^+)^{t_1}(x)L_{12}(y-x) \ . \end{aligned} \tag{6.84}$$

In fact all previous results can be applied to $U^\pm$.

It is easy to prove that for any x and y

$$[\Delta\{U^-(x)\}, U^-(y)] = 0 \ , \tag{6.85}$$

where

$$\Delta\{U^-(x)\} = \mathrm{tr}_{12}\, P_{12}^- U_1^-(x - i(c/2))L_{12}(2x - ic)U_2^-(x + i(c/2)) \ . \tag{6.86}$$

The latter quantity can be considered as the "Casimir operator" of the algebra U^-. The reader can check that

$$\begin{aligned} \Delta\{U^-(x)\} &= \delta\{T^-(x)\}\delta\{T^-(-x)\}\Delta\{K(x - i(c/2), \xi_-)\} \\ &= -(2x - 2ic)(x + \xi_-)(x - \xi_-)\prod_{j=1}^{L} \delta\{L_j(x)\}\delta\{L_j(-x)\} \ . \end{aligned} \tag{6.87}$$

Consider now $U^-(\lambda)$ in the auxiliary space as the 2×2 matrix

$$U^-(\lambda) = \begin{pmatrix} \mathcal{A} & \mathcal{B} \\ \mathcal{C} & \mathcal{D} \end{pmatrix} \ . \tag{6.88}$$

One can write $(U^-)^{-1}(x) = \tilde{U}^-(x - ic)/\Delta\{U^-(x - i(c/2))\}$, where $\tilde{U}_1^-(x) = 2\,\mathrm{tr}_{12}\, P_{12}^- U_2^-(x)L_{12}(x)$ is usually called the *algebraic adjunct* of $U^-(x)$. This quantity can be written in the auxiliary space as 2×2 matrix

$$\tilde{U}^-(\lambda) = \begin{pmatrix} \tilde{\mathcal{D}} & -\tilde{\mathcal{B}} \\ -\tilde{\mathcal{C}} & \tilde{\mathcal{A}} \end{pmatrix}$$
$$= -\begin{pmatrix} b(2\lambda)\mathcal{A} - c(2\lambda)\mathcal{D} & \mathcal{B} \\ \mathcal{C} & -c(2\lambda)\mathcal{A} + b(2\lambda)\mathcal{D} \end{pmatrix} . \qquad (6.89)$$

Then

$$\begin{aligned} \Delta\{U^-(x)\} &= U^-(x + i(c/2))\tilde{U}^-(x - i(c/2)) \\ &= \tilde{U}^-(x - i(c/2)U^-(x + i(c/2)) \\ &= \tilde{\mathcal{D}}(x - i(c/2))\mathcal{D}(x + i(c/2)) \\ &\quad - \tilde{\mathcal{B}}(x - i(c/2))\mathcal{C}(x + i(c/2)) . \end{aligned} \qquad (6.90)$$

It follows from these facts that

$$\begin{aligned} \hat{\tau}(\lambda) &= \operatorname{tr} U^+(\lambda)U^-(\lambda) \\ &= \operatorname{tr} K(\lambda + i(c/2), \xi_+)T(\lambda)K(\lambda - i(c/2), \xi_-)\sigma^y T^t(-\lambda)\sigma^y \end{aligned} \qquad (6.91)$$

is the even function of the spectral parameter.

Let us introduce the mathematical vacuum as $\mathcal{C}|0\rangle = 0$, and with the diagonal action of the operators $\mathcal{A}$ and $\mathcal{D}$: $\mathcal{A}(x)|0\rangle = \alpha(x)|0\rangle$ and $\mathcal{D}(x)|0\rangle = \delta(x)|0\rangle$ (do not confuse with a delta-function). It is easy to see that

$$\Delta_+(x + i(c/2))\Delta_-(x - i(c/2)) = \Delta\{U^-(x)\} , \qquad (6.92)$$

where $\Delta_+(x) = \alpha(x)$ and $\Delta_-(x) = 2x\delta(x) - ic\alpha(x)$. Again, we construct the vector

$$|\lambda_1, \ldots, \lambda_M\rangle = \prod_{j=1}^{M} \mathcal{B}(\lambda_j)|0\rangle . \qquad (6.93)$$

Let us apply the operator

$$\hat{\tau}(\lambda) = (\lambda + \xi_+ + i(c/2))\mathcal{A}(\lambda) - (\lambda - \xi_+ + i(c/2))\mathcal{D}(\lambda) \qquad (6.94)$$

to the vector $|\lambda_1, \ldots, \lambda_M\rangle$. The following commutation relations can be used:

$$\begin{aligned}(x-y)(x+y)\mathcal{D}(x)\mathcal{B}(y) &= (x-y+ic)(x+y+ic)\mathcal{B}(y)\mathcal{D}(x)\\ &\quad +2c^2\mathcal{B}(x)\mathcal{A}(y)-ic(x+y+ic)\mathcal{B}(x)\mathcal{D}(y)\\ &\quad +ic(x-y+2ic)\mathcal{B}(x)\mathcal{D}(y)\ ,\\ (x-y)(x+y)\mathcal{A}(x)\mathcal{B}(y) &= (x-y-ic)(x+y-ic)\mathcal{B}(y)\mathcal{A}(x)\\ &\quad +ic(x+y-ic)\mathcal{B}(x)\mathcal{A}(y)\\ &\quad -ic(x-y)\mathcal{B}(x)\mathcal{D}(y)\ ,\end{aligned} \tag{6.95}$$

which stem from intertwining relations for monodromies. It is important to emphasize that these relations differ from the ones for periodic boundary conditions not only by the presence of coefficients with sums $(x+y)$, but also due to onsets of the operators $\mathcal{A}$ in the first relation and $\mathcal{D}$ in the second one. The vector $|\lambda_1,\ldots,\lambda_M\rangle$ is an eigenstate of the transfer matrix of an open chain if

$$-\frac{\lambda_j+\xi_+-i(c/2)}{\lambda_j-\xi_++i(c/2)}\frac{\Delta_+(\lambda_j)}{\Delta_-(\lambda_j)}(2\lambda_j-ic)=\prod_{\substack{l=1\\ l\neq j}}^{M}\frac{\lambda_j-\lambda_l-ic}{\lambda_j-\lambda_l+ic}\frac{\lambda_j+\lambda_l-ic}{\lambda_j+\lambda_l+ic}\ , \tag{6.96}$$

where $j=1,\ldots,M$. The eigenvalue of the transfer matrix is

$$\begin{aligned}\Lambda(\lambda) &= \frac{2\lambda+ic}{2\lambda}(\lambda+\xi_+-i(c/2))\Delta_+(\lambda)\prod_{j=1}^{M}\frac{\lambda-\lambda_j-ic}{\lambda-\lambda_j}\frac{\lambda+\lambda_j-ic}{\lambda+\lambda_j}\\ &\quad -\frac{1}{2\lambda}(\lambda-\xi_-+i(c/2))\Delta_-(\lambda)\prod_{j=1}^{M}\frac{\lambda-\lambda_j+ic}{\lambda-\lambda_j}\frac{\lambda+\lambda_j+ic}{\lambda+\lambda_j}\ .\end{aligned} \tag{6.97}$$

Let us now put $T^-(\lambda)=T(\lambda)$ into the definition of $U^-(\lambda)$. In the auxiliary 2×2 space the monodromy of a periodic chain can be written as

$$T(\lambda)=\begin{pmatrix}\hat{A} & \hat{B}\\ \hat{C} & \hat{D}\end{pmatrix}\ . \tag{6.98}$$

Its components act on the mathematical vacuum as

$$\begin{aligned}\hat{C}|0\rangle = 0\ ,\ \hat{A}|0\rangle &= \delta_+(\lambda)|0\rangle\ ,\\ \hat{D}|0\rangle &= \delta_-(\lambda)|0\rangle\ ,\end{aligned} \tag{6.99}$$

where $\delta_+(x+i(c/2))\delta_-(x-i(c/2)) = \delta\{T(x)\}$. Then after some tedious but straightforward calculations one gets

$$\begin{aligned}\Delta_+(x) &= (x+\xi_- - i(c/2))\delta_+(x)\delta_-(-x) \ , \\ \Delta_-(x) &= -(2x-ic)(x-\xi_- + i(c/2))\delta_+(-x)\delta_-(x) \ . \end{aligned} \tag{6.100}$$

These expressions can be introduced into Eqs. (6.96) and (6.97). Then taking the logarithmic derivative of Eq. (6.97) with respect to the spectral parameter λ, equating $\lambda = 0$, and shifting $\lambda_j \to \lambda_j - i(c/2)$ we finally obtain Bethe ansatz equations and the expression for the energy for a Heisenberg chain with open boundaries (with $c = 1$), presented in the previous section.

6.4 Open Hubbard Chain

So far in this chapter we considered quantum spin chains with open boundary conditions. Let us now see how open boundaries affect the behaviour of correlated electron chains with the possible charge and spin dynamics. Let us start with the open Hubbard chain, in which Hamiltonian Eq. (4.1) we exclude terms $-\sum_\sigma(a^\dagger_{N,\sigma}a_{1,\sigma} + \mathrm{H.c.})$, but, instead, introduce the term $-\sum_{j=1,L}\sum_\sigma p_{j\sigma}n_{j\sigma}$, describing local boundary potentials and magnetic fields, which act on edge electrons. We assume the following wave function to be the eigenfunction of the Hubbard Hamiltonian of an open chain:

$$\begin{aligned}&\psi(x_1,\dots,x_N,\sigma_1,\dots,\sigma_N) \\ &\quad = \sum_P \epsilon_P A_{\sigma_{Q_1},\dots,\sigma_{Q_N}}(k_{P_1},\dots,k_{P_N})\exp\left(i\sum_{j=1}^N k_{P_j}x_{Q_j}\right) \ , \end{aligned} \tag{6.101}$$

where the sum extends over all permutations and negations of of $k_1,\dots,k_N$ and ϵ_P changes sign at each such "mutation". Our strategy is the same as in Chapter 4, but taking into account negations as in the second section of the present chapter. By using the co-ordinate representation of the wave function in a stationary Schrödinger equation we define scattering and reflection matrices as follows

$$\begin{aligned}&A_{\dots\sigma_j,\sigma_{j+1},\dots}(\dots,k_{P_j},k_{P_{j+1}},\dots) \\ &\qquad = Y(k_{P_j},k_{P_{j+1}})A_{\dots\sigma'_j,\sigma'_{j+1},\dots}(\dots,k_{P_{j+1}},k_{P_j},\dots) \ , \\ &A_{\sigma_{Q_1},\dots}(k_{P_1},\dots) = U_{\sigma_{Q_1}}(k_{P_1})A_{\sigma_{Q_1},\dots}(-k_{P_1},\dots) \ , \\ &A_{\dots\sigma_{Q_N}}(\dots,k_{P_N}) = V_{\sigma_{Q_N}}(-k_{P_N})A_{\dots\sigma_{Q_N}}(\dots,-k_{P_N}) \ , \end{aligned} \tag{6.102}$$

where

$$Y(k_j, k_l) = \frac{(\sin k_l - \sin k_j)\Pi + i(U/2)I}{\sin k_l - \sin k_j + i(U/2)} , \tag{6.103}$$

(I and Π are the identity and permutation operators), and boundary matrices $U_\sigma(k)$ and $V_\sigma(k)$ are

$$U(k) = \begin{pmatrix} \frac{\alpha_\uparrow(k)}{\alpha_\uparrow(-k)} & 0 \\ 0 & \frac{\alpha_\downarrow(k)}{\alpha_\downarrow(-k)} \end{pmatrix} , \qquad V(k) = \begin{pmatrix} \frac{\beta_\uparrow(k)}{\beta_\uparrow(-k)} & 0 \\ 0 & \frac{\beta_\downarrow(k)}{\beta_\downarrow(-k)} \end{pmatrix} ,$$
$$\alpha_\sigma(k) = 1 - p_{1\sigma}\exp(-ik) ,$$
$$\beta_\sigma(k) = [1 - p_{L\sigma}\exp(-ik)]\exp[ik(L+1)] . \tag{6.104}$$

They have to satisfy the following relations:

$$U_\uparrow(k)U_\downarrow(-k) = \frac{\xi_1 + \sin k}{\xi_1 - \sin k} , \qquad V_\uparrow(k)V_\downarrow(-k) = \frac{\xi_L + \sin k}{\xi_L - \sin k} , \tag{6.105}$$

where $\xi_{1,L} = \infty$ for $p_{1,L\uparrow} = p_{1,L\downarrow}$, and $\xi_{1,L} = (1 - p^2_{1,L\uparrow})/2ip_{1,L\uparrow}$ for $p_{1,L\uparrow} = -p_{1,L\downarrow}$. The fulfillment of these relations is necessary to apply the algebraic Bethe ansatz for open chains, described in the previous section. Hence, it is possible to obtain a Bethe ansatz solution for an open Hubbard chain only for these two sets of boundary potentials, *i.e.*, if one has only boundary potentials or only boundary magnetic fields. We point out that, as for an open spin chain, we can obtain all other similar relations by negations and permutations of k_j.

By using these matrices we obtain the following relation

$$A_{\dots\sigma_{Q_j},\dots}(\dots, k_{P_j}, \dots) = T_j A_{\dots\sigma_{Q_j},\dots}(\dots, k_{P_j}, \dots) , \tag{6.106}$$

where

$$T_j = \mathrm{tr}_0\, K_0^+(k_j)L_{01}(k_j, -k_1) \times \cdots \times L_{0N}(k_j, -k_N) \times K_0^-(k_j)L_{0N}(k_j, k_N) \times \cdots \times L_{01}(k_j, k_1) , \tag{6.107}$$

where

$$L_{0n}(k_j, k_n) = \frac{(\sin k_j - \sin k_n)I_{0n} + i(U/2)\Pi_{0n}}{\sin k_j - \sin k_n + i(U/2)} , \tag{6.108}$$

with I_{0n} and P_{0n} are the identity and permutation operators acting in the quantum subspace V_n and auxiliary subspace V_0, and

$$K_0^-(k) = V(k) \ ,$$

$$K_0^+(k) = \frac{2\sin k + i(U/2)}{2\sin k(2\sin k + iU)} \begin{pmatrix} (2\sin k + i(U/2))\frac{\alpha_\uparrow(k)}{\alpha_\uparrow(-k)} & \\ -i(U/2)\frac{\alpha_\downarrow(k)}{\alpha_\downarrow(-k)} & 0 \\ 0 & (2\sin k + i(U/2))\frac{\alpha_\downarrow(k)}{\alpha_\downarrow(-k)} \\ & -i(U/2)\frac{\alpha_\uparrow(k)}{\alpha_\uparrow(-k)} \end{pmatrix} . \tag{6.109}$$

To diagonalize the matrix T_j we proceed along the lines of Chapter 4 by using the fundamental spin problem for an open spin-$\frac{1}{2}$ chain from the previous section with introduced inhomogeneities (charge rapidities). Then one obtains Bethe ansatz equations for an open Hubbard chain (this diagonalization was first performed by H. Schulz for $p = 0$) for the sets k_j ($j = 1, \ldots, N$, N is the number of electrons) and λ_γ ($\gamma = 1, \ldots, M$, M is the number of electrons with spins directed downward)

$$\begin{aligned} \prod_{1,L} e_{1+i(4\xi_{1,L}/U)}(\lambda_\gamma) &= \prod_{\pm}\prod_{j=1}^{N} e_1^{-1}(\lambda_\gamma \pm \sin k_j) \prod_{\substack{\beta=1 \\ \beta\neq\gamma}}^{M} e_2(\lambda_\gamma \pm \lambda_\beta) \ , \\ \frac{\alpha_\uparrow(k_j)}{\alpha_\uparrow(-k_j)}\frac{\beta_\uparrow(k_j)}{\beta_\uparrow(-k_j)} &= \prod_{\beta=1}^{M} e_1(\sin k_j - \lambda_\beta) e_1(\sin k_j + \lambda_\beta) \ , \end{aligned} \tag{6.110}$$

where $e_n(x) = (4x + iUn)/(4x - iUn)$. The energy is

$$E = -2\sum_{j=1}^{N} \cos k_j \ . \tag{6.111}$$

In the framework of the string hypothesis we can write the thermodynamic Bethe ansatz equations for the open Hubbard chain. Let us perform it first for the values of boundary fields or potentials, at which there are no boundary bound states, see the analysis below.

In the limit of large L, N, N with N/L and M/L kept fixed we can consider three main classes of solutions of Eq. (6.110): one again considers real charge rapidities, spin-singlet pairs and bound states of them, and spin bound states, with the same notations as in Chapter 4. By using straightforward procedures, similar to the case of a Heisenberg open chain, we obtain thermodynamic Bethe ansatz equations for densities

$$
\begin{aligned}
2[\rho(k)+\rho_h(k)] &= \frac{2}{\pi}+\frac{2}{L}P(k)+\cos k\sum_{n=1}^{\infty}[a_{Un/4}(\sin k-\lambda) \\
&\quad + a_{Un/4}(\sin k+\lambda)] * [\sigma_n(\lambda)+\sigma'_n(\lambda)] \ , \\
2\sigma_{nh}(\lambda) &= \frac{1}{L}Q_n(\lambda)+[a_{Un/4}(\lambda-\sin k)+a_{Un/4}(\lambda+\sin k)] * \rho(k) \\
&\quad - \sum_{m=1}^{\infty}[A_{nm}(\lambda-\lambda')+A_{nm}(\lambda+\lambda')] * \sigma_m(\lambda') \ , \qquad (6.112) \\
2\sigma'_{nh}(\lambda) &= \frac{4}{\pi}\mathrm{Re}\frac{1}{\sqrt{1-[\lambda-in(U/4)]^2}}+\frac{2}{L}Q'_n(\lambda) \\
&\quad - \sum_{m=1}^{\infty}[A_{nm}(\lambda-\lambda')+A_{nm}(\lambda+\lambda')] * \sigma'_m(\lambda') \\
&\quad - [a_{Un/4}(\lambda-\sin k)+a_{Un/4}(\lambda-\sin k)] * \rho(k) \ ,
\end{aligned}
$$

where

$$
\begin{aligned}
P(k) &= \frac{1-p_{1\uparrow}p_{L\uparrow}}{\pi}\frac{1+p_{1\uparrow}p_{L\uparrow}-(p_{1\uparrow}+p_{L\uparrow})\cos k}{(1+p_{1\uparrow}^2-2p_{1\uparrow}\cos k)(1+p_{L\uparrow}^2-2p_{L\uparrow}\cos k)} \ , \\
Q_n(\lambda) &= \sum_{j=1}^{n}[a_{x_{j1}}(\lambda)+a_{x_{jL}}(\lambda)] \ , \qquad (6.113) \\
Q'_n(\lambda) &= -\sum_{j=1}^{2n}P(k^{n,j})-Q_n(\lambda) \ ,
\end{aligned}
$$

with

$$
x_{j1,L} = \frac{1-p_{1,L\uparrow}^2}{2p_{1,L\uparrow}}+(n-2j)\frac{U}{4} \ , \qquad (6.114)
$$

and

$$
\begin{aligned}
k^{n,2j+1} &= \pi-\sin^{-1}[\lambda+i(n-2j)(U/4)] \ , \ j=1,\ldots,n-1 \ , \\
k^{n,2j} &= \sin^{-1}[\lambda+i(n-2j)(U/4)] \ , \ j=1,\ldots,n-1 \ .
\end{aligned} \qquad (6.115)
$$

Thermodynamic Bethe ansatz equations for dressed energies have the form

$$
\begin{aligned}
\varepsilon(k) &= -\left(2\cos k + \frac{H}{2} + \mu\right) + \frac{T}{2}\sum_{n=1}^{\infty}[a_{Un/4}(\sin k - \lambda) \\
&\quad + a_{Un/4}(\sin k + \lambda)] * \ln\frac{1+\kappa_n^{-1}(\lambda)}{1+\eta_n^{-1}(\lambda)} , \\
T\ln[1+\eta_n(\lambda)] &= nH - \frac{T}{2}\cos k[a_{Un/4}(\lambda - \sin k) \\
&\quad + a_{Un/4}(\lambda + \sin k)] * \ln[1+\xi^{-1}(k)] \\
&\quad + \frac{T}{2}\sum_{m=1}^{\infty}[A_{nm}(\lambda - \lambda') \\
&\quad + A_{nm}(\lambda + \lambda')] * \ln[1+\eta_m^{-1}(\lambda')] , \\
T\ln[1+\kappa_n(\lambda)] &= -4\,\mathrm{Re}\sqrt{1-[\lambda - in(U/4)]^2} - 2n\mu \\
&\quad - \frac{T}{2}\cos k[a_{Un/4}(\lambda - \sin k) \\
&\quad + a_{Un/4}(\lambda + \sin k)] * \ln[1+\xi^{-1}(k)] \\
&\quad + \frac{T}{2}\sum_{m=1}^{\infty}[A_{nm}(\lambda - \lambda') \\
&\quad + A_{nm}(\lambda - \lambda')] * \ln[1+\kappa_m^{-1}(\lambda')] .
\end{aligned}
\tag{6.116}
$$

To remind the notations: $a_{Un/4}(x) = (nU/4)/\pi[x^2 + (nU/4)^2]$, the Fourier transform of $A_{nm}(x)$ is $\coth(|\omega U|/8)[\exp(-|n-m||\omega U|/8) + \exp(-(n-m)|\omega U|/8)]$. The internal energy, the number of electrons and the total magnetic moment are given by

$$
\begin{aligned}
E &= -2L\int_0^{\pi} dk\cos k\rho(k) \\
&\quad - 4L\sum_{n=1}^{\infty}\mathrm{Re}\int_0^{\infty} d\lambda\sqrt{1-[\lambda - in(U/4)]^2}\sigma_n'(\lambda) , \\
N &= L\int_0^{\pi} dk\rho(k) + 2L\sum_{n=1}^{\infty} n\int_0^{\infty} d\lambda\sigma_n'(\lambda) , \\
M^z &= \frac{L}{2}\int_0^{\pi} dk\rho(k) - L\sum_{n=1}^{\infty} n\int_0^{\infty} d\lambda\sigma_n(\lambda) .
\end{aligned}
\tag{6.117}
$$

Thermodynamic Bethe ansatz equations for densities of an open attractive Hubbard chain follow from Eq. (6.112) with the change of the signs of the third term in the right hand side of the first equation. Thermodynamic Bethe ansatz equations for dressed energies of an open attractive Hubbard chain follow from Eq. (6.116) with the change of the signs of $\ln(1+\xi^{-1})$ and the sign of the driving term for dressed energies of pair excitations, while $2n\mu$ keeps the same sign.

Noting that thermodynamic Bethe ansatz equations for densities are linear integral equations, one can divide them into two sets: one of the scale 1 for the main (of order of L) contribution to the energy, magnetization, number of electrons, *etc.*, and the other one, of order of L^{-1}. The former describes thermodynamics of the bulk of an open Hubbard chain, while the latter reveals the contribution from electrons sitting at edges. The set of equations for dressed energies does not have terms of order of L^{-1} explicitly.

Since all dressed energies, densities and kernels of integrals in Eqs. (6.112) and (6.116) are even functions, we can re-write those equations in the form, similar to a periodic Hubbard chain, permitting distributions of k and λ over total intervals instead of half-intervals. It is important, however, to exclude double counting of states related to k_j and λ_γ with $j=\gamma=0$ and to count the possibility of $\lambda_\gamma=\lambda_\beta$ in Eq. (6.110), which implies subsequent changes in the equations for densities of order of L^{-1}, *cf.* the results for an open Heisenberg chain. After this procedure we see that the main contribution in L is, naturally, the same for a Hubbard chain with periodic and open boundary conditions. The difference appears in behaviours of energies, magnetic moments, valences (the average number of electrons per site) of edge sites. By performing this program we obtain the set of equations for densities (with the superscript (1) we again denote the values of order of L^{-1}), while the set of thermodynamic Bethe ansatz equations for dressed energies of an open Hubbard chain formally coincides with Eq. (4.58)

$$\rho^{(1)}(k)+\rho_h^{(1)}(k)=\hat{P}(k)+\cos k$$

$$\times\sum_{n=1}^{\infty}a_{Un/4}(\sin k-\lambda)*[\sigma_n^{(1)}(\lambda)+(\sigma'_n)^{(1)}(\lambda)]\ ,$$

$$\sigma_{nh}^{(1)}(\lambda)=\hat{Q}_n(\lambda)+a_{Un/4}(\lambda-\sin k)*\rho^{(1)}(k)$$

$$-\sum_{m=1}^{\infty}A_{nm}(\lambda-\lambda')*\sigma_m^{(1)}(\lambda')\ ,$$

$$(\sigma'_{nh})^{(1)}(\lambda) = \hat{Q}'_n(\lambda) - \sum_{m=1}^{\infty} A_{nm}(\lambda - \lambda') * (\sigma'_m)^{(1)}(\lambda') \\ - a_{Un/4}(\lambda - \sin k) * \rho^{(1)}(k) \ , \tag{6.118}$$

where

$$\begin{aligned} \hat{P}(k) &= P(k) - \sum_{m=1}^{\infty} a_{Um/4}(\sin k)(2 - \delta_{M'_m,0} - \delta_{M_m,0}) \ , \\ \hat{Q}_n(\lambda) &= Q_n(\lambda) + \sum_{m=1}^{\infty} A_{nm}(\lambda)(1 - \delta_{M_m,0}) \ , \\ \hat{Q}'_n(\lambda) &= Q'_n(\lambda) + \sum_{m=1}^{\infty} A_{nm}(\lambda)(1 - \delta_{M'_m,0}) \ . \end{aligned} \tag{6.119}$$

The internal energy, valences and magnetic moments of edge sites of an open Hubbard chain are then

$$\begin{aligned} e_{edge} &= -2 \int_0^{\pi} dk \cos k \rho^{(1)}(k) \\ &\quad - 4 \sum_{n=1}^{\infty} \mathrm{Re} \int_0^{\infty} d\lambda \sqrt{1 - [\lambda - in(U/4)]^2} (\sigma'_n)^{(1)}(\lambda) \ , \\ n_{edge} &= \int_0^{\pi} dk \rho^{(1)}(k) + 2 \sum_{n=1}^{\infty} n \int_0^{\infty} d\lambda (\sigma'_n)^{(1)}(\lambda) \ , \\ m^z_{edge} &= \frac{1}{2} \int_0^{\pi} dk \rho^{(1)}(k) - \sum_{n=1}^{\infty} n \int_0^{\infty} d\lambda \sigma_n^{(1)}(\lambda) \ . \end{aligned} \tag{6.120}$$

These corrections are nonzero even for $p_{1,L\sigma} = 0$, which is the contribution from free edges themselves.

It is interesting to compare, *e.g.*, the behaviours of a local specific heat (connected to edge sites) and the one for bulk sites of an open Hubbard chain. Denote the specific heat of bulk sites per site as c_∞ and the local specific heat of edges as c_{edge} (notice that we do not now distinguish contributions from edges themselves and boundary potentials or fields). Then after some straightforward but tedious calculations one obtains, for example, that at low temperatures (with $H \gg T$)

$$c_{edge} = c_\infty \delta_f \tag{6.121}$$

for $-\mu \leq -2 - (H/2)$, where $\delta_f = 0$ for $p_{1,L\uparrow} = 1$, and $\delta_f = (1 + 2p_{1\uparrow} - p_{1\uparrow}p_{L\uparrow})/(1 - p_{1\uparrow})(1 - p_{L\uparrow})$ elsewhere. In this region of parameters the low temperature specific heat is proportional to $\sqrt{T}$. Observe that in the absence of boundary potentials or fields, $p_{1,L\uparrow} = 0$, the local specific heat of edges coincides with the bulk one per site. For $-\mu > -2 - (H/2)$ the specific heat is proportional to T (*i.e.*, the Sommerfeld coefficient is constant). Calculations yield

$$c_{edge} = c_{\infty}\pi\hat{P}(Q) , \tag{6.122}$$

where Q is the Fermi point of unbound electron excitations. For $-\mu \geq 2 - (H/2)$, $H \geq (\sqrt{16+U^2} - U)$ the specific heat is also linear in T and one gets

$$c_{edge} = c_{\infty}\frac{\pi\hat{P}(\pi)c_c + \Gamma c_s}{c_c + c_s} , \tag{6.123}$$

where $c_{c,s}$ are contributions to the bulk specific heat due to charged and spin low-lying excitations, respectively ($c_{\infty} = c_c + c_s$) and

$$\Gamma = \frac{\pi\sqrt{1+(U/4)^2}}{2}\left(\int_{-\pi}^{\pi} dk a_{U/4}(\sin k)\hat{P}(k) + \hat{Q}_1(0)\right) . \tag{6.124}$$

Now let us consider the effect of boundary bound states. In principle one is free to leave the boundary bound states empty. This gives rise to another continuum of states, which are important if one studies, *e.g.*, multiple Fermi edge singularities in the presence of those boundary bound states. For simplicity we shall discuss the case with $p_{L\sigma} = 0$ and $p_{1\uparrow} = p_{1\downarrow} = p$, *i.e.*, without boundary magnetic fields but with only one boundary potential applied to the left edge of the chain. Other cases can be studied analogously. We can consider four situations. For $p < 1$ there are no boundary bound states, and we can use the previous analysis. For $1 < p < p_1 = (U/4) + \sqrt{1+(U/4)^2}$ there is a complex solution to the Bethe ansatz equations $k_N = i\ln p$. To take into account this solution, but to keep the total number of roots of Bethe ansatz equations fixed, we renormalize Eq. (6.110) as

$$\begin{aligned} e^{2ik_j(L+1)}s_p(k_j) &= \prod_{\pm}\prod_{\beta=1}^{M} e_1(\sin k_j \pm \lambda_\beta) , \\ \prod_{\pm} e_{1\pm(8f/U)}\prod_{j=1}^{N-1} e_1(\lambda_\gamma - \sin k_j) &= \prod_{\pm}\prod_{\substack{\beta=1 \\ \beta\neq\gamma}}^{M} e_2(\lambda_\gamma \pm \lambda_\beta) , \end{aligned} \tag{6.125}$$

where $j = 1, \ldots, N-1$, $\gamma = 1, \ldots, M$, $s_p(k) = [1 - p\exp(-ik)]/[1 - p\exp(ik)]$ and $f = -i\sin k_N = (p^2-1)/2p < (U/4)$. The contribution of this bound state to the energy is $E_{b1} = -(p^2+1)/p + \mu - (H/2)$. For $p_1 < p < p_2 = (U/2) + \sqrt{1+U^2}$ there appears the complex solution $\lambda_M = i[f - (U/4)]$ (here $f > U/4$). We have to renormalize Bethe ansatz equations to

$$\begin{aligned} & e^{2ik_j(L+1)} s_p(k_j) \\ & \quad = e_{2-(8f/U)}(\sin k_j) e_{8f/U}(\sin k_j) \prod_{\pm} \prod_{\beta=1}^{M-1} e_1(\sin k_j \pm \lambda_\beta) , \\ & e_{1-(8f/U)}(\lambda_\gamma) e_{3-(8f/U)}(\lambda_\gamma) \prod_{\pm} \prod_{j=1}^{N-1} e_1(\lambda_\gamma \pm \sin k_j) \\ & \quad = \prod_{\pm} \prod_{\substack{\beta=1 \\ \beta \neq \gamma}}^{M-1} e_2(\lambda_\gamma \pm \lambda_\beta) , \end{aligned} \tag{6.126}$$

where $j = 1, \ldots, N-1$, $\gamma = 1, \ldots, M-1$. Finally, for $p > p_2$ a boundary singlet bound state with $\lambda_M = \sin k_N - i(U/4) = \sin k_N + i(U/4) = i[f - (U/4)]$ appears. In this case Bethe ansatz equations are modified as

$$\begin{aligned} & e^{2ik_j(L+1)} s_p(k_j) = e_{4-(8f/U)}(\lambda_\gamma) e_{8f/U}(\lambda_\gamma) \prod_{\pm} \prod_{\beta=1}^{M-1} e_1(\sin k_j \pm \lambda_\beta) , \\ & \prod_{\pm} \prod_{j=1}^{N-2} e_1(\lambda_\gamma \pm \sin k_j) = \prod_{\pm} \prod_{\substack{\beta=1 \\ \beta \neq \gamma}}^{M-1} e_2(\lambda_\gamma \pm \lambda_\beta) , \end{aligned} \tag{6.127}$$

where $j = 1, \ldots, N-2$, $\gamma = 1, \ldots, M-1$. The energy of this singlet boundary bound state is $E_{b2} = -2\sqrt{1 + [f - (U/2)]^2} + \mu - (H/2)$.

The energy of edges for a repulsive Hubbard open chain in the ground state can be calculated as

$$\begin{aligned} e_{edge} = & \int_0^Q dk \varepsilon(k) \tilde{P}(k) + \int_0^B d\lambda \phi_1(\lambda) \tilde{Q}_1(\lambda) - \frac{\mu}{2} - \frac{H}{4} \\ & + 1 + \theta(p-1) E_{b1} + H\theta(p - p_1) + \theta(p - p_2) E_{b2} , \end{aligned} \tag{6.128}$$

where $\theta(x)$ is the Heaviside step function,

$$\begin{aligned} \tilde{P}(x) = & \frac{1}{\pi} - \cos k a_{U/4}(\sin k) + \frac{p\cos k - p^2}{\pi(1 + p^2 - 2p\cos k)} \\ & + \theta(p - p_1) \cos k [a_{4f}(\sin k) + a_{(U/2)-4f}(\sin k)] , \end{aligned} \tag{6.129}$$

and

$$\tilde{Q}_1(\lambda) = a_{U/2}(\lambda) + \theta(p-1)\theta(p_1-p)[a_{(U/4)-4f}(\lambda) + a_{(U/4)+4f}(\lambda)] + \theta(p-p_1)\theta(p_2-p)[a_{4f-(U/4)}(\lambda) + a_{(3U/4)-4f}(\lambda)] \,. \qquad (6.130)$$

The dressed energies are the solution of the ground state equations Eq. (4.69). In the absence of the homogeneous magnetic field $H = 0$ we can simplify those equations, which become

$$\varepsilon(k) = \mu - 2\cos k + \int_{-Q}^{Q} dk' G_{(U/2),(U/2)}(\sin k - \sin k') \cos k' \varepsilon(k') \,, \qquad (6.131)$$

where $G_{a,b}(x)$ is the Fourier transform of $\exp(-a|\omega|/2)/2\cosh(b\omega/2)$. Then the valence of the first site is

$$n_1 = -\frac{\partial e_{edge}}{\partial p} = \theta(p-1)(1-p^{-2}) + \theta(p-p_2)\frac{[f-(U/2)](1+p^2)}{2p^2\sqrt{1+[f-(U/2)]^2}} - \frac{1}{2}\int_{-Q}^{Q} dk\varepsilon(k)\Bigg(\gamma_p(k) + \theta(p-1)\theta(p_2-p) \times \cos k \frac{\partial}{\partial p}[G_{(U/2)-2f,(U/2)}(\sin k) + G_{(U/2)+2f,(U/2)}(\sin k)] + \theta(p-p_2)\cos k\frac{\partial}{\partial p}[a_{4f}(\sin k) + a_{4f-(U/2)}(\sin k)]\Bigg) \,, \qquad (6.132)$$

where

$$\gamma_p(k) = \frac{p^2\cos k + \cos k - 2p}{\pi(1+p^2-2p\cos k)} \,. \qquad (6.133)$$

Notice that θ-functions in Eq. (6.128) must not be differentiated with respect to p to obtain Eq. (6.132). In the limit of large p only the first two terms survive and the expected answer $n_1|_{p\to\infty} = 2$ results. Depending on U and the total number of electrons (or μ) the local charge stiffness of the first site of an open Hubbard chain reveals one, two or no features, which are related to the values of p at which n_1 is close to zero, 1 and 2. This behaviour is very different from the behaviour of bulk electrons in a Hubbard chain.

6.5 Open Supersymmetric *t*-*J* Chain

Let us now consider the behaviour of a supersymmetric t-J chain with open boundary conditions. In this section we shall limit ourselves with the case $V = -J/4$ and $J = 2$.

The analysis of reflection equations of an open supersymmetric t-J chain in the framework of the algebraic Bethe ansatz shows that there are several solutions of those equations. Here we shall consider two of them. In the first case boundary potentials are applied as

$$\mathcal{H}_{b1} = -p_1' n_1 - p_L' n_L \tag{6.134}$$

and in the second case edge sites are affected by the mixed influence of boundary potentials and fields which are equal to each other

$$\mathcal{H}_{b2} = -\frac{1}{2}[p_1(2S_1^z - n_1) + p_L(2S_L^z - n_L)] \ . \tag{6.135}$$

Let us denote

$$S_{1,L}' = 2 + \frac{2}{p_{1,L}'} \tag{6.136}$$

for the first case and

$$S_{1,L} = 1 + \frac{2}{p_{1,L}} \tag{6.137}$$

for the second case.

The construction of Bethe ansatz equations of an open supersymmetric t-J chain is similar to the one of a Hubbard chain of the previous section. We, however, shall use a slightly different Bethe ansatz description of this problem. Let us start with Bethe ansatz equations for the sets of spin $\{\lambda_\alpha\}_{\alpha=1}^M$ and charge $\{p_j\}_{j=1}^N$ rapidities, where N and M denote the number of electrons and the number of electrons with spins down, respectively. The Bethe ansatz equations are (for simplicity we now study the case with $p_{1,L} = p_{1,L}' = 0$; the case with nonzero boundary potentials can be treated analogously):

$$\begin{aligned} \prod_{\pm}\prod_{j=1}^{N} e_1(\lambda_\alpha \pm p_j) &= \prod_{\pm}\prod_{\beta=1}^{M} e_2(\lambda_\alpha \pm \lambda_\beta) \\ e_1^{2L}(p_j) &= \prod_{\pm}\prod_{\beta=1}^{M} e_1(p_j \pm \lambda_\beta) \ , \end{aligned} \tag{6.138}$$

where $e_n(x) = (2x + in)/(2x - in)$. Equations (6.138) are written for the FFB grading. This form of Bethe ansatz equations for a periodic model was introduced by C. K. Lai and P. Schlottmann. However, for some purposes it will be more convenient to use the BFF grading scheme of the algebraic Bethe ansatz with the Grassmann parities $\epsilon_1 = 0$ (this state is related to a hole at the site) and $\epsilon_2 = \epsilon_3 = 1$. We re-write Eq. (6.138) as:

$$\begin{aligned} e_1^{2L}(u_j) &= \prod_{\pm} \prod_{k=1}^{N^h+M} e_2(u_j \pm u_k) \prod_{\beta=1}^{N^h} e_1^{-1}(u_j \pm \nu_\beta) \\ 1 &= \prod_{\pm} \prod_{k=1}^{N^h+M} e_1(\nu_\alpha \pm u_k) \ , \end{aligned} \tag{6.139}$$

where $N^h = L - N$ is the number of holes (non-occupied sites). This form of Bethe ansatz equations for a supersymmetric t-J chain with periodic boundary conditions was first introduced by B. Sutherland. It is easy to show that the two forms, Eqs. (6.138) and Eq. (6.139), are equivalent. For this purpose one can consider the second set of Eq. (6.139) as the root of some polynomial $P(\nu_\alpha) = 0$ with

$$P(x) = \prod_{\pm} \prod_{k=1}^{N^h+M} \left(x \pm u_k - \frac{i}{2}\right) - \prod_{\pm} \prod_{k=1}^{N^h+M} \left(x \pm u_k + \frac{i}{2}\right) . \tag{6.140}$$

We separate the first N^h roots ν_α of the $N^h + M$ roots of $P(x)$ and label the remaining M roots by λ_α. Then we have the factorization

$$P(x) = \text{const} \cdot \prod_{\pm} \prod_{\alpha=1}^{M} (x \pm \lambda_\alpha) \prod_{\beta=1}^{N_h} (x \pm \nu_\beta) \tag{6.141}$$

from which it follows

$$\prod_{\pm} \prod_{\alpha=1}^{M} e_1(u_j \pm \lambda_\alpha) \prod_{\beta=1}^{N^h} e_1(u_j \pm \nu_\beta) = \frac{P(u_j + i/2)}{P(u_j - i/2)} = \prod_{\pm} \prod_{k=1}^{N^h+M} e_2(u_j \pm u_k) . \tag{6.142}$$

Then using this relation and the first set of Eq. (6.139) we obtain the second set of Eq. (6.138), with $u_j = p_j$.

Next, the second equation of (6.138) can be re-written as $Q(p_j) = 0$ with the definition

$$Q(x) = \left(x + \frac{i}{2}\right)^{2L} \prod_{\pm} \prod_{\beta=1}^{M} \left(x \pm \lambda_\beta - \frac{i}{2}\right)$$
$$- \left(x - \frac{i}{2}\right)^{2L} \prod_{\pm} \prod_{\beta=1}^{M} \left(x \pm \lambda_\beta + \frac{i}{2}\right) . \tag{6.143}$$

As above, separating first the N roots p_j of this polynomial and labeling the remaining $N^h + M$ roots by u_k we obtain the factorization

$$Q(x) = \text{const} \cdot \prod_{\pm} \prod_{j=1}^{N} (x \pm p_j) \prod_{k=1}^{N_h+M} (x \pm u_k). \tag{6.144}$$

From this we get

$$\prod_{\pm} \prod_{j=1}^{N} e_1(\lambda_\alpha \pm p_j) \prod_{k=1}^{N^h+M} e_1(\lambda_\alpha \pm u_k) = \frac{Q(\lambda_\alpha + i/2)}{Q(\lambda_\alpha - i/2)}$$
$$= \prod_{\pm} \prod_{\beta=1}^{M} e_2(\lambda_\alpha \pm \lambda_\beta) . \tag{6.145}$$

Together with the second set of Eq. (6.139) for λ_α in place of ν_α it gives the first set of Eq. (6.138).

For nonzero boundary potentials we have

$$\eta_{1,2}(u_j) e_1^{2L}(u_j) = \prod_{\pm} \prod_{k=1}^{N^h+M} e_2(u_j \pm u_k) \prod_{\beta=1}^{N^h} e_1^{-1}(u_j \pm \nu_\beta)$$
$$1 = \zeta_{1,2}(\nu_\alpha) \prod_{\pm} \prod_{k=1}^{N^h+M} e_1(\nu_\alpha \pm u_k) , \tag{6.146}$$

where

$$\eta_1 = 1 , \qquad \zeta_1(x) = e_{-S_1'}(x) e_{-S_L'}(x) ,$$
$$\eta_2(x) = e_{-S_1}(x) e_{-S_L}(x) , \qquad \zeta_2 = 1 . \tag{6.147}$$

The energy corresponding to the solution of Bethe ansatz equations is

$$E = -(\mu - 2)(L - N_h) - \frac{H}{2}(L - N_h - 2M) - E_{p1,2} - \sum_{j=1}^{N_h+M} \frac{4}{4u_j^2 + 1} , \tag{6.148}$$

where $E_{p1} = p'_1 + p'_L$ and $E_{p2} = (p_1 + p_L)/2$. The number of down spins must be smaller than or equal to the number of up spins, *i.e.*, $2M \leq L - N_h$ in this construction of the Bethe ansatz. The states which do not satisfy this constraint must be constructed by switching the mathematical vacuum to the state with down spin electrons at each site. This change formally leads to $p'_{1,L}, p_{1,L} \to -p'_{1,L}, -p_{1,L}$ in the definition of $S'_{1,L}$ and $S_{1,L}$. The advantage of this approach is the simplicity of the limiting case $N = L$ (*i.e.*, $N_h = 0$). The reader can see that in this case Bethe ansatz equations for a supersymmetric t-J model coincide with the ones of a Heisenberg spin-$\frac{1}{2}$ chain.

The analysis of thermodynamics of a supersymmetric open t-J chain is similar to the one of the periodic t-J chain and to the analysis of an open Hubbard chain from the previous section. We shall not present it here to save space. The interested reader can perform such calculations without difficulty, using the knowledge of methods of the previous section. We limit ourselves with the most interesting case $T = 0$. The other advantage of the Bethe ansatz equations (6.146) is that the ground state corresponds to only real u_j and ν_α. In the limit of large L, M and N_h, but with the ratios N_h/L and $(N_h + M)/L$ fixed, the ground state Bethe ansatz equations for dressed energies (with obvious notations) are

$$\varepsilon_s(x) = H - 2\pi a_1(x) - \int_{-B}^{B} dy a_2(x-y)\varepsilon_s(y) + \int_{-Q}^{Q} dy a_1(x-y)\varepsilon_c(y) \,,$$
$$\varepsilon_c(x) = \mu - 2 - \frac{H}{2} + \int_{-B}^{B} dy a_1(x-y)\varepsilon_s(y) \,, \tag{6.149}$$

where $a_n(x) = 2n/\pi(4x^2 + n^2)$, and $\varepsilon_s(u)$ is minus the energy of an elementary excitation with the real rapidity u, while $\varepsilon_c(\nu)$ is minus the energy of an elementary excitation with the real rapidity ν (*i.e.*, they rather describe holes, than quasiparticles). The Fermi points, as usual, are determined as $\varepsilon_s(\pm B) = 0$ and $\varepsilon_c(\pm Q) = 0$, which are related in such a way to the values of the chemical potential μ and the homogeneous magnetic field H. The ground state Bethe ansatz equations for densities are

$$\rho_{ch}(x) + \rho_c(x) = \frac{1}{L}[Y_{1,2}(x) - a_1(x)] + \int_{-B}^{B} dy a_1(x-y)\rho_{sh}(y) \,,$$
$$\rho_{sh}(x) + \rho_s(x) = 2a_1(x) + \frac{1}{L}[a_2(x) + X_{1,2}(x)] \tag{6.150}$$
$$- \int_{-B}^{B} dy a_2(x-y)\rho_{sh}(y) + \int_{-Q}^{Q} dy a_1(x-y)\rho_{ch}(y) \,,$$

where

$$X_1(x) = 0 \ , \qquad X_2(x) = a_{-S_1}(x) + a_{-S_L}(x) \ ,$$
$$Y_1(x) = a_{-S_1'}(x) + a_{-S_L'}(x) \ , \qquad Y_2(x) = 0 \ . \tag{6.151}$$

The ground state energy is equal to

$$\begin{aligned} E &= (\mu - 2)N + HM^z + L\varepsilon_c(0) - 2\mu L + 4L \\ &+ \frac{1}{4\pi}\int_{-B}^{B} dx X_{1,2}(x)\varepsilon_s(x) + \frac{1}{4\pi}\int_{-Q}^{Q} dx Y_{1,2}(x)\varepsilon_c(x) \\ &- \frac{1}{2}\varepsilon_s(0) - \frac{\mu}{2} + 1 + \frac{H}{4} + E_{p1,2} + 0(L^{-1}) \ . \end{aligned} \tag{6.152}$$

It is worthwhile to present results here for the ground state properties of bulk electrons, of edges, and contributions from boundary potentials. The internal ground state energy for bulk electrons per site is

$$e = -\mu + 2 - 2\ln 2 - 2a - \frac{2\ln 2\zeta(3)}{\pi} A^3 \ , \tag{6.153}$$

where

$$\begin{aligned} a &= \frac{H^2}{8\pi^2}\left(1 - \frac{1}{2\ln H} + \cdots\right) \ , \\ A^2 &= \frac{2}{2\zeta(3)}\left(\bar{\mu} + 2a + \frac{8\ln 2(\bar{\mu}+2a)^{3/2}}{3\pi\sqrt{6\zeta(3)}}\right) \ , \\ \bar{\mu} &= 2\ln 2 - \mu + 2 \ . \end{aligned} \tag{6.154}$$

The magnetic moment per bulk site is

$$m^z = \frac{H}{2\pi^2}\left(1 + \frac{\ln 2}{\pi}\sqrt{\frac{8(\bar{\mu}+2a)}{3\zeta(3)}}\right)\left(1 - \frac{1}{2\ln H} + \cdots\right) + \cdots \ . \tag{6.155}$$

The magnetic susceptibility per bulk site at $H = 0$ is $\chi = 1/2\pi^2 + \ldots$. The average valence of bulk sites is

$$\frac{N}{L} = 1 - \frac{\ln 2}{\pi}\sqrt{\frac{8(\bar{\mu}+2a)}{3\zeta(3)}} + \cdots \tag{6.156}$$

and the charge stiffness per bulk site is

$$\chi_c = \frac{2\ln 2}{\pi}\frac{1}{\sqrt{6\zeta(3)(\bar{\mu}+2a)}} + \ldots \ . \tag{6.157}$$

These answers coincide with the ones for periodic boundary conditions, naturally. At half filling the charge susceptibility is divergent, as the consequence of a one-dimensional van Hove singularity.

The contribution from edges themselves is

$$\begin{aligned}
e_{edge} &= -\frac{1}{2}\left[-\frac{H}{2\ln H}\left(1+\frac{\ln\ln H}{2\ln H}\right)+\mu-2-\pi-\sqrt{\frac{8(\bar{\mu}+2a)^3}{27\zeta(3)}}+\cdots\right],\\
m^z_{edge} &= -\frac{H}{4\ln H}\left(1+\frac{\ln\ln H}{2\ln H}\right)+\cdots,\\
\chi_{edge} &= \frac{1}{4H\ln^2 H}\left(1+\frac{\ln\ln H}{2\ln H}\right)+\cdots,\\
n_{edge} &= \frac{1}{2}\left(1+\sqrt{\frac{2(\bar{\mu}+2a)}{3\zeta(3)}}\right)+\cdots,\\
\chi_{c,edge} &= -\frac{4\ln 2+\pi}{2\pi\sqrt{6\zeta(3)(\bar{\mu}+2a)}}+\cdots.
\end{aligned} \tag{6.158}$$

We see that the ground state magnetic susceptibility of open edges themselves of a supersymmetric t-J chain is divergent. The boundary charge stiffness (charge susceptibility, or compressibility) of edges is negative and diverges as one approaches half-filling.

Now we want to consider contributions from boundary potentials. For small boundary potentials p' $(-2[1+(p')^{-1}] \gg B)$ the contributions due to each boundary potential to the magnetic moment, magnetic susceptibility, valence and compressibility are

$$\begin{aligned}
m^z_{p'} &= \frac{H}{\pi^3}F\left(1-\frac{1}{2\ln H}\right)+\ldots,\\
\chi_{p'} &= \frac{F}{\pi^3}\left(1+\frac{H^2}{4\pi^2(\bar{\mu}+2a)}\right)\left(1-\frac{1}{2\ln H}\right)+\ldots\\
n_{p'} &= -\frac{2F}{\pi}+\ldots, \qquad \chi_{c,p'} = \frac{F}{\pi(\bar{\mu}+2a)}+\ldots,
\end{aligned} \tag{6.159}$$

where

$$F = -\frac{p'}{1+p'}\sqrt{\frac{(\bar{\mu}+2a)}{6\zeta(3)}}, \tag{6.160}$$

we determined the valence and the charge stiffness as derivatives with respect to the chemical potential μ. [The case in which boundary potentials and μ are connected to each other has to be considered numerically.] The reader can see that valences of edges, caused by boundary potentials, are larger than in the bulk. For large boundary potentials of the first type one gets

$$
\begin{aligned}
m_{p'}^z &= \frac{H}{4\pi^2}\left(1-\frac{1}{2\ln H}\right)\left(1-\frac{1}{\pi F}\right)+\cdots\ ,\\
\chi_{p'} &= \frac{1}{4\pi^2}\left(1-\frac{1}{\pi F}+\frac{H^2}{4\pi^3 F(\bar\mu+2a)}\right)\left(1-\frac{1}{2\ln H}\right)+\cdots\\
n_{p'} &= -\frac{1}{2}\left(1+\frac{\ln 2}{\pi}\sqrt{\frac{2(\bar\mu+2a)}{3\zeta(3)}}\right)+\frac{1}{2\pi F}+\cdots\ ,\\
\chi_{c,p'} &= \frac{3\ln 2}{\pi\sqrt{6\zeta(3)(\bar\mu+2a)}}+\frac{1}{2\pi F(\bar\mu+2a)}+\cdots\ .
\end{aligned}
\tag{6.161}
$$

The reader can see that the valence at edges is less than the one in the bulk due to the strong nonzero boundary potential of the first type.

Now let us consider the action of boundary potentials/magnetic fields of the second kind. For each large boundary potential/magnetic field $p \gg -2\pi/|\ln H|$ we can find (here we write the first terms in series)

$$
\begin{aligned}
m_p^z &= -\frac{1}{4}+\frac{p+1}{2p\ln H}+\frac{HA}{4\pi^3}\left(\psi[(p-1)/p]-\psi(-1/2p)\right)\ ,\\
\chi_p &= -\frac{1+h}{2hH\ln^2 H}+\frac{1}{4\pi^3}\left(\psi[(p-1)/p]-\psi(-1/2p)\right)\left(A+\frac{H^2}{6\pi^2\zeta(3)A}\right)\ ,\\
n_p &= -\frac{1}{2\pi}\left(\psi[(p-1)/p]-\psi(-1/2p)\right)\sqrt{\frac{2(\bar\mu+2a)}{3\zeta(3)}}\ ,\\
\chi_{c,p} &= \frac{1}{2\pi}\left(\psi[(p-1)/p]-\psi(-1/2p)\right)\sqrt{\frac{1}{6(\bar\mu+2a)\zeta(3)}}\ ,
\end{aligned}
\tag{6.162}
$$

where we again considered a boundary potential as an independent parameter and determined characteristics as derivatives with respect to H and μ. Each small boundary potential/magnetic field yields (we again present the first terms in series)

$$
\begin{aligned}
m_p^z &= -\frac{p\ln H}{2\pi^2}\ , \qquad \chi_p = -\frac{p}{2\pi^2 H}\ ,\\
n_p &= \frac{p}{2\pi}\sqrt{\frac{2(\bar\mu+2a)}{3\zeta(3)}}\ , \qquad \chi_{c,p} = -\frac{p}{2\pi\sqrt{6(\bar\mu+2a)\zeta(3)}}\ .
\end{aligned}
\tag{6.163}
$$

It turns out that it is impossible to take $H \to 0$ without taking $p \to 0$ first. This is why, the magnetic moment in this situation is small. However, depending on how fast one takes $p \to 0$, as compared to the homogeneous magnetic field H, the contribution to the magnetic susceptibility, caused by a weak boundary potential of the second kind, may diverge or not.

To summarize, in this chapter we introduced the reader to the main results of exact solutions of many-body quantum systems with open boundary conditions. We started with simple XY and Ising chains, and then presented Bethe ansatz solution to the problem in the frameworks of co-ordinate and algebraic Bethe ansatz. For the latter the reflection equations (complimentary to Yang–Baxter relations) are derived. Finally, we compared the behaviours of correlated electron models with open and periodic boundary conditions.

Open XY and Ising spin-$\frac{1}{2}$ chains were studied in [Lieb, Schulz and Mattis (1961); Pfeuty (1970)]. Studies of open Bethe ansatz integrable chains was pioneered by M. Gaudin [Gaudin (1971)], see also [Gaudin (1983)]. The co-ordinate Bethe ansatz ground state calculations for an open spin-$\frac{1}{2}$ Heisenberg–Ising chain can be found in [Alcaraz, Barber, Batchelor, Baxter and Quispel (1987)]. The reader can find calculations of the contributions from free edges themselves and boundary potentials of an open Heisenberg chain in [Frahm and Zvyagin (1997b)], and thermodynamics of an open Heisenberg–Ising chain in [de Sa and Tsvelik (1995)]. Reflection equations were proposed in [Cherednik (1984)], and the algebraic Bethe ansatz for open integrable chains was developed in [Sklyanin (1988)] (we closely follow this work in the derivation of the algebraic Bethe ansatz for open chains). The first Bethe ansatz solution of an open Hubbard chain was performed in [Schulz (1985)]. Role of boundary potentials in the behaviour of open spin and correlated electron chains was studied (and in this chapter we follow those studies) in [Asakawa and Suzuki (1995); Frahm and Zvyagin (1997a); Bedürftig and Frahm (1997); Yue and Deguchi (1997)] for Hubbard and quantum spin chains and in [Eßler (1996)] for a supersymmetric t-J chain.

Chapter 7

Correlated Quantum Chains with Isolated Impurities

In this chapter we shall present exact results for thermodynamic characteristics of impurities in quantum spin and correlated electron chains and compare their behaviours with those of free edges of homogeneous quantum chains, and with magnetic and hybridization impurities in three-dimensional metals (the *Kondo and Anderson* impurities).

7.1 Impurities in XY Chains

In previous chapters we considered mostly homogeneous quantum spin and correlated electron chains. The only inhomogeneity, considered so far, was the possibility of cutting a periodic chain. In that case the behaviour of edges of open quantum chains was different from the behaviour of bulk sites of that chain. This situation (with open chains) can be realized if one introduces a nonmagnetic impurity into a periodic spin chain. However, more generic situation with an impurity in a quantum correlated chain is the following:

- An impurity can have different local characteristics (*e.g.*, an effective magnetic moment or local potential energy) from those of other sites of a chain;
- an impurity can be coupled to other sites of a chain with an interaction, different in its strength (and, generally speaking, in the way of coupling) from interactions between other sites of a chain.

Actually, an open boundary, studied in the previous chapter, can be considered as a special impurity. The strength of an interaction of the link between the first and the last sites of a periodic chain is zero for an open chain, and, hence, it is different from couplings in the bulk. Also, an effective magnetic moment of such an "impurity" is modeled by a local boundary magnetic field, and a local potential energy of such an impurity is modeled by a local boundary potential. It is important, nevertheless, to study the behaviour of a more general impurity, which is coupled to a host chain with the coupling, different from the one of a host, but when this coupling does not cut that chain.

Let us consider first a simple isotropic spin-$\frac{1}{2}$ XY chain with a single (isolated) impurity. The Hamiltonian of such a model can be written as

$$\mathcal{H}_{XYi} = -\sum_{j\neq j_0, j_0-1} J(S_j^x S_{j+1}^x + S_j^y S_{j+1}^y) - \sum_{j\neq j_0} \mu H S_j^z$$
$$- J'(S_{j_0}^x S_{j_0+1}^x + S_{j_0}^y S_{j_0+1}^y + S_{j_0-1}^x S_{j_0}^x + S_{j_0-1}^y S_{j_0}^y) - \mu' H S_{j_0}^z \ , \quad (7.1)$$

where we situated an impurity in the site number j_0 (nothing actually depends on the number of the impurity site, see below). The impurity is different from other sites of the chain by its effective magneton μ' (with μ denoting magnetons of host sites), and by the exchange constant J', which describes the coupling of the impurity site with the right and left nearest neighbors (with J being the homogeneous coupling of the host), *i.e.*, in this case the local field, acting on the impurity is $h = \mu' H$.

To diagonalize the Hamiltonian $\mathcal{H}_{XYi}$ we first use the Jordan–Wigner transformation Eq. (2.10), which exactly relates the spin-$\frac{1}{2}$ Hamiltonian to the quadratic form of spinless Fermi operators as

$$\mathcal{H}_{XYi} = -\frac{J}{2}\left[\sum_{\substack{j=1,\\ j\neq j_0, j_0-1}}^{L-1} (a_j^\dagger a_{j+1} + a_{j+1}^\dagger a_j) - \nu_{L+1} a_L^\dagger a_1 - \nu_{L+1} a_1^\dagger a_L\right]$$
$$- \frac{\mu H}{2}\sum_{\substack{j=1\\ j\neq j_0}}^{L} (1 - 2a_j^\dagger a_j) - \frac{\mu' H}{2}(1 - 2a_{j_0}^\dagger a_{j_0})$$
$$- \frac{J'}{2}(a_{j_0}^\dagger a_{j_0+1} + a_{j_0+1}^\dagger a_{j_0} + a_{j_0-1}^\dagger a_{j_0} + a_{j_0}^\dagger a_{j_0-1}) \ . \quad (7.2)$$

Suppose L is even. Dividing the Hamiltonian Eq. (7.2) into $\mathcal{H}^\pm$ by using the projection operators $\frac{1}{2}(1 \pm \nu_L)$ as in Chapter 2, we can study those parts separately.

$$\mathcal{H}_{XYi} = \frac{1}{2}(1 + \nu_L)\mathcal{H}_{XYi} + \frac{1}{2}(1 - \nu_L)\mathcal{H}_{XYi}$$
$$= \frac{1}{2}(1 + \nu_L)\mathcal{H}_i^+ + \frac{1}{2}(1 - \nu_L)\mathcal{H}_i^- \ , \quad (7.3)$$

where

$$\mathcal{H}_i^+ = -\frac{J}{2}\left[\sum_{\substack{j=1\\ j\neq j_0, j_0-1}}^{L-1} (a_j^\dagger a_{j+1} + a_{j+1}^\dagger a_j) - a_L^\dagger a_1 - a_1^\dagger a_L\right]$$
$$- \frac{\mu H}{2}\sum_{\substack{j=1\\ j\neq j_0}}^{L} (1 - 2a_j^\dagger a_j) - \frac{\mu' H}{2}(1 - 2a_{j_0}^\dagger a_{j_0})$$
$$- \frac{J'}{2}(a_{j_0}^\dagger a_{j_0+1} + a_{j_0+1}^\dagger a_{j_0} + a_{j_0-1}^\dagger a_{j_0} + a_{j_0}^\dagger a_{j_0-1}) , \qquad (7.4)$$

and

$$\mathcal{H}_i^- = -\frac{J}{2}\sum_{\substack{j=1\\ j\neq j_0, j_0-1}}^{L-1} (a_j^\dagger a_{j+1} + a_{j+1}^\dagger a_j) + \mu H \sum_{\substack{j=1\\ j\neq j_0}}^{L} a_j^\dagger a_j - \frac{(\mu' + \mu L)H}{2}$$
$$+ \mu' H a_{j_0}^\dagger a_{j_0} - \frac{J'}{2}(a_{j_0}^\dagger a_{j_0+1} + a_{j_0+1}^\dagger a_{j_0} + a_{j_0-1}^\dagger a_{j_0} + a_{j_0}^\dagger a_{j_0-1}) . \qquad (7.5)$$

One can find eigenvalues for each of $\mathcal{H}_i^\pm$ separately and then take into account the effect of factors $\frac{1}{2}(1 \pm \nu_{L+1})$ by selecting half of eigenvalues of $\mathcal{H}_i^+$ and half of those of $\mathcal{H}_i^-$.

Consider, *e.g.*, the Hamiltonian $\mathcal{H}_i^-$ ($\mathcal{H}_i^+$ can be studied in a similar way). It is a quadratic form of Fermi operators, and, hence, it can be diagonalized with the help of the unitary transformation

$$a_j = \sum_\lambda u_j(\lambda) a_\lambda , \qquad i\hbar\frac{\partial a_\lambda}{\partial t} = \varepsilon_\lambda a_\lambda . \qquad (7.6)$$

The coefficients $u_j(\lambda)$ satisfy the following equations (we omit the explicit dependence on λ here)

$$(\varepsilon - \mu H)u_j + \frac{J}{2}(u_{j+1} + u_{j-1}) = 0 , \qquad j \neq j_0, j_0 \pm 1 ,$$
$$(\varepsilon - \mu H)u_{j_0\pm1} + \frac{1}{2}(J u_{j_0\pm2} + J' u_{j_0}) = 0 , \qquad (7.7)$$
$$(\varepsilon - \mu' H)u_{j_0} + \frac{1}{2}J'(u_{j_0+1} + u_{j_0-1}) = 0 .$$

The last three equations can be considered as the boundary condition for the first homogeneous linear equation in finite differences. The standard solution of these equations describes a gas of fermions with the dispersion law

$$\varepsilon_k = \mu H - J \cos k \ . \tag{7.8}$$

However, if

$$I^2 > 1 \pm x \ , \tag{7.9}$$

where $I = J'/J$, $x = (\mu H/J)(\alpha - 1)$, $\alpha = \mu'/\mu$, impurity bound states are split off from the continuous-like spectrum with the energies of these local levels

$$\varepsilon_{1,2} = \mu H - J \frac{x(1 - I^2) \pm I^2 \sqrt{x^2 + 2I^2 - 1}}{2I^2 - 1} \ . \tag{7.10}$$

If just one of the inequalities is satisfied, then either the level ε_1 below the band of the continuous spectrum, or the level ε_2 above the band of the continuous spectrum, respectively, are split off. Notice that for $2I^2 = 1$ the bound state level is $\varepsilon = \mu' H + J/2x$ for $|x| > \frac{1}{2}$.

It is not difficult to calculate the magnetic moment of an impurity in thermal equilibrium for large L (notice that in order to keep the total number of states in a finite system one has to remove two states from the bands of extended states)

$$m_{j_0}^z = \frac{\mu'}{2} - \mu' I^2 \sum_k \frac{\sin^2 k}{|x + \cos k - I^2 \exp(ik)|^2} \frac{1}{1 + \exp(\varepsilon_k/T)}$$
$$- \mu' \sum_{j=1,2} \frac{(1 - r_{1,2}^2)}{1 + r_{1,2}^2 (2I^2 - 1)} \frac{1}{1 + \exp(\varepsilon_{1,2}/T)} \theta(I^2 - 1 \mp x) \ , \tag{7.11}$$

where $\theta(x)$ is the Heaviside step function, and

$$r_{1,2} = \frac{x \pm \sqrt{x^2 + 2I^2 - 1}}{2I^2 - 1} \ . \tag{7.12}$$

It turns out that we are interested only in $|r_{1,2}| < 1$, *i.e.*, in decaying solutions. In what follows we shall consider only the case of large enough L.

In the ground state the expression for the impurity magnetic moment reduces to

$$m_{j_0}^z = \frac{\mu'}{2} - \frac{\mu' I^2}{\pi} \int_0^{k_0} \frac{dk \sin^2 k}{|x + \cos k - I^2 \exp(ik)|^2}$$
$$- \mu' \frac{(1 - r_1^2)}{1 + r_1^2 (2I^2 - 1)} \theta(I^2 - 1 - x) \theta(-\varepsilon_1) \ , \tag{7.13}$$

where $k_0 = \cos^{-1}(\mu H/J)$. The reader can see that the magnetic moment of an impurity can have a jump when arguments of the Heaviside functions vanish. However, for $I^2 = x + 1$ we have $r_1 = 1$ and there is no jump. Hence, the jump exists only at the field value, at which $\varepsilon_1 = 0$. The necessary condition for such an onset of a jump is the inequality

$$I^2 > \frac{\mu\alpha}{2(1+\alpha)} = \frac{\mu\mu'}{2(\mu+\mu')} . \tag{7.14}$$

The critical value of the field at which a jump can exist is

$$H_j = \frac{JI^2}{\mu\sqrt{\alpha(2I^2-\alpha)}} . \tag{7.15}$$

The stronger inequality follows from the condition $r_1|_{H=H_j} \leq 1$, which reads $I^2 \geq \alpha$. The magnitude of a jump is equal to

$$\delta m_{j_0}^z = \mu' \frac{I^2 - \alpha}{I^2(\alpha+1) - \alpha} , \tag{7.16}$$

which has the maximum value $\mu\mu'/(\mu+\mu')$. As I^2 tends to α, the jump vanishes and H_j becomes equal to $H_s = J/\mu$, *i.e.*, the value at which the quantum phase transition in the homogeneous isotropic XY model into the spin-polarized phase takes place. The local susceptibility of the impurity is

$$\begin{aligned}\chi_{j_0} = & \frac{\mu\mu' I^2(\alpha-1)r_1^2}{J\sqrt{x^2+2I^2-1}[1+r_1^2(2I^2-1)]^2} \\ & + \frac{\mu' I^2\sqrt{1-(H/H_s)^2}}{2\pi J[(I^2-\alpha)^2 - I^4](H/H_s)^2 + I^4}\theta(J-\mu H) \\ & + \frac{2\mu'(\mu'-\mu)}{\pi J}\int_0^{k_0} \frac{dk \sin^2 k(x+\cos k - I\cos k)}{|x+\cos k - I^2\exp(ik)|^4} \\ & + \mu' \frac{\sqrt{H_s^2 - H^2}}{H^2(1-2I^2) + H_s^2(x^2+I^4) + 2HH_s(x-I^2)} .\end{aligned} \tag{7.17}$$

If $H > H_s$, then the second, third and fourth terms in the above formula vanish and the sign of the first term is determined by the sign of $\alpha - 1$, *i.e.*, for $\alpha < 1$ the susceptibility in this domain of values of field (adjoining H_j) is negative. A decrease of the magnetic moment of an impurity in this domain is then compensated by a positive jump. For $I^2 = \alpha$ the second term in Eq. (7.17) has the same square root singularity as in the homogeneous isotropic XY chain. For $I^2 > \alpha$ both the moment and the local susceptibility of an impurity are regular for all values of the magnetic field.

The presence of an impurity affects the distribution of magnetic moments of a host. For example, in the ground state we obtain

$$\frac{m_j^z}{\mu} = \frac{1}{2} - \frac{1}{\pi}\left(k_0 - \frac{\sin 2k_0|j-j_0|}{2|j-j_0|}\right) - \frac{I^2(1-r_1)^2 r_1^{2|j-j_0|}\theta(I^2-1-x)}{1+(2I^2-1)r_1^2}$$
$$- \frac{I^2}{\pi}\int_0^{k_0} dk \frac{\sin^2 k[(x+\cos k)\sin 2k|j-j_0| - I^2 \sin k(2|j-j_0|-1)]}{|x+\cos k - I^2\exp(ik)|^2} , \tag{7.18}$$

from which the reader can see that magnetic moments of the host also undergo jumps for $I^2 > \alpha$ at $H = H_j$, which magnitudes decay exponentially as distances from the impurity $|j-j_0|$ grow (naturally, then, they grow due to the periodicity). The total jump of the total magnetic moment is equal to

$$\Delta M^z = \mu' \frac{2I^2 - \alpha}{I^2(1+\alpha) - \alpha} . \tag{7.19}$$

It is important to emphasize that the total jump of the total spin moment is equal to 1.

It is interesting that results for a semi-infinite XY chain with an impurity at the edge follow from the above results using the change $I^2 \to I^2/2$. Also, results for a (non-magnetic) impurity, which renormalizes a coupling constant of only one link can be obtained with the change $I^2 \to I^2/2$ for $\mu' = \mu$ $(\alpha = 1)$.

Now, let us consider the behaviour of an impurity in the XY chain if the local field h is directed, *e.g.*, along x. This problem can be solved explicitly only for the semi-infinite chain with an impurity at the edge and only for the case $H = 0$. The Hamiltonian is

$$\mathcal{H}_{XYi\perp} = -\sum_{j\neq 0}^{\infty} J(S_j^x S_{j+1}^x + S_j^y S_{j+1}^y) - hS_0^x - J'(S_0^x S_1^x + S_0^y S_1^y) . \tag{7.20}$$

Here we can use the following trick. The average magnetic moment of an impurity with the above Hamiltonian is equal to

$$\langle S_0^x \rangle = Z^{-1}\, \mathrm{tr}[S_0^z \exp(-\beta \mathcal{H}_{XYi\perp})] , \tag{7.21}$$

where the partition function is $Z = \mathrm{tr}\exp(-\beta\mathcal{H}_{XYi\perp})$. Let us introduce the auxiliary Hamiltonian $\mathcal{H}_a = \mathcal{H}_{XYi\perp} + hS_0^x - 2hS_0^x S_{-1}^x$, where we formally added one spin $j = -1$ to the semi-infinite chain. Since the operator S^x_{-1} has the eigenstates $\pm\frac{1}{2}$ the reader can see that

$$\langle S_0^x \rangle = \mathrm{tr}[\rho_a S_0^x(1 + 2S^x_{-1})] = 2\,\mathrm{tr}[S_0^x S^x_{-1}\rho_a] \ , \tag{7.22}$$

where $\rho_a = \exp(-\beta\mathcal{H}_a)/\mathrm{tr}[\exp(-\beta\mathcal{H}_a)]$.

The diagonalization of the Hamiltonian $\mathcal{H}_a$ is obtained with the help of the Jordan–Wigner transformation and following unitary transformation of Fermi operators

$$a_j = \sum_\lambda [u_j(\lambda)a_\lambda + v_j(\lambda)a_\lambda^\dagger] \ , \tag{7.23}$$

the coefficients of which satisfy the following set of equations (we again drop the explicit dependence on λ)

$$\begin{gathered} -J(u_{j+1} + u_{j-1}) = 2\varepsilon u_j \ , \qquad J(v_{j+1} + v_{j-1}) = 2\varepsilon v_j \ , \qquad j = 1, \ldots, \infty \ , \\ -Ju_2 - J'u_0 = 2\varepsilon u_1 \ , \qquad Jv_2 + J'v_0 = 2\varepsilon v_1 \ , \\ -J'u_1 - h(u_{-1} - v_{-1}) = 2\varepsilon u_0 \ , \qquad -Jv_2 + h(u_{-1} - v_{-1}) = -2\varepsilon v_0 \ , \\ -h(u_0 - v_0) = 2\varepsilon u_{-1} \ , \qquad h(u_0 - v_0) = 2\varepsilon v_{-1} \ . \end{gathered} \tag{7.24}$$

Again, as for the previous case, the reader can see that there are two types of solutions. One of them describes a continuous-like spectrum of extended states, and the second describes local levels of impurity bound states. It is also clear that solutions can be divided into two classes: the one, which is dependent on h, and the other one, which is h-independent. Using this solution we obtain the value of the magnetic moment of an impurity for large L limit

$$\begin{aligned} m_0^x = {} & \frac{2\nu I^2}{\pi} \int \frac{dk \sin^2 k \cos k}{|2(\cos^2 k - \nu^2) - I^2 \cos k \exp(ik)|^2} \tanh(J\cos k/2T)) \\ & + \frac{\nu r(1 - r^2)}{1 + r^4(I^2 - 1)} \tanh[J(r^2 + 1)/4rT]\theta(I^2 - 2 + 2\nu^2) \ , \end{aligned} \tag{7.25}$$

where $\nu = h/J$ and

$$r = \sqrt{\frac{2\nu^2 + I^2 - 2 - \sqrt{(4\nu^2 + I^2)^2 - 16\nu^2}}{2(1 - I^2)}} \ . \tag{7.26}$$

It is interesting to consider some limiting cases. In the ground state for small h we have

$$m_0^z \sim (h/J)\ln(h/J) \ , \tag{7.27}$$

while for $h \gg J$ we obtain

$$m_0^z = \frac{1}{2} - \frac{I^2}{\nu^2} . \tag{7.28}$$

It means that the magnetic susceptibility of an impurity spin is divergent in the ground state for small fields, while the saturation value $\frac{1}{2}$ is achieved only in the infinitely large field h.

Now let us study the behaviour of an impurity in the dimerized XY chain of spins $\frac{1}{2}$. The Hamiltonian of this system has the form

$$\begin{aligned}\mathcal{H}_{dimp} = &\sum_{j \neq j_0} [J_1(S_{j,1}^x S_{j,2}^x + S_{j,1}^y S_{j,2}^y) - \mu_1 H S_{j,1}^z - \mu_2 H S_{j,2}^z] \\ &+ \sum_{j \neq j_0, j_0 - 1} J_2(S_{j,2}^x S_{j+1,1}^x + S_{j,2}^y S_{j+1,1}^y) - \mu_1' H S_{j_0,1}^z \\ &+ J_1'(S_{j_0,1}^x S_{j_0,2}^x + S_{j_0,1}^y S_{j_0,2}^y) + J_2'(S_{j_0,2}^x S_{j_0+1,1}^x + S_{j_0,2}^y S_{j_0+1,1}^y) ,\end{aligned} \tag{7.29}$$

where we used the same notations as in Chapter 2. An impurity is defined by couplings $J_{1,2}'$ and the effective moment μ_1', which can be different from the values in a host.

The Hamiltonian $\mathcal{H}_{dimp}$ can be exactly diagonalized using the generalized Jordan–Wigner transformation from Chapter 2 and the procedure, similar to the one, described above for the isotropic XY chain with an impurity. Eigenstates of the system are divided into:

- Two bands of a continuous-like spectrum with energies

$$\begin{aligned}\varepsilon_{1,2}(k) = &\frac{(\mu_1 + \mu_2)H}{2} \\ &\pm \frac{1}{2}\sqrt{(\mu_1 - \mu_2)^2 H^2 + J_1^2 + J_2^2 + 2J_1 J_2 \cos k} ;\end{aligned} \tag{7.30}$$

- Discrete levels are split off the edges of the bands with energies

$$\begin{aligned}\varepsilon_{1,2}^j = &\frac{(\mu_1 + \mu_2)H}{2} \\ &\pm \frac{1}{2}\sqrt{(\mu_1 - \mu_2)^2 H^2 + (J_1 + J_2 r_j)(J_1 + J_2 r_j^{-1})} .\end{aligned} \tag{7.31}$$

Here $|r_j| \le 1$, $j = 1,2$ are real solutions of the equation

$$r^2[J_2^2 J_1^2 - (J_1')^2 J_2^2 - (J_2')^2 J_1^2] + rJ_1J_2[J_1^2 + J_2^2 - (J_1')^2 - (J_2')^2]$$
$$+ J_1^2 J_2^2 = 2J_1J_2(\mu_1' - \mu_1)Hr\big[(\mu_1 - \mu_2)H$$
$$\pm \sqrt{(\mu_1 - \mu_2)^2H^2 + J_1^2 + J_2^2 + J_1J_2(r + r^{-1})}\,\big] \,. \tag{7.32}$$

We are interested only in solutions of this equation, which satisfy the condition $|r| < 1$. This equation is of the fourth order, and, hence, there can be, generally speaking, up to four solutions (and, hence, up to four localized levels, each being split off the upper or lower edges of two branches of the continuous-like spectrum). The reader can can see that for $J_{1,2}' = J_{1,2}$ and $\mu_1' = \mu_1$ the only solutions are $r = \pm 1$, *i.e.*, there are no local levels of impurity bound states in the homogeneously dimerized XY chain, as it must be.

It is straightforward to calculate the specific heat of the system in the limit of large L:

$$c = \frac{1}{4T^2} \sum_{m=1,2} \left[\int \frac{dk[\varepsilon_m(k)]^2}{\pi \cosh^2(\varepsilon_m(k)/2T)} + \sum_{l=1,2} \frac{(\varepsilon_m^l)^2 \theta(1 - |r_l|)}{\cosh^2(\varepsilon_m^l(k)/2T)} \right] , \tag{7.33}$$

where $\theta(x)$ is the Heaviside function. We can also calculate for large enough L the total magnetic susceptibility of the model:

$$\chi = \frac{1}{2} \sum_{m=1,2} \left[\int \frac{dk}{\pi} \left(\frac{\partial^2 \varepsilon_m(k)}{\partial H^2} \tanh(\varepsilon_m(k)/2T) + \frac{1}{2T} \left(\frac{\partial \varepsilon_m(k)}{\partial H} \right)^2 \right.\right.$$
$$\left. \times \frac{1}{\cosh^2(\varepsilon_m(k)/2T)} \right) + \sum_{l=1,2} \theta(1 - |r_l|) \left(\frac{\partial^2 \varepsilon_m^l}{\partial H^2} \tanh(\varepsilon_m^l/2T) \right.$$
$$\left.\left. + \frac{1}{2T} \left(\frac{\partial \varepsilon_m^l}{\partial H} \right)^2 \frac{1}{\cosh^2(\varepsilon_m^l/2T)} \right) \right] . \tag{7.34}$$

It is clear from the above expressions that contributions from extended states (bands) to the specific heat and magnetic susceptibility are similar to the ones for the homogeneous dimerized XY chain (notice that in order to keep the total number of states in a finite system one has to remove four states from extended states), but there appear additional contributions from local levels of bound states.

By using eigenvalues and eigenvectors we can obtain the magnetic moment of an impurity

$$
\begin{aligned}
m_{j_0}^z = \mu_1' \Bigg[\frac{1}{2} - \sum_{m=1,2} \int \frac{dk}{\pi} & \frac{2J_2^2(J_1'J_2 + J_2J_1')^2 \sin^2 k}{[\varepsilon_1(k) - \varepsilon_2(k)]} n_{mk} \\
& \times [\varepsilon_m(k) - \mu_2 H](|2J_2(\mu_1' - \mu_1)H[\varepsilon_m(k) - \mu_2 H] - J_2'(J_1'J_2 + J_2'J_1) \\
& + J_2 \exp(ik)[J_1^2 + J_2^2 + 2J_1J_2\cos k - (J_1')^2 - (J_2')^2]|^2)^{-1} \\
& - \sum_{l,m=1,2} \theta(1 - |r_l|) n_{l,m} 4(1 - r_l^2) J_1^2 J_2^2 (\varepsilon_m^l - \mu_2 H)^2 ((1 - r_l^2) J_2^2 \\
& \times [2J_1(\varepsilon_m^l - \mu_2 H) + J_1'(J_1 + J_2 r_l)]^2 + r_l^2 4(\varepsilon_m^l - \mu_2 H)^2 [(J_1')^2 J_2^2 \\
& + (J_2')^2 J_1^2] + (J_1')^2 J_2^2 (J_1 + J_2 r_l)^2 + (J_2')^2 J_1^2 (J_1 + J_2 r_l^{-1})^2)^{-1} \Bigg] ,
\end{aligned}
\tag{7.35}
$$

where $n_{1,2k} = [1 + \exp(\varepsilon_{1,2}(k)/T)]^{-1}$ and $n_{lm} = [1 + \exp(\varepsilon_m^l/T)]^{-1}$. In the ground state, since $\varepsilon_1(k)$ and $\varepsilon_1^{1,2}$ are always non-negative, $n_{1k} = n_{1m} = 0$. Also one has to replace $n_{2k} \to \theta[-\varepsilon_2(k)]$ and $n_{2m} \to \theta[-\varepsilon_2^{1,2}]$.

It turns out that for $\mu_1' = \mu_1$ we can write the explicit formula for r:

$$
\begin{aligned}
r_{1,2} = (2[(J_1')^2 J_2^2 + (J_2')^2 J_1^2 - J_1^2 J_2^2])^{-1} J_1 J_2 [J_1^2 + J_2^2 - (J_1')^2 - (J_2')^2 \\
\pm \sqrt{[J_1^2 + J_2^2 - (J_1')^2 - (J_2')^2]^2 + 4[(J_1')^2 J_2^2 + (J_2')^2 J_1^2 - J_2^2 J_1^2]}] .
\end{aligned}
\tag{7.36}
$$

In this case eigenvalues of local impurity levels are:

$$
\begin{aligned}
\varepsilon_{1,2}^{1,2} = \frac{(\mu_1 + \mu_2)H}{2} & \pm \frac{1}{2} \Bigg[(\mu_1 - \mu_2)^2 H^2 + (J_1')^2 + (J_2')^2 \\
& - \frac{[(J_1')^2 J_2^2 + (J_2')^2 J_1^2]}{2[(J_1')^2 J_2^2 + (J_2')^2 J_1^2 - J_1^2 J_2^2]} (J_1^2 + J_2^2 - (J_1')^2 - (J_2')^2 \\
& \pm \sqrt{[J_1^2 + J_2^2 - (J_1')^2 - (J_2')^2]^2 + 4[(J_1')^2 J_2^2 + (J_2')^2 J_1^2 - J_2^2 J_1^2]}) \Bigg]^{1/2} .
\end{aligned}
\tag{7.37}
$$

It is interesting to note that for $\mu_1' = \mu_1 = \mu_2$ the magnetic moment of an impurity in the ground state depends on the magnetic field only *via* the limit of integration for extended states (the contribution from local levels of bound states and integrand do not depend on H). Hence, in this limit the local magnetic susceptibility of an impurity is zero in the phases $H \leq H_c$

and $H \geq H_s$, where the critical values of the homogeneous dimerized XY chain, *cf.* Chapter 2, are

$$H_{c,s} = \frac{|J_1 \mp J_2|}{2\sqrt{\mu_1 \mu_2}} \ . \tag{7.38}$$

In the ground state for $\mu_1' = \mu_1$ the Heaviside functions $\theta(-\varepsilon_2^l)$ imply the onset of critical values

$$\begin{aligned} H_{j1,2} &= \frac{\sqrt{J_1^2 + J_2^2 + J_1 J_2 (r_{1,2} + r_{1,2}^{-1})}}{2\sqrt{\mu_1 \mu_2}} = (2\sqrt{\mu_1 \mu_2})^{-1} \Bigg[(J_1')^2 + (J_2')^2 \\ &- \frac{[(J_1')^2 J_2^2 + (J_2')^2 J_1^2]}{2[(J_1')^2 J_2^2 + (J_2')^2 J_1^2 - J_1^2 J_2^2]} (J_1^2 + J_2^2 - (J_1')^2 - (J_2')^2 \\ &\pm \sqrt{[J_1^2 + J_2^2 - (J_1')^2 - (J_2')^2]^2 + 4[(J_1')^2 J_2^2 + (J_2')^2 J_1^2 - J_2^2 J_1^2]}) \Bigg]^{1/2} . \end{aligned} \tag{7.39}$$

At these values of an external field a jump of the magnetic moment of an impurity can take place. If $0 < r_{1,2} \leq 1$, the corresponding critical field $H_{j1,2}$ exceeds critical values of the homogeneous dimerized XY chain, *i.e.*, $H_c \leq H_s \leq H_{j1,2}$. On the other hand, if $-1 \leq r_{1,2} < 0$, we have $H_{j1,2} \leq H_c \leq H_s$. Depending on the relation between exchange constants the following situations are possible:

- (a) For $J_1^2 + J_2^2 > (J_1')^2 + (J_2')^2$ and $(J_1')^2 J_2^2 + (J_2')^2 J_1^2 > J_1^2 J_2^2$, or for $J_1^2 + J_2^2 < (J_1')^2 + (J_2')^2$ and $(J_1')^2 J_2^2 + (J_2')^2 J_1^2 > J_1^2 J_2^2$, one has $r_1 > 0$, $r_2 < 0$ and, hence, $H_{j2} \leq H_c$ and $H_s \leq H_{j1}$;
- (b) For $J_1^2 + J_2^2 > (J_1')^2 + (J_2')^2$ and $(J_1')^2 J_2^2 + (J_2')^2 J_1^2 < J_1^2 J_2^2$, one has $r_1 < 0$, $r_2 < 0$, and $H_{j1,2} \leq H_c$;
- (c) Finally, for $J_1^2 + J_2^2 < (J_1')^2 + (J_2')^2$ and $(J_1')^2 J_2^2 + (J_2')^2 J_1^2 < J_1^2 J_2^2$, one has $r_1 > 0$, $r_2 > 0$, and $H_{j1,2} \geq H_s$.

In the case (a) local levels of bound states $\varepsilon_{1,2}^1$ emerge when the following conditions are satisfied

$$2J_1^2 J_2^2 < (J_1')^2 J_2^2 + (J_2')^2 J_1^2 - J_1 J_2 [J_1^2 + J_2^2 - (J_1')^2 - (J_2')^2] \ , \tag{7.40}$$

while local levels $\varepsilon_{1,2}^2$ emerge when

$$2J_1^2 J_2^2 < (J_1')^2 J_2^2 + (J_2')^2 J_1^2 + J_1 J_2 [J_1^2 + J_2^2 - (J_1')^2 - (J_2')^2] \ . \tag{7.41}$$

In the case (b) local levels $\varepsilon_{1,2}^{1,2}$ are formed, if one has

$$2J_1^2J_2^2 > (J_1')^2J_2^2 + (J_2')^2J_1^2 - J_1J_2[J_1^2 + J_2^2 - (J_1')^2 - (J_2')^2] . \quad (7.42)$$

Finally, in the case (c) local levels $\varepsilon_{1,2}^{1,2}$ are formed if

$$2J_1^2J_2^2 > (J_1')^2J_2^2 + (J_2')^2J_1^2 + J_1J_2[J_1^2 + J_2^2 - (J_1')^2 - (J_2')^2] . \quad (7.43)$$

Jumps of host magnetic moments exponentially decrease with increasing the distance between host spins and an impurity (again, one has to take into account periodic boundary conditions).

7.2 Impurities in Spin Chains: Bethe Ansatz

In the previous section we considered characteristics of magnetic impurities in isotropic XY spin-$\frac{1}{2}$ chains. Those studies are relatively easy because the reader knows that Hamiltonians of spin-$\frac{1}{2}$ XY chains in the transverse magnetic field can be exactly mapped onto Hamiltonians of quadratic forms of Fermi operators (by using the Jordan–Wigner transformation). However, a nonzero Ising component, $J_z \neq 0$, as the reader knows from Chapter 2, introduces an interaction between Jordan–Wigner fermions, and we needed a more sophisticated approach, the Bethe ansatz. Is it possible to study exactly, *e.g.*, a Heisenberg–Ising chain with an impurity, similar to the one of the previous section? It is easy to check that the Hamiltonian

$$\begin{aligned}\mathcal{H}_i = &\sum_{j\neq j_0, j_0-1} [J(S_j^xS_{j+1}^x + S_j^yS_{j+1}^y) + J_zS_j^zS_{j+1}^x] - \sum_{j\neq j_0} \mu H S_j^z \\ &+ J'(S_{j_0}^xS_{j_0+1}^x + S_{j_0}^yS_{j_0+1}^y + S_{j_0-1}^xS_{j_0}^x + S_{j_0-1}^yS_{j_0}^y) \\ &+ J_z'(S_{j_0}^zS_{j_0+1}^z + S_{j_0-1}^zS_{j_0}^z) - \mu' H S_{j_0}^z \qquad (7.44)\end{aligned}$$

cannot be diagonalized by using the Bethe ansatz. The reason is very simple: the two-particle scattering matrix of the system with the Hamiltonian Eq. (7.44) does not satisfy Yang–Baxter relations, which are necessary conditions to apply the Bethe ansatz scheme, as we showed in Chapter 5. Moreover, the reader saw in Chapter 6 that the only possibility to introduce a local magnetic field (a local potential) is to apply it to edges of an open quantum chain, to preserve the Bethe ansatz integrability. Hence, at least from this perspective, the reader already knows how impurities, which can be described only by the action of a local magnetic field (or a local potential), behave. Nonetheless, the question appears: can one consider inhomogeneous quantum chains, in which a coupling between an impurity and the host is different from couplings between other sites of the host, but such that their Hamiltonians can be diagonalized by using the Bethe's ansatz? The answer is affirmative. To find such Hamiltonians, we shall follow the pioneering idea of N. Andrei and H. Johannesson, look for some transfer matrices composed with "defect" L-operators (which will define an impurity), but constructed in such a way that those impurity L-operators and, hence, monodromy operators with impurity L-operators included satisfy intertwining relations with R-matrices (which satisfy Yang–Baxter relations). The idea was to use the fact, already known to the reader from Chapter 5. Namely, we know that L-operators of higher spin values, Eq. (5.39), which describe the Takhtajan–Babujian model, satisfy intertwining relations with the R-matrix of the Heisenberg chain, *cf.* Fig 7.1.

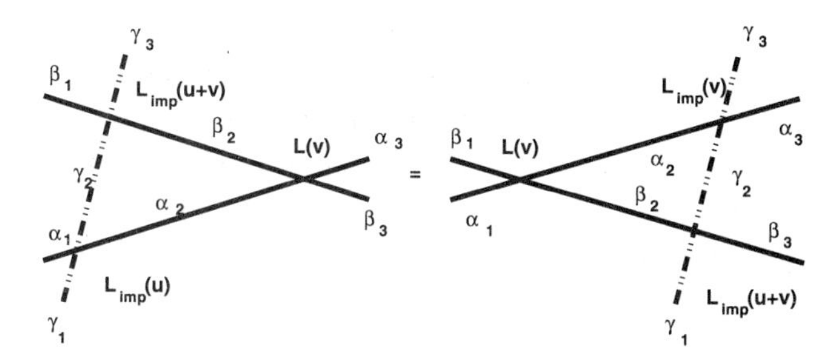

Fig. 7.1 Illustration of the Yang–Baxter relations for a host's and impurity's L-operator.

Let us construct the monodromy matrix

$$T_{imp}(\lambda) = L_{01}(\lambda) \cdots L_{0L}(\lambda) L^{S'}_{0imp}(\lambda - \theta) \ , \tag{7.45}$$

where

$$L_{0n}(\lambda) = \frac{1}{\lambda + i(c/2)} I_0 \otimes I_n + ic\frac{1}{2}\vec{\sigma}_0 \otimes \vec{\sigma}_n \ , \tag{7.46}$$

and

$$L^{S'}_{0imp}(\lambda) = \frac{1}{\lambda - \theta + icS'} \begin{pmatrix} \lambda - \theta + ic(S')^z_{imp} & ic(S')^-_{imp} \\ ic(S')^+_{imp} & \lambda - \theta - ic(S')^z_{imp} \end{pmatrix} , \tag{7.47}$$

see Fig. 7.2.

The subscript *imp* denotes the co-ordinate of an impurity. The impurity L-operator differs from other L-operators by two parameters. The first

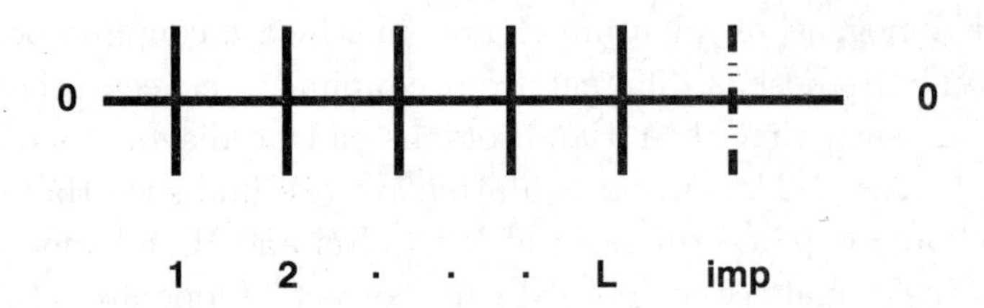

Fig. 7.2 Illustration of a monodromy operator of an integrable model with an impurity, situated at the last site of the lattice.

one, S' determines the value of the spin of an impurity, while the role of the other parameter, θ, (absent in the analysis of Andrei and Johannesson) we shall clarify below. Naturally, for $\theta = 0$ and $S' = \frac{1}{2}$ the impurity L-operator coincides with the ones of the host. It is easy to check that the L-operators Eqs. (7.46) and (7.47) satisfy Yang–Baxter relations mutually and intertwining relations with R-matrices of the Heisenberg chain for any S' and θ. It is also easy to check that the monodromy operators Eq. (7.45) satisfy intertwining relations with the R-matrices of the Heisenberg chain, and, moreover, this fact does not depend on the position of an impurity L-operator, *cf.* Fig. 7.3.

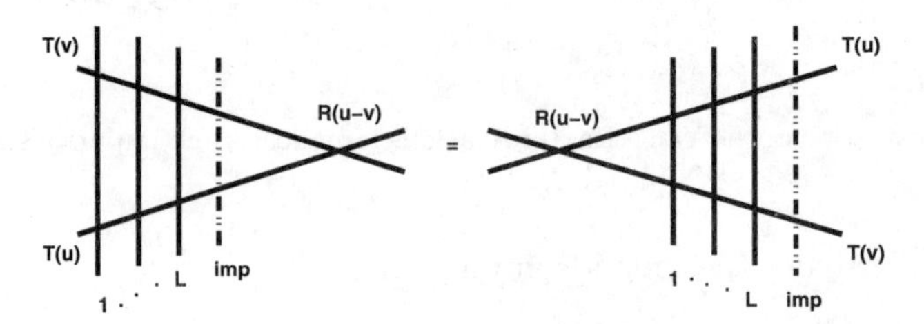

Fig. 7.3 Illustration of intertwining relations for monodromy operators of an integrable model with an impurity.

Then, it follows that the transfer matrices with different spectral parameters, constructed as traces of Eq. (7.45) over the auxiliary subspace, commute. This constitutes the exact integrability of the problem. Now it is necessary to construct the Hamiltonian of the Bethe ansatz solvable Heisenberg spin chain with an impurity. To do it, we shall follow the procedure, described in Chapter 5, *i.e.*, we shall use as the Hamiltonian the logarithmic derivative of the transfer matrix (with embedded impurity L-operator) with respect to the spectral parameter λ, putting then $\lambda = 0$. The Hamiltonian is $\mathcal{H} = \mathcal{H}_H + \mathcal{H}_{imp}$, where $\mathcal{H}_H = J\sum_{j=1}^{L} \vec{S}_j\vec{S}_{j+1}$ ($S = \frac{1}{2}$, and periodic boundary conditions are assumed). Suppose that the impurity L-operator is situated between mth and $(m+1)$th L-operators of the host, then

$$\mathcal{H}_{imp} = J_{imp}\big((\vec{S}_m + \vec{S}_{m+1})\vec{S}'_{imp} + \{\vec{S}_m\vec{S}'_{imp}, \vec{S}_{m+1}\vec{S}'_{imp}\} \\ - 2i\theta[\vec{S}_m\vec{S}'_{imp}, \vec{S}_{m+1}\vec{S}'_{imp}] + (\theta^2 - 2S'(S'+1))\vec{S}_m\vec{S}_{m+1}\big) \ , \quad (7.48)$$

where

$$J_{imp} = \frac{4J}{4\theta^2 + (2S'+1)^2} \quad (7.49)$$

plays the role of a coupling constant between the impurity site and two neighboring sites of the host Heisenberg chain, and $[.,.]$ ($\{.,.\}$) denotes a commutator (anticommutator), for the illustration see Fig. 7.4. The term with the commutator can be re-written as $-i\theta[\vec{S}_m\vec{S}'_{imp}, \vec{S}_{m+1}\vec{S}'_{imp}] = \theta\vec{S}_m(\vec{S}'_{imp} \times \vec{S}_{m+1})$, where $(\vec{a} \times \vec{b})$ denotes the vector product. The reader can see that θ actually determines the coupling between the impurity and the host, *i.e.*, distinguishes the impurity site even for $S' = \frac{1}{2}$. It is important to notice that for $S' = \frac{1}{2}$ and $\theta = 0$ the model reduces to the $L+1$-long periodic Heisenberg chain. On the other hand, for $\theta = \infty$ the impurity term is totally decoupled from the host Heisenberg Hamiltonian. The fact that the Hamiltonian is blind to the position of an impurity is, naturally, the artifact of the Bethe ansatz construction, used here.

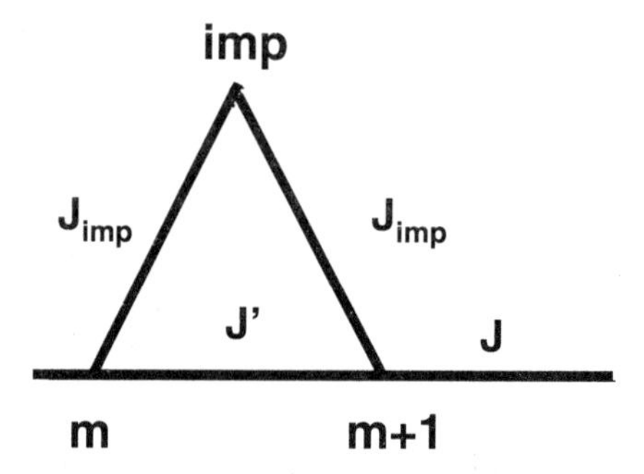

Fig. 7.4 Illustration of interactions in a Hamiltonian with a Bethe ansatz integrable impurity. In the simplest case of the isotropic Heisenberg spin-$\frac{1}{2}$ chain the local impurity-host exchange constant is $J_{imp} = 4J/[4\theta_j^2 + (2S'+1)^2]$.

Bethe ansatz equations, which solutions determine eigenfunctions and eigenvalues of the Schrödinger equation with the Heisenberg Hamiltonian with an integrable impurity are

$$e_1^L(\lambda_j)e_{2S'}(\lambda_j-\theta)=\prod_{\substack{l=1\\l\neq j}}^{M}e_2(\lambda_j-\lambda_l)\ ,\qquad j=1,\ldots,M\ ,\tag{7.50}$$

where $e_n(x)=(2x+in)/(2x-in)$, and M is the number of spins down (the total magnetic moment is $M^z=(L/2)+S'-M$), and the energy is

$$\begin{aligned}E&=E_0-\sum_{j=1}^{M}\left(\frac{2J}{4\lambda_j^2+1}-H\right)\ ,\\E_0&=\frac{(L-1)J}{4}-\frac{H(2S'+L)}{2}+\frac{J_{imp}(\theta^2+2S')}{4}\ .\end{aligned}\tag{7.51}$$

Notice, that except for the trivial contribution to the energy of the ferromagnetic state E_0, the expression for the energy does not depend on the parameters of an impurity explicitly.

It is straightforward to generalize the above construction for the Heisenberg–Ising chain with an integrable impurity. For $J_z\neq J$ we again introduce the value $\cos\eta=J_z/J$ (real values of η are related to the "easy-plane" magnetic anisotropy $|J_z|\leq J$, while the "easy-axis" magnetic anisotropy with $|J_z|\geq J$ is described by imaginary values of η). The Bethe ansatz equations and the energy can be written as

$$\left(\frac{\sin[\lambda_j+(\eta/2)]}{\sin[\lambda_j-(\eta/2)]}\right)^L\frac{\sin[\lambda_j-\theta+S'\eta]}{\sin[\lambda_j-\theta-S'\eta]}=\prod_{\substack{l=1,\\l\neq j}}^{M}\frac{\sin[\lambda_j-\lambda_l+\eta]}{\sin[\lambda_j-\lambda_l-\eta]}\tag{7.52}$$

and the energy is

$$\begin{aligned}E=&-\frac{H(L+2S')}{2}+\frac{(L-1)J_z}{4}+\frac{J_{imp}^{an}J_z(\cosh\theta+2S')}{4J}\\&-\sum_{j=1}^{M}\left(J_z-H-\frac{J\sin[\lambda_j+(\eta/2)]}{2\sin[\lambda_j-(\eta/2)]}-\frac{J\sin[\lambda_j-(\eta/2)]}{2\sin[\lambda_j+(\eta/2)]}\right)\ ,\end{aligned}\tag{7.53}$$

where $J_{imp}^{an}=J\sin^2\eta/(\sinh^2\theta+\sin^2\eta)$.

For the most interesting case of $S'=\frac{1}{2}$ the Hamiltonian of the Heisenberg–Ising ring with an integrable impurity, *e.g.*, for the "easy-plane" magnetic anisotropy has the form $\mathcal{H}=\sum_j\mathcal{H}_{j,j+1}+\mathcal{H}_{imp}$, where the impurity site is supposed to be situated between mth and $(m+1)$th sites of the host,

$$\mathcal{H}_{j,j+1} = J\vec{S}_j\vec{S}_{j+1} + (J_z - J)S_j^z S_{j+1}^z \ , \tag{7.54}$$

and

$$\begin{aligned}\mathcal{H}_{imp} = J_{imp}^{an}(\hat{B}_1(\mathcal{H}_{m,imp} + \mathcal{H}_{imp,m+1}) - \mathcal{H}_{m,m+1} \\ - 2iB_2[\mathcal{H}_{m,imp}, \mathcal{H}_{imp,m+1}]) \ ,\end{aligned} \tag{7.55}$$

where the operator $\hat{B}_1$ modifies the Heisenberg-like interaction by multiplying transverse terms (with x and y components) with $\cosh\theta$, and $B_2 = \tanh\theta/\sin\eta$. The isotropic Heisenberg limit is obtained by the rescaling $\theta, \eta \to 0$ with $\theta/\eta \to \theta$ being fixed.

The reader can see that an impurity acts threefold. First, it is coupled to two neighboring sites of the host chain. Second, it renormalizes the coupling between neighboring sites of the host. And, finally, it introduces three-site terms. All these terms are determined by nonzero θ and the value of the spin of an impurity, S', being nonequal to the value of host spins. There is no other free parameter of an impurity, using one, one can remove the second and third parts of the impurity Hamiltonian. Three-site terms violate time-reversal (T) and parity (P) symmetries separately, but TP, naturally, holds, so that the CPT-theorem works. These three-spin terms introduce the topological *spin current* (*spin chirality*) around the elementary triangular cell (formed by the impurity spin and spins of two neighboring sites of the host). One can check that these three-site terms are only important in a quantum mechanical description. If one replaces quantum spins by classical vectors, then a three-spin term is a total time derivative, and, hence, does not change classical equations of motion. One can speak about three-spin terms as about local Noether spin topological currents, induced by an impurity (they are also similar to the Pontriagin indices, or winding numbers). Naturally, the Hamiltonian of an integrable impurity is different from what was expected, Eq. (7.44) for $S' = \frac{1}{2}$. The question appears, whether one can avoid the action of these three-spin terms and the renormalization of an interaction between host sites. The answer comes if one considers a chain with open boundary conditions. If an impurity is situated in the bulk of an open chain, the only difference is the presence in the Hamiltonian terms with boundary fields and the absence of interactions between the first and the last sites of the chain. However, suppose an impurity is situated at the edge (*e.g.*, first) site of an open chain.

Then the impurity Hamiltonian with an impurity situated to the left from the first host site has the form

$$\mathcal{H}_{impop} = J_{imp}\Bigg(\vec{S}_1\vec{S}'_{imp} + h_1\Big[(S')^z_{imp} + 2\theta(S^x_1(S')^y_{imp} - S^y_1(S')^x_{imp}) + \{(S')^z_{imp}, \vec{S}'_{imp}\vec{S}_1\} + \left(\theta^2 - S'(S'+1) + \frac{1}{4}\right)S^z_1\Big]\Bigg) , \quad (7.56)$$

where h_1 is the boundary local magnetic field acting on the impurity site. If we put $h_1 = 0$, we obtain the deserved Hamiltonian $\mathcal{H}_{impop} = J_{imp}\vec{S}_1\vec{S}'_{imp}$, in which the action of an impurity is only in the renormalization of the coupling of an impurity to the neighboring host site, *i.e.*, what we wanted. Then, by using the methods of the previous chapter, we can write the Bethe ansatz equations and the expression for the energy of the open Heisenberg chain with an impurity

$$e_1^{2L}(\lambda_j)e_{2S'}(\lambda_j - \theta))e_{2S'}(\lambda_j + \theta)e_{2S_1}(\lambda_j)e_{2S_L}(\lambda_j) = \prod_{\pm}\prod_{\substack{l=1,\\ l\neq j}}^{M} e_2(\lambda_j \pm \lambda_l) \quad (7.57)$$

and

$$E = -\frac{H(L - 2M + 2S') - h_1 - h_2}{2} + \frac{J_{imp}S'}{2} + \frac{(L-1)J}{4} - 2J\sum_{j=1}^{M}(4\lambda_j^2 + 1)^{-1} , \quad (7.58)$$

where

$$2S_{1,L} = \frac{J}{h_{1,L}} - 1 , \quad (7.59)$$

for an impurity in the bulk, and $h_1 \to 4h_1[4\theta^2 + (2S'+1)^2]$ for an impurity situated at the left edge of an open chain. Notice that it is the only change related to the position of the impurity. Actually for $h_1 = h_L = 0$ Bethe ansatz equations for an open Heisenberg chain with an impurity do not depend on the position of an impurity. These equations for the interesting for us case $h_1 = h_L = 0$ are similar to Bethe ansatz equations of a periodic Heisenberg chain with an integrable impurity, but with several changes. First, there is a renormalization $L \to 2L$. The reader already knows what does this change produce from the previous chapter. As for the impurity effect: for an open chain we have two multipliers, related to the impurity, with $\pm\theta$, instead of one term in Eq. (7.50). Physics of this change is clear: it originated from the fact that a reflection from a boundary wave also scatters off an impurity, but that reflected wave has the opposite sign of its wave vector, and, thus, of the rapidity, while the sign of θ remains the same. In what follows we shall analyze the behaviour of an integrable impurity in an open Heisenberg chain, and shall only point out differences which appear in the periodic case. Also, we mostly limit ourselves to real θ.

In the limit of large L (results will be given for the generic case L odd) in the framework of the string hypothesis thermodynamic Bethe ansatz equations for an open chain with an impurity

$$\rho_{mh}(\lambda) + \frac{1}{2}\sum_{n=1}^{\infty}[A_{m,n}(\lambda-\lambda') + A_{m,n}(\lambda+\lambda')] * [\rho_n(\lambda') - p(\lambda')\delta_{m,1}]$$
$$= \frac{1}{2L}\sum_{n=1}^{\infty}[A_{m,n}(\lambda-\lambda') + A_{m,n}(\lambda+\lambda')] * p(\lambda'-\theta)(\delta_{m,2S'}) \ , \quad (7.60)$$

where $p(\lambda) = 1/4\cosh(\pi\lambda/2)$, $*$ denotes the convolution,

$$A_{m,n}(x) = a_{|m-n|}(x) + 2\sum_{l=1}^{\min(n,m)-1} a_{m+n-2l}(x) + a_{m+n}(x) \ , \quad (7.61)$$

and $a_m(x) = 2m/[\pi(4x^2+m^2)]$. Then the internal energy E and the total magnetic moment M^z are given as

$$E = E_0 - J\sum_{m=1}^{\infty}\int_0^{\infty} d\lambda\theta'_{m,1}(\lambda)\rho_m(\lambda) \ ,$$
$$M^z = \frac{L}{2} + S' - L\sum_{m=1}^{\infty} m\int_0^{\infty} d\lambda\rho_m(\lambda) \ . \quad (7.62)$$

The set of thermodynamic equations for dressed energies $\varepsilon_n(\lambda) = T\ln[\rho_{nh}(\lambda)/\rho_n(\lambda)] = \eta_n(\lambda)$ is

$$Hm - J\theta'_{m,1}(\lambda) = T\ln[1+\eta_m(\lambda)]$$
$$-\frac{T}{2}\sum_n[A_{n,m}(\lambda-\lambda') + A_{n,m}(\lambda+\lambda')] * \ln[1+\eta_n^{-1}(\lambda')] \ , \quad (7.63)$$

which completes the set Eq. (7.60). We see that equations for dressed energies do not depend on the parameter of an impurity explicitly.

Thermodynamic Bethe ansatz equations for densities are linear integral equations. There are two kinds of driving terms: the ones of order of 1, and the ones of order of L^{-1}. This is why, we can divide the densities as $\rho_n(\lambda) = \rho_n^{(0)}(\lambda) + L^{-1}\rho_n^{(1)}(\lambda)$ (and the same for densities of holes). Then one can separate Bethe ansatz equations for densities into two sets: one of the scale 1 for the main (of order of L) contribution to the energy, magnetization, *etc.*, *i.e.*, for $\rho_n^{(0)}(\lambda)$ only, and the other one of the scale L^{-1} for the finite contribution (of order of 1) to the energy, magnetic moment, *etc.*, *i.e.*, for $\rho_n^{(1)}(\lambda)$ only. The former describes thermodynamics of the bulk, while the latter reveals the contribution from edges of open chain and from an impurity.

The most interesting behaviour of the one-dimensional quantum system is in the ground state and at low temperatures. For the spin-$\frac{1}{2}$ Heisenberg chain only spinons have a Dirac sea. The latter is defined as the solution of the equation

$$\varepsilon_1(\lambda) + \frac{1}{2}[A_{1,1}(\lambda-\lambda') + A_{1,1}(\lambda+\lambda')] * \varepsilon_1^-(\lambda') = H - J\theta'_{1,1}(\lambda) \,. \quad (7.64)$$

The Fermi point (related to the limit of integration) is determined from the condition $\varepsilon_1(B) = 0$. The equations for densities in the ground state are

$$\begin{aligned}\rho_1(\lambda) + \rho_{1h}(\lambda) &= a_1(\lambda) + \frac{1}{2L}\left(a_{2S'}(\lambda+\theta) + a_{2S'}(\lambda-\theta)\right) \\ &\quad - \frac{1}{2}\int_0^B d\lambda'[a_2(\lambda-\lambda') + a_2(\lambda+\lambda')]\rho_1(\lambda') \,. \end{aligned} \quad (7.65)$$

The ground state internal energy can be written as

$$E_{T=0} = E_0 + \int_0^B d\lambda[H - J\theta'_{1,1}(\lambda)]\rho_1(\lambda) \quad (7.66)$$

and the ground state magnetization is equal to

$$M^z = \frac{L}{2} + S - L\int_0^B d\lambda \rho_1(\lambda) \,. \quad (7.67)$$

Two additional terms in Eq. (7.65) comparing to the homogeneous case describe the behaviour of an impurity.

For the antiferromagnetic case large values of the external magnetic field $|H| > H_s = 2J$ the system is in the ferromagnetic state and $B = 0$. In these regions of values of H the ground state energy is equal to E_0, the magnetic moments of all spins have their nominal values, $\pm\frac{1}{2}$ for the host and $\pm S'$ for an impurity, and the magnetic susceptibility is zero. For zero magnetic field in the antiferromagnetic situation at $H = 0$ we have $B = \infty$.

Since $\epsilon_1(\lambda)$ and $\rho_1(\lambda)$ are even functions, we can re-write the equation for dressed energies as

$$\varepsilon_1(\lambda) + A_{1,1}(\lambda - \lambda') * \varepsilon_1^-(\lambda') = H - J\theta'_{1,1}(\lambda) . \tag{7.68}$$

This equation, naturally, coincides with the one for a periodic Heisenberg chain with an impurity. The main contribution to equations of densities, which describes the behaviour of the bulk, can be written as

$$\rho_1^{(0)}(\lambda) + \rho_{1h}^{(0)}(\lambda) = a_1(\lambda) - \int_B^B d\lambda' a_2(\lambda - \lambda')\rho_1(\lambda') . \tag{7.69}$$

It is easy to check that the answers for the main contribution for an open chain coincide with those for a periodic chain, as expected.

Let us then concentrate on finite size corrections (considering the case L odd), for which we have the equation for dressed energies:

$$\begin{aligned}\rho_1^{(1)}(\lambda) + \rho_{1h}^{(1)}(\lambda) &= \frac{1}{2}\left[a_2(\lambda) + a_1(\lambda) + a_{2S'}(\lambda + \theta) + a_{2S'}(\lambda - \theta)\right] \\ &\quad - \int_B^B d\lambda' a_2(\lambda - \lambda')\rho_1^{(1)}(\lambda') ,\end{aligned} \tag{7.70}$$

where we introduced the term $(1/2L)[a_1(\lambda) + a_2(\lambda)]$ to avoid the double counting due to the symmetrization of functions (with $\lambda = 0$) and to take into account the term with $\lambda_\alpha = \lambda_\beta$ in the right hand side of Eq. (7.57). The limits of integration are determined by the host. For periodic boundary conditions we have to change the driving term as

$$\frac{1}{2}\left[a_2(\lambda) + a_1(\lambda) + a_{2S'}(\lambda + \theta) + a_{2S'}(\lambda - \theta)\right] \to a_{2S'}(\lambda - \theta) . \tag{7.71}$$

Combining all contributions we obtain for the vanishing homogeneous magnetic field $H = 0$, where $B = \infty$:

$$\begin{aligned}E = E_0 - \frac{2L+1}{2}J\ln 2 + \frac{\pi J}{4} - \frac{J}{4}\Bigg[&\psi(3/4) - \psi(1/4) \\ &+ \sum_{\pm}(\psi[(2S'+3)/4 \pm i\theta] - \psi[(2S'+1)/4 \pm i\theta])\Bigg] ,\end{aligned} \tag{7.72}$$

where $\psi(x)$ are digamma functions. For small values of the homogeneous magnetic field H we can apply the Wiener–Hopf technique. For a periodic system we have to replace the terms of order of 1 by

$$-\frac{J}{2}(\psi[(2S'+3)/4 - i\theta] - \psi[(2S'+1)/4 - i\theta]) . \tag{7.73}$$

The magnetic moment of free edges themselves is given in the previous chapter. The magnetic moment of an impurity $S' \neq \frac{1}{2}$ in an open chain is

$$m_{imp}^z = \mu_i \Bigg[1 + \sum_{\pm} \Bigg(\pm \frac{1}{2|\ln \sqrt{e}H/\sqrt{\pi^3}JA_\pm|} - \frac{\ln \frac{1}{2}|\ln \sqrt{e}H/\sqrt{\pi^3}JA_\pm|}{4(\ln \sqrt{e}H/\sqrt{\pi^3}JA_\pm)^2} + \cdots \Bigg) \Bigg] . \tag{7.74}$$

Here $A_\pm = \exp(\pm\pi\theta)$. There is a resonance at $|\ln \sqrt{e}H/\sqrt{\pi^3}J| = \pi|\theta|$, and, hence, $T_K = \sqrt{\pi^3/e}J\exp(-\pi|\theta|)$ can be considered as the usual Kondo temperature for a magnetic impurity. This is why, the reader can see that the parameter θ in fact determines the resonance shift of the Abrikosov–Suhl (Kondo) resonance of a magnetic impurity in a quantum spin chain. As we saw, it is related to the coupling between an impurity and the host. For a periodic chain with an impurity we obtain

$$m_{imp}^z = \mu_i \left[1 \pm \frac{1}{2|\ln H/T_K|} - \frac{\ln \frac{1}{2}|\ln H/T_K|}{4(\ln H/T_K)^2} + \cdots \right] . \tag{7.75}$$

In the limit of small H the difference in answers for the behaviour of the impurity magnetization in open and periodic chains is negligible; it can be essential for high enough values of the field. For $H \ll T_K$ we take $\mu_i = S' - \frac{1}{2}$, and the upper sign, *i.e.*, the impurity spin S' is underscreened by the low-lying excitations of the chain to the value $S' - \frac{1}{2}$, and the latter behaves asymptotically free. On the other hand, for $H \gg T_K$ we choose $\mu_i = S'$ and the lower sign, which means that at higher values of the magnetic field the non-screened spin of an impurity S' behaves asymptotically free. For $\theta = 0$, which pertains to the impurity spin coupled to the host with the maximal strength, the Kondo screening is maximum (T_K is maximum). On the other hand, $T_K \to 0$ for $\theta \to \infty$, and the spin of an impurity is not screened. This case is also obvious physically, because it corresponds to the impurity spin totally decoupled from the host. If we have $S' = \frac{1}{2}$ in the region $H \ll T_K$ we have to use $\mu = 2H/\pi^2 J \cosh \pi\theta$ for an open chain and $\mu = 4H\exp(\pi|\theta|)/\pi^2 J$ for a periodic chain. This implies that the ground state of an impurity is singlet, *i.e.*, it is totally screened by low-lying excitations of the host with the magnetic susceptibility of an impurity

$\chi_{imp} = 4/\sqrt{\pi e}T_K$, *i.e.*, inverse proportional to the Kondo temperature, as expected. It is renormalized by a factor of T_K^{-1} with respect to the host susceptibility.

As it is clear from Eqs. (7.48) and (7.56), an imaginary θ implies the non-Hermitian Hamiltonian for an impurity situated in the bulk. The only possibility to consider imaginary θ with the Hermitian Hamiltonian is to study an impurity at the edge of an open spin chain with the zero boundary field. One can see that imaginary θ with $\theta > S + \frac{1}{2}$ pertain to the ferromagnetic coupling of the impurity spin to the host (other cases correspond to an antiferromagnetic impurity-host interaction). For imaginary θ we can divide it into its integer and fractional part, $2|\theta| = [2|\theta|] + \{2|\theta|\}$. In this case the fractional part can define the "Kondo temperature" of an impurity $T_K \sim J[\cos(\pi\{2|\theta|\}/2)]^{-1}$. Hence, this "crossover scale" is larger than the characteristic energy of spinons, which, in turn, defines the critical field H_s of a quantum phase transition to the spin-saturated (ferromagnetic) phase. It follows that the only special point in the behaviour of such an impurity is H_s. Another feature of an imaginary θ is that incident and reflected waves effectively scatter off different effective "impurity spins", $S \pm \frac{[2|\theta|]}{2}$. Negative effective spins signal the onset of local levels (related to bound states caused by the impurity spin situated at the edge, however their appearance is not connected with the boundary field, as in Chapter 6, but only with the ferromagnetic coupling of the impurity spin to the host).

For low temperatures $T \ll T_K$ we can use the Sommerfeld expansion and calculate $H = 0$ contributions to the magnetic susceptibility and the specific heat of the $S' = \frac{1}{2}$ impurity. The entropy of such an impurity is zero at $T = 0$. The Sommerfeld coefficient of an impurity is equal to $\gamma_{imp} = 8\pi^{3/2}/3\sqrt{e}T_K$, which implies the Wilson ratio $\gamma_{imp}/\chi_{imp} = 2\pi^2/3$, *i.e.*, the universal Fermi liquid-like behaviour. On the other hand, at high temperatures $T \gg T_K$, we have a Curie-like behaviour of the magnetic susceptibility $\chi_{imp} \sim 1/12T$ and a Schottky-like behaviour of the specific heat. For $S' \neq \frac{1}{2}$ for both high and low temperatures we have a Curie-like behaviour of the magnetic susceptibility of an impurity, but with the coefficient, proportional to $[(S')^2 - \frac{1}{4}]/3$ at $T \ll T_K$ and $[S'(S'+1)]/3$ for $T \gg T_K$. The remnant entropy is equal to $\mathcal{S}_{imp} = \ln(2S' - 1)/2$ in this case, and we have a Schottky-like maximum in the behaviour of the specific heat of an impurity.

It is important to emphasize that the finite value of the magnetic susceptibility of an impurity in the case of an open chain is very small comparing to the divergent magnetic susceptibility of free edges, *cf.* the previous chapter.

At high temperatures $T \gg J$ with H/T finite the impurity spin behaves as a free spin S'.

We can generalize the approach considering the behaviour of a spin S' impurity in an anisotropic spin-S Takhtajan–Babujian chain. To construct the Hamiltonian we start with $R^{\mu_i \mu_{i+1}}_{\alpha_i \beta_i}(\lambda)$, the standard R-matrix of a spin S chain with the uniaxial "easy-plane" anisotropy. Indices α_i and β_i denote states of the spin at site i (acting in the Hilbert space V_i), and μ denotes states in the auxiliary space (Hilbert space V_0). The R-matrix has the form

$$R = P \sum_{j=0}^{2S} \prod_{l=0}^{j-1} \frac{\sinh \eta[i2(2S-l) - \lambda]}{\sinh \eta[i2(2S-l)]} \prod_{l=j}^{2S-1} \frac{\sinh \eta[i2(2S-l) + \lambda]}{\sinh \eta[i2(2S-l)]}$$
$$\times \prod_{\substack{p=0 \\ p \neq j}}^{2S} \frac{2 \sin^2 \eta \hat{X}_{0i} - \sin \eta p \sin \eta (p+1)}{\sin \eta (j-p) \sin \eta (j+p+1)} , \tag{7.76}$$

where λ is the spectral parameter, η is the parameter of the ("easy-plane") magnetic anisotropy, the operator P permutes the spaces V_i and V_0 and

$$\hat{X}_{0i} = e^{i\eta S_i^z} \left(\frac{1}{2}[S_i^+ S_0^- + S_i^- S_0^+] + \frac{\cos \eta S \cos \eta (S+1)}{\sin^2 \eta} \sin \eta\, S_i^z \sin \eta S_0^z \right.$$
$$\left. + \frac{\sin \eta S \sin \eta (S+1)}{\sin^2 \eta} \cos \eta S_i^z \cos \eta S_0^z \right) e^{-i\eta S_0^z} , \tag{7.77}$$

which in the limit of the SU(2)-symmetric system ($\eta \to 0$) simplifies to $\vec{S}_i \vec{S}_0 + S(S+1)$. R-matrices satisfy the Yang–Baxter relation. The transfer matrix $\hat{\tau}^{\beta}_{\alpha}(\lambda)$ has the form of the trace over the auxiliary space of the product of L-operators (constructed similar to the ones in Chapter 5) with the same values of spins S in sites of the host and the L-operator of a spin S' with its spectral parameter shifted by θ in the impurity site. L-operators satisfy intertwining relations, hence, transfer matrices with different spectral parameters commute and the problem is exactly integrable. The Hamiltonian of the uniaxial quantum spin S chain with an impurity with the spin S' is obtained as the derivative of the logarithm of the transfer matrix with respect to the spectral parameter (taken at $\lambda = 0$). It has the form $\mathcal{H} = \sum_j J \mathcal{H}_{j,j+1} + \mathcal{H}_{imp}$. In general, the form of the lattice Hamiltonian is very complicated; it depends on S, S', θ and the anisotropy η. For example, for the isotropic SU(2)-symmetric spin S host the structure of the Hamiltonian with an impurity is (without an impurity it corresponds to a

Takhtajan–Babujian chain, *cf.* Chapter 5)

$$\mathcal{H}_{imp} = J\Big(\mathcal{H}_{m,imp} + \mathcal{H}_{imp,m+1} + \{\mathcal{H}_{m,imp}, \mathcal{H}_{imp,m+1}\} - 2i\theta[\mathcal{H}_{m,imp}, \mathcal{H}_{imp,m+1}] + (\theta^2 - 2S'(S'+1))\mathcal{H}_{m,m+1}\Big) , \quad (7.78)$$

where

$$\mathcal{H}_{a,b} = \sum_{j=|S-S'|+1}^{S+S'} \sum_{k=|S-S'|+1}^{j} \frac{k}{k^2 + \delta_{a,b,imp}\theta^2} \prod_{l=|S-S'|}^{S+S'} \frac{x - x_l}{x_j - x_l} , \quad (7.79)$$

$x = \vec{S}_a\vec{S}_b$ $(a, b = m, m+1, imp)$, and $2x_j = j(j+1) - S(S+1) - S'(S'+1)$. Note that in this case the multiplier at the impurity term is $[\theta^2 + (S+S')^2]^{-1}$ and the coefficient in front of $\mathcal{H}_{m,m+1}$ becomes $-2S'(S'+1) - (S'+S)^2$. For an anisotropic case one has to replace x by $\hat{X}_{m,m+1}$, *cf.* Eq. (7.77) and x_j by appropriate coefficients from Eq. (7.76).

Bethe ansatz equations, the solutions of which determine eigenfunctions and eigenvalues of the Schrödinger equation with the Takhtajan–Babujian Hamiltonian with an integrable impurity in the case of periodic boundary conditions are

$$e_{2S}^L(\lambda_j)e_{2S'}(\lambda_j - \theta) = \prod_{\substack{l=1 \\ l\neq j}}^{M} e_2(\lambda_j - \lambda_l) , \qquad j = 1, \ldots, M , \quad (7.80)$$

where $e_n(x) = (2x + in)/(2x - in)$, the total magnetic moment is $M^z = LS + S' - M)$, and the energy is

$$E = E_0 - J\sum_{j=1}^{M}\left(\frac{S}{\lambda_j^2 + S^2} - H\right) . \quad (7.81)$$

Again, the Hamiltonian is simplified in the case of an impurity situated at the edge of an open chain with boundary fields equal to zero. The reader already saw above that the difference in the behaviour of a magnetic impurity itself in an open and periodic chain is small in the most interesting case of a weak magnetic field, and, therefore, we shall present results for the periodic case below. Results for the behaviour of a magnetic impurity with open boundary conditions can be straightforwardly obtained from the ones for a periodic situation. We emphasize again, that for open boundary conditions there exists a contribution of free edges themselves, which is of the same order of magnitude as the contribution from a single impurity, but its features can be stronger than the ones from an impurity.

We shall present our results for the generic case L odd. Let us first study the ground state behaviour of the considered system. In the absence of a magnetic field the ground state energy of an impurity is

$$[e_0(\theta)]_{imp} = -\frac{\pi J \sin(2\eta S)}{4\eta S} \times \int d\omega e^{i\frac{\pi\omega\theta}{\eta}} \frac{\sinh[\frac{\pi\omega}{\eta}\min(S,S')]\sinh[(\frac{\pi^2}{2\eta}-\pi\max(S,S'))\omega]}{\sinh(\pi\omega)\sinh(\frac{\pi\omega}{2})} . \quad (7.82)$$

Consider now the ground state behaviour of an impurity in a small magnetic field H. The ground state energy of an impurity is equal to (we shall consider small enough $\eta < \pi/2S$)

$$[e_0(\theta,H)]_{imp} = [e_0(\theta)]_{imp} - \int \frac{d\omega}{2\eta} e^{\frac{i\omega\pi\theta}{\eta}} \frac{y^+(\frac{\pi\omega}{\eta})\sinh(\omega\pi S')}{2\cosh(\frac{\pi\omega}{2})\sinh(\omega\pi S)} \quad (7.83)$$

for $S' \leq S$ and

$$[e_0(\theta,H)]_{imp} = [e_0(\theta)]_{imp} - \frac{\pi(S'-S)H}{\pi - 2S\eta} - \int \frac{d\omega}{2\eta} e^{\frac{i\omega\pi\theta}{\eta}} \frac{y^+(\frac{\pi\omega}{\eta})\sinh[(\omega\frac{\pi^2}{2\eta}-\pi S')]}{2\cosh(\frac{\pi\omega}{2})\sinh[(\omega\frac{\pi^2}{2\eta}-\pi S)]} \quad (7.84)$$

for $S' \geq S$. Here $y^+(\omega)$ is the positive part of the solution of the equation

$$y(u) + \int_0^\infty du' y(u')J(u-u') - HS + \frac{\pi J \sin(2\eta S)}{4\eta S \cosh[\frac{\pi(u+B)}{\eta}]} = -\int_0^\infty du' y(u')J(u+u'+2B) , \quad (7.85)$$

where the Fourier transform of $J(x)$ is

$$J(\omega) = \frac{\sinh(\frac{\eta\omega}{2})\sinh(\frac{\pi\omega}{2})}{2\cosh(\frac{\eta\omega}{2})\sinh(\eta\omega S)\sinh[\omega(\frac{\pi}{2}-\eta S)]} \quad (7.86)$$

and B is connected with the value of the external magnetic field. Equation (7.85) for small fields can be solved as the sequence of Wiener–

Hopf equations. It also gives the connection between H and B:

$$H = \frac{\pi^2 J \sin(2\eta S)}{2\eta S} e^{-\frac{(B+a)\pi}{\eta}} \frac{\Gamma(1+\frac{\pi}{2\eta})}{\Gamma(1+S)\Gamma(1-S+\frac{\pi}{2\eta})} + \dots , \tag{7.87}$$

where a is a constant.

For $S' = S$ we close the contour of integration in Eq. (7.84) through the upper half-plane (the main pole is of $\cosh(\pi\omega/2)$) and obtain

$$[e_0(\theta, H)]_{imp} = [e_0(\theta)]_{imp} - \frac{2S^2\eta(\pi - 2\eta S)H^2}{2\pi^3 \sin(2\eta S) T_K} + \frac{AH^{2+4\eta/(\pi-2\eta S)}}{T_K} + \dots , \tag{7.88}$$

for $H \ll T_K$, where A is a constant and for small η, $T_K = v^F \exp(-\pi|\theta|/\eta)$ ($v^F = \pi J \frac{\sin(2\eta S)}{4\eta S} \to \pi J/2$ is the Fermi velocity of low-lying excitations, *i.e.*, strings of the length $2S$). It plays the role of the Kondo temperature, similar to the crossover scale in the behaviour of a magnetic impurity in a metal. The spin of a magnetic impurity is totally compensated for $H \le T_K$. The magnetic susceptibility of an impurity is finite as $H \to 0$ and it is renormalized by a factor of T_K^{-1} with respect to the host susceptibility. For $H \gg T_K$ the impurity spin is not screened.

For $S' > S$ the main contribution to the integral arises from the poles at $\omega = i\pi/\eta$ (and then $\omega = 2\pi/(\pi - 2\eta S)$) which produces for $H \ll T_K$

$$[e_0(\theta, H)]_{imp} = [e_0(\theta)]_{imp} - \frac{(S'-S)\pi H}{(\pi - 2\eta S)} - CH\left(\frac{H}{T_K}\right)^{2\eta/(\pi-2\eta S)} + \dots , \tag{7.89}$$

where C is a constant. The reader can see that for $H \to 0$ the spin of an impurity is underscreened to the value $S' - S$ by host low-lying excitations. For $H \gg T_K$ the spin of an impurity S' is not screened and behaves with the known asymptotic freedom. It turns out that some authors connect the multiplier $(1 - 2\eta S/\pi)$ with the renormalization of the effective g-factor of spins, while other works relate such a change to the non Fermi liquid critical behaviour caused by the magnetic anisotropy.

Finally, for $S' < S$ and $H \ll T_K$ we get

$$[e_0(\theta, H)]_{imp} = [e_0(\theta)]_{imp} - C'H\left(\frac{H}{T_K}\right)^{1/S} + \dots , \tag{7.90}$$

for $S > 1$, where C' is a constant, and for $S = 1$, $S' = \frac{1}{2}$ we have

$$[e_0(\theta, H)]_{imp} = [e_0(\theta)]_{imp} - \frac{2\eta(\pi - 2\eta)H^2}{4\pi^4 \sin(2\eta)T_K} \ln(T_K/H) + \dots \,. \qquad (7.91)$$

Hence, for $S' < S$ the spin of an impurity is overscreened, which produces the critical, non Fermi liquid behaviour.

For low T the temperature behaviour of the magnetic susceptibility and specific heat of an impurity also strongly depends on relative values of host spins S and the impurity spin S'. For $S > S'$ the impurity is underscreened by low-lying excitations of the chain. The magnetic susceptibility χ_{imp} of such an impurity is divergent at $H = 0$ for $T \to 0$. The specific heat c_{imp} exhibits a Schottky anomaly, related to the undercompensated spin of an impurity. The entropy of an impurity at $T = H = 0$ becomes nonzero, $\mathcal{S}_{imp} = \ln[1 + 2\pi(S' - S)/(\pi - 2\eta S)]$. A finite magnetic field lifts the degeneracy and the remnant entropy becomes zero. On the other hand, for $S' < S$ the spins of low-lying excitations of the antiferromagnetic critical chain overscreen the spin of an impurity. This yields the critical behaviour, which reveals itself in divergences of the $T \to 0$ magnetic susceptibility of an impurity and of the low-T Sommerfeld coefficient of the specific heat for $H = 0$. In this case one has a remnant $T = H = 0$ entropy of an impurity $\mathcal{S}_{imp} = \ln(\sin[\pi(2S' + 1)/(2S' + 2)]/\sin[\pi/(2S + 2)])$, which is removed by a finite magnetic field that lifts the spin degeneracy of the system. It is not difficult to show that at low T one has $c_{imp} \propto \chi_{imp} \sim (T/T_K)^{2/(S+1)}$ for $S > 1$, and $\gamma_{imp} \propto \chi_{imp} \sim T_K^{-1} \ln(T_K/T)$ at zero magnetic field. For the case $S' = S$ we obtain the low temperature behaviour of the free energy of an impurity (for $H = 0$)

$$f(\theta)_{imp} = [e_0(\theta)]_{imp} - \frac{\pi S T^2}{2(S+1)T_K} \left[1 + \frac{3S^3}{[\ln(\alpha T_K/T)]^3}\right] + \dots \,, \qquad (7.92)$$

where α is a constant. In the presence of a weak magnetic field $H \ll T$ we can calculate the temperature corrections to the free energy of an impurity

$$f(\theta)_{imp} = e_0(\theta, H)_{imp} - \frac{\pi S T^2}{2(S+1)T_K} - \frac{S H^2}{2\pi T_K} \times \left[1 + \frac{S}{\ln(\alpha T_K/T)} + \frac{S^2 \ln|\ln(\alpha T_K/T)|}{\ln^2(\alpha T_K/T)}\right] + O(T^2) \,, \qquad (7.93)$$

which is the famous Kondo behaviour of the asymptotically free spin (characteristic for a Kondo impurity in a free electron host).

At high temperatures $T \gg J$ with H/T finite the impurity spin behaves as a free spin S'.

We see that the behaviour of an impurity in an antiferromagnetic quantum spin chain is similar to the behaviour of a magnetic impurity in a metal (the Kondo impurity). This similarity is not occasional. Let us consider the behaviour of a Kondo impurity in a metal. The Hamiltonian, which describes the Kondo problem, is

$$\mathcal{H}_K = \sum_{\mathbf{k},\sigma} \epsilon_k \psi^\dagger_{\mathbf{k},\sigma}\psi_{\mathbf{k},\sigma} + (I/2) \sum_{\mathbf{k},\mathbf{k}',\sigma,\sigma'} \vec{S}'\psi^\dagger_{\mathbf{k},\sigma}\vec{\sigma}_{\sigma\sigma'}\psi_{\mathbf{k}'\sigma'} \ , \tag{7.94}$$

where $\psi^\dagger_{\mathbf{k},\sigma}$ creates an electron with spin σ and quasimomentum $\mathbf{k}$, $\vec{S}$ is the operator of the impurity spin ($\langle(\vec{S}')^2\rangle = S'(S'+1)$) situated at x_0, ϵ_k is the energy of the free electron gas and I is the local exchange constant between an impurity and the free electron host. It is easy to prove that the problem is effectively one-dimensional. One expands the electron wave in spherical harmonics about an impurity

$$\psi^\dagger_{\mathbf{k},\sigma} = \sum_l \sum_m Y_{lm}(\mathbf{k}/k)\psi^\dagger_{k,l,m,\sigma} \ . \tag{7.95}$$

Then the Hamiltonian obtains the form

$$\begin{aligned}\mathcal{H}_K = &\sum_{k,l,m,\sigma} \epsilon_k \psi^\dagger_{k,l,m,\sigma}\psi_{k,l,m,\sigma} \\ &+ (I/2) \sum_{k,k',\sigma,\sigma'} \vec{S}'\psi^\dagger_{k,0,0,\sigma}\vec{\sigma}_{\sigma\sigma'}\psi_{k',0,0,\sigma'} \ ,\end{aligned} \tag{7.96}$$

where only s-waves interact with the impurity. Fourier transforming the Hamiltonian yields the effectively one-dimensional Hamiltonian. Then usually the relativistic dispersion law for electrons (linearized about Fermi points) is considered (we put the Fermi velocity equal to 1 below).

In the framework of the Bethe ansatz the behaviour of the Kondo model is described by the solution of Bethe ansatz equations. They determine the sets of quantum numbers, charge ($\{k_j\}_{j=1}^N$, N is the number of electrons) and spin ($\{\lambda_\alpha\}_{\alpha=1}^M$, M being the number of down spins) rapidities, which parametrize eigenvalues and eigenfunctions of the Schrödinger equation of the Kondo Hamiltonian. There are two types of scattering processes in the problem. Let us look for the two-particle scattering matrix between a magnetic impurity and an electron in the form:

$$S^i = \hat{S}^{\sigma,\sigma'}_{s,s'}(\alpha) = \frac{1}{2}[a'(\alpha)+c'(\alpha)]\delta_{\sigma,\sigma'}\delta_{s,s'} + [a'(\alpha)-c'(\alpha)]\vec{S}'_{s,s'}\vec{\sigma}_{\sigma,\sigma'} \ , \tag{7.97}$$

where α is the spectral parameter, s and s' denote z-projections of the impurity's spin before and after scattering, (while σ and σ' denote z-projections of the spin of an electron, respectively), $c'(\alpha) = \alpha a'(\alpha)/(\alpha + ig)$, and g is related to the impurity-electron exchange constant I. We emphasize that a dynamical magnetic Kondo impurity produces only elastic scattering, without any reflection. For the exact integrability of the problem two-particle scattering matrices have to satisfy the Yang–Baxter relations:

$$S_{i,j}(\alpha)S^i_{i,x_0}(\alpha+\alpha')S^i_{j,x_0}(\alpha') = S^i_{j,x_0}(\alpha')S^i_{i,x_0}(\alpha+\alpha')S_{i,j}(\alpha) \tag{7.98}$$

and

$$S_{i,j}(\alpha)S_{i,k}(\alpha+\alpha')S_{j,k}(\alpha') = S_{j,k}(\alpha')S_{i,k}(\alpha+\alpha')S_{i,j}(\alpha) \ , \tag{7.99}$$

where i, j, k enumerate positions of electrons and x_0 denotes the position of a magnetic impurity. Then two-particle scattering matrices between electrons dynamically yield the form, similar to Eq. (7.97) due to the Kondo interaction with the magnetic impurity:

$$\begin{aligned} S_{1,2} = S^{\sigma_2,\sigma_2'}_{\sigma_1,\sigma_1'}(\alpha) &= \frac{1}{2}[a(\alpha)+c(\alpha)]\delta_{\sigma_1,\sigma_1'}\delta_{\sigma_2,\sigma_2'} \\ &+ \frac{1}{2}[a(\alpha)-c(\alpha)]\vec{\sigma}_{\sigma_1,\sigma_1'}\vec{\sigma}_{\sigma_2,\sigma_2'} \ , \end{aligned} \tag{7.100}$$

where $c(\alpha) = \alpha a(\alpha)/(\alpha + ig)$ and $a(0) = 1$. Those electron-electron scatterings also do not produce any reflection. It is easy to check that Yang–Baxter relations are satisfied provided $h(\alpha) = h'(\alpha)$ and $h(\alpha) + h(\alpha') = h(\alpha+\alpha')$, where $h(\alpha) = c(\alpha)/[a(\alpha)-c(\alpha)]$. The Hamiltonian $\mathcal{H}_K$ yields the electron-impurity scattering matrix of the form Eq. (7.97) with $a'(\alpha_0) = (\alpha_0 + ig)[e^{iIS'/2} + e^{-i(S'+1)I/2}]/2\alpha_0$ and $(2S'+1)g = \alpha_0 \tan[(2S'+1)I/4]$. [Notice, that as the reader knows from Chapter 5, Yang–Baxter relations can be satisfied up to an arbitrary factor in $a(\alpha)$, $b(\alpha)$, $a'(\alpha)$ and $b'(\alpha)$, *i.e.*, not in the unique way.] Then, proceeding as it was described in Chapter 4, we obtain Bethe ansatz equations for the Kondo problem with periodic boundary conditions in a box of length L

$$\begin{aligned} e_{2S'}(\lambda_\alpha + \alpha_0/g)e_1^N(\lambda_\alpha) &= \prod_{\beta=1,\beta\neq\alpha}^{M} e_2(\lambda_\alpha - \lambda_\beta) \ , \\ \exp(ik_jL) = \exp(iIS'/2)\prod_{\alpha=1}^{M} e_1(\lambda_\alpha) \ , & \ E = \sum_{j=1}^{N} k_j \ , \end{aligned} \tag{7.101}$$

where $\alpha = 1, \ldots, M$, $j = 1, \ldots, N$. Taking into account the second set of Eqs. (7.101) we can re-write the expression for the energy as

$$E = \frac{2\pi}{L}\sum_{j=1}^{N} I_j + \frac{N(IS' - 2\pi M)}{2L} - \frac{2N}{L}\sum_{\alpha=1}^{M} \tan^{-1} 2\lambda_\alpha \ , \tag{7.102}$$

where I_j appear because the logarithm is the multi-valued function. The total magnetic moment is $M^Z = (L/2) + S' - M$. We see that actually the Bethe ansatz equations for an impurity in a quantum spin-$\frac{1}{2}$ chain coincide with the ones for a Kondo impurity in a metal, up to re-definitions of eigenvalues and θ. It is also easy to show that the Bethe ansatz equations for the n-channel Kondo impurity in a metal coincide with the ones for a spin-S' impurity in a $S = n/2$ SU(2)-symmetric quantum antiferromagnetic chain, up to similar re-definitions. Hence, our analogies become transparent. The Kondo temperature for the Kondo case is defined as $T_K = (2N/L)\exp(-\pi\alpha_0/g)$, *i.e.*, it is also related to the shift in Bethe ansatz equations, as for a spin chain.

7.3 Impurity in Correlated Electron Chains

So far in this chapter we considered the behaviour of magnetic impurities in insulating systems, in which only spin degrees of freedom possessed dynamics. It is interesting to investigate how impurities behave in correlated electron chains using exact methods. From the previous chapter the reader already knows one possible model of an impurity: it pertains to a local field or potential. For the Bethe ansatz integrable models this kind of impurity can be considered only when it is situated at the edge of an open chain. Now we shall study the behaviour of an integrable impurity of another kind: the one, studied in the previous section, which is introduced into the Bethe ansatz scheme *via* a special L-operator, which, though, satisfies Yang–Baxter (intertwining) relations with R-matrices of the host model.

Let us start our consideration with the supersymmetric antiferromagnetic t-J model for the most popular case $V = -J/2$ and $J = 2$. We shall first work in the framework of the graded scheme of the algebraic Bethe ansatz, introduced in Chapter 5. We begin with the gl(1|2) invariant R-matrix, which satisfies the Yang–Baxter equation. We already proved in Chapter 5 that one can construct monodromy operators, which traces, transfer matrices, with different spectral parameters mutually commute. This constitutes the exact integrability. The only condition we demanded from a monodromy matrix was the following. The action of diagonal matrix elements of the monodromy $T_{\alpha\alpha}$ on the mathematical vacuum in the auxiliary 3×3 space has to produce c-numbers, $T_{\alpha\alpha}(\lambda)|0\rangle = a_\alpha(\lambda)|0\rangle$, *i.e.*, the mathematical vacuum is the eigenstate for these diagonal components, and the action of all upper elements $T_{\alpha\beta}$ with $\alpha < \beta$ is zero ($T_{\alpha\beta}(\lambda)|0\rangle = 0$ for $\alpha < \beta$). Then the eigenvalue of the transfer matrix for the eigenstate

$$|\lambda_1^0, \dots, \lambda_N^0|F\rangle = F_{a_1\cdots a_N} \prod_{j=1}^{N} C_{a_j}(\lambda_j^0)|0\rangle \ , \tag{7.103}$$

where C plays the role of "creation operators" is

$$\Lambda(\lambda) = a_3(\lambda)\prod_{j=1}^{N} c^{-1}(\lambda_j^0 - \lambda) - a_2(\lambda)\prod_{j=1}^{N} c^{-1}(\lambda_j^0 - \lambda)\prod_{\gamma=1}^{M} c^{-1}(\lambda - \lambda_\gamma)$$
$$- a_1(\lambda)\prod_{\gamma=1}^{M} c^{-1}(\lambda_\gamma - \lambda) \ . \tag{7.104}$$

Taking the logarithmic derivative of the eigenvalue $\Lambda(\lambda)$ at $\lambda = 0$ we get

$$E = \sum_{j=1}^{N}\left[A\frac{ic}{(\lambda_j^0 + ic)\lambda_j^0} - 2\right] + Aa_3^{-1}(0)\frac{da_3(\lambda)}{d\lambda}\bigg|_{\lambda=0} \ , \tag{7.105}$$

where A is a constant. The Bethe ansatz equations were the conditions on rapidities $\{\lambda_j^0\}_{j=1}^N$ and $\{\lambda_\gamma\}_{\gamma=1}^M$:

$$\begin{aligned} \frac{a_2(\lambda_\gamma)}{a_1(\lambda_\gamma)}\prod_{j=1}^{N} c^{-1}(\lambda_j^0 - \lambda_\gamma) &= \prod_{\substack{\beta=1\\ \beta\neq\gamma}}^{M} \frac{c(\lambda_\gamma - \lambda_\beta)}{c(\lambda_\beta - \lambda_\gamma)} \ , \qquad \gamma = 1, \dots, M \ , \\ \frac{a_3(\lambda_j^0)}{a_2(\lambda_j^0)} &= \prod_{\gamma=1}^{M} c^{-1}(\lambda_j^0 - \lambda_\gamma) \ , \qquad j = 1, \dots, N \ . \end{aligned} \tag{7.106}$$

Let us now study the representation of diagonal matrix elements of the monodromy matrix for a supersymmetric t-J model with an impurity. Consider the unity operator I_j, the operator of the number of electrons per site n_j, and three operators of projections of the total spin of the system, $S_j^{\pm,z}$, respectively. They form U(1) and SU(2) subalgebras ($[S_j^z, S_j^\pm] = \pm S_j^\pm$, $[S_j^+, S_j^-] = 2S_j^z$) of gl(2|1). Fermion operators $(Q_{1,2}^\pm)_j$ satisfy anticommutation relations

$$\{(Q_1^\pm)_j, (Q_2^\pm)_j\} = \pm\frac{S_j^\pm}{2} \ , \qquad \{(Q_1^\pm)_j, (Q_2^\mp)_j\} = \pm\frac{-S_j^z \pm n_j}{2} \ . \tag{7.107}$$

with other mutual anticommutators being zero. They satisfy commutation relations with the bosonic generators

$$[S_j^z, (Q_l^{\pm})_j] = \pm \frac{(Q_l^{\pm})_j}{2} , \qquad [n_j, (Q_l^{\pm})_j] = (-1)^{l+1} \frac{(Q_l^{\pm})_j}{2} ,$$
$$[S_j^{\mp}, (Q_l^{\pm})_j] = (Q_l^{\mp})_j , \qquad [S_j^{\pm}, (Q_l^{\pm})_j] = 0 , \tag{7.108}$$

with $l = 1, 2$. To remind, we denote $[.,.]$ ($\{.,.\}$) a commutator (anticommutator). In the basis, where n_j, $\mathbf{S}_j^2$ and S_j^z are diagonal, non-vanishing matrix elements of $(Q_{1,2}^{\pm})_j$ are

$$\left\langle S + \frac{1}{2}, S - \frac{1}{2}, \sigma \pm \frac{1}{2} |(Q_1^{\pm})_j| S, S, \sigma \right\rangle = \pm \sqrt{\frac{S \mp \sigma}{2}} ,$$
$$\left\langle S, S, \sigma |(Q_2^{\pm})_j| S + \frac{1}{2}, S - \frac{1}{2}, \sigma \mp \frac{1}{2} \right\rangle = \sqrt{\frac{S \pm \sigma}{2}} . \tag{7.109}$$

Actually these operators are sums of local operators of the same structure at each site of the system. For $S = \frac{1}{2}$ one can express these operators in terms of standard electron creation and annihilation operators as $n_j = n_{j,\uparrow} + n_{j,\downarrow}$, $2S_j^z = n_{j,\uparrow} - n_{j,\downarrow}$, $S_j^{\pm} = c_{j,\uparrow,\downarrow}^{\dagger} c_{j,\downarrow,\uparrow}$, $(Q_1^+)_j = (1 - n_{j,\downarrow}) c_{j,\uparrow}^{\dagger}$, $(Q_2^+)_j = (1 - n_{j,\uparrow}) c_{j,\downarrow}$, and $(Q_{1,2}^-)_j = (Q_{1,2}^+)_j^+$. The multipliers $(1 - n_{j,\sigma})$ of fermionic operators Q exclude double occupations of each site, as it must be for a t-J model. Notice that in Chapter 5 we used definitions, in which, for $S = \frac{1}{2}$, $(Q_1^+)_j = Q_{j\uparrow}$, $(Q_2^+)_j = Q_{j\downarrow}$, $(Q_1^-)_j = Q_{j\uparrow}^{\dagger}$, $(Q_2^-)_j = Q_{j\downarrow}^{\dagger}$, and we used the operator $N_j = I_j - (1/2) n_j$.

Let us consider the L-operator of host sites of a supersymmetric t-J chain for $V = -J/4$ and $J = 2$ as

$$L_j(\lambda) = c(\lambda) I_j^{(1|2)} - b(\lambda)$$
$$\times \begin{pmatrix} (N_j + S_j^z)(I_j - N_j + S_j^z) & -S_j^+ & -Q_{j\uparrow} \\ S_j^- & (N_j - S_j^z)(I_j - N_j - S_j^z) & -Q_{j\downarrow} \\ -Q_{j\uparrow}^{\dagger} & -Q_{j\downarrow}^{\dagger} & -(N_j - S_j^z)(N_j + S_j^z) \end{pmatrix} . \tag{7.110}$$

For the impurity site we introduce the operator with its spin equal to S' and the spectral parameter being shifted by θ, *i.e.*,

$$L_{imp}(\lambda) = c(\lambda - \theta) I_{imp}^{(1|2)} - b(\lambda - \theta)$$
$$\times \begin{pmatrix} (N_j + S_j^z)(I_j - N_j + S_j^z) & -S_j^+ & -(Q_1^+)_j \\ S_j^- & (N_j - S_j^z)(I_j - N_j - S_j^z) & -(Q_2^+)_j \\ -(Q_1^-)_j & -(Q_2^-)_j & -(N_j - S_j^z)(N_j + S_j^z) \end{pmatrix} . \tag{7.111}$$

In these formulas we used the same values for $b(x)$ and $c(x)$ as in Chapter 5 for a t-J chain. It is easy to check that these L-operators satisfy graded intertwining relations (Yang–Baxter relations) for L-operators with a graded R-matrix, Eq. (5.112), from which we started. Corresponding monodromy matrix of a supersymmetric t-J chain with an impurity $T(\lambda) = L_{imp}(\lambda)L_L(\lambda) \otimes \cdots \otimes L_1(\lambda)$ also satisfies graded intertwining relations. One can check that the action of this monodromy on the mathematical vacuum $|0\rangle$ is

$$T(\lambda)|0\rangle = \begin{pmatrix} c^L(\lambda)c_{S'}(\lambda-\theta) & 0 & 0 \\ 0 & c^L(\lambda)c_{S'}(\lambda-\theta)Z(\lambda-\theta) & 0 \\ C_1(\lambda) & C_2(\lambda) & 1 \end{pmatrix}, \quad (7.112)$$

where $c_{S'}(x) = (x+icS')/[x+ic(1+S')]$ and $Z(x) = (x-icS')/(x+icS')$. Equation (7.112) means the triangular action of the monodromy matrix on the mathematical vacuum, *i.e.*, this choice of the L-operator can be used for the above described scheme with $a_1(\lambda) = c^L(\lambda)c_{S'}(\lambda-\theta)$, $a_2(\lambda) = c^L(\lambda)c_{S'}(\lambda-\theta)Z(\lambda-\theta)$ and $a_3(\lambda) = 1$. The Hamiltonian of the supersymmetric t-J chain with an impurity for $V = -J/4$ and $J = 2$ can be obtained (up to constants) as

$$\mathcal{H}_{tJimp} = -icA\frac{\partial}{\partial\lambda}\ln[\mathrm{str}\,\hat{\tau}(\lambda)]|_{\lambda=0} \quad (7.113)$$

for shifted rapidities $\lambda_j^0 \to \lambda_j^0 - ic/2$ and $c=1$. It consists of two parts, the host Hamiltonian, $\mathcal{H}_{host}$, and the impurity Hamiltonian, $\mathcal{H}_{imp}$. The host Hamiltonian is $\mathcal{H}_{host} = \sum_j \mathcal{H}_{j,j+1}$, where

$$\mathcal{H}_{j,j+1} = -\sum_\sigma \mathcal{P}(c^\dagger_{j,\sigma}c_{j+1,\sigma} + c^\dagger_{j+1,\sigma}c_{j,\sigma})\mathcal{P} + c^\dagger_{j,\downarrow}c_{j,\uparrow}c^\dagger_{j+1,\uparrow}c_{j+1,\downarrow}$$
$$+ c^\dagger_{j,\uparrow}c_{j,\downarrow}c^\dagger_{j+1,\downarrow}c_{j+1,\uparrow} - n_{j,\uparrow}n_{j+1,\downarrow} - n_{j,\downarrow}n_{j+1,\uparrow}\,, \quad (7.114)$$

which is the standard Hamiltonian of a t-J chain, studied in previous chapters. The impurity's part of the Hamiltonian (for an impurity situated between sites m and $m+1$) is

$$\mathcal{H}_{imp} = \frac{(M,\sigma|M+\sigma)}{\theta^2 + (S+\frac{1}{2})^2}(\mathcal{H}_{m,imp} + \mathcal{H}_{imp,m+1} - 2S(S-1)\mathcal{H}_{m,m+1}$$
$$+ \{\mathcal{H}_{m,imp}, \mathcal{H}_{imp,m+1}\} - 2i\theta[\mathcal{H}_{m,imp}, \mathcal{H}_{imp,m+1}])\,, \quad (7.115)$$

where $\{.,.\}$ ($[.,.]$) denote anticommutator (commutator) and $(M,\sigma|M+\sigma)$ denotes the Clebsch–Gordan coefficient $(\frac{1}{2}\sigma, S'M|\frac{1}{2}S'SM+\sigma)$ with $S = S' + \frac{1}{2}$. An integrable impurity embedded in a host lattice is (as for the spin case) located on a link of the chain and interacts with electrons on both sites joined by the link. All the coupling constants of the impurity Hamiltonian depend on two parameters: S, determining the spin of the impurity, and the off-resonance shift θ (here we mostly limit ourselves with real θ), determining the impurity-host coupling even for $S = \frac{1}{2}$. From Eq. (7.115) it is clear, that the "impurity" of spin $S = \frac{1}{2}$ and $\theta = 0$ is, in fact, an addition of one more site to the host. On the other hand, the case $\theta \to \infty$ defines an impurity, totally decoupled from the host ring. Three-site terms of the impurity Hamiltonian violate the T and P symmetries separately, while their product PT is of course invariant. These terms are total time derivatives in the classical sense and are only important in quantum mechanical aspects. Although the reflection amplitude is zero as a consequence of the integrability, an impurity interacts with both partial waves (forward and backward moving electrons). Three-site terms can be avoided by placing the impurity site at the open end of the host chain. This considerably simplifies the impurity Hamiltonian, since one of the neighboring host sites is absent. Bethe ansatz equations for the supersymmetric t-J model for $V = -J/4$ and $J = 2$ with an impurity for periodic boundary conditions are:

$$
\begin{aligned}
e_{2S'}(\lambda_\alpha - \theta) \prod_{j=1}^{N} e_1(\lambda_\alpha - p_j) &= \prod_{\beta=1}^{M} e_2(\lambda_\alpha - \lambda_\beta) \ , \qquad \alpha = 1, \ldots, M \ , \\
e_{2S'+1}(p_j - \theta) e_1^L(p_j) &= \prod_{\alpha=1}^{M} e_1(p_j - \lambda_\alpha) \ , \qquad j = 1, \ldots, N \ .
\end{aligned}
\tag{7.116}
$$

The total magnetization is $M^z = (N/2) + S - M$ and the energy of the system is given by

$$
E = -JN + J \sum_{j=1}^{N} \frac{(1/2)}{(1/4) + p_j^2} \ . \tag{7.117}
$$

The reader already knows how to generalize Bethe ansatz equations (and following results) for an impurity with open boundary conditions:

$$
\begin{aligned}
\prod_{\pm} e_{2S'}(\lambda_\alpha \pm \theta) \prod_{j=1}^{N} e_1(\lambda_\alpha \pm p_j) \prod_{\beta=1}^{M} e_2^{-1}(\lambda_\alpha \pm \lambda_\beta) &= 1 \ , \qquad \alpha = 1, \ldots, M \ , \\
e_1^{2L}(p_j) = \prod_{\pm} e_{2S'+1}^{-1}(p_j \pm \theta) \prod_{\alpha=1}^{M} e_1(p_j \pm \lambda_\alpha) & \ , \qquad j = 1, \ldots, N \ .
\end{aligned}
\tag{7.118}
$$

It is instructive to obtain the Bethe ansatz description of the same model using a different Bethe ansatz scheme. We can start from the two-particle scattering matrices of the host

$$\hat{X}(\lambda) = \frac{\hat{I}_s\lambda \pm i\hat{P}_s}{\lambda \pm i} , \qquad (7.119)$$

where $\hat{I}_s$ is the identity and $\hat{P}_s$ the two-particle permutation operator in the spin subspace. The impurity scattering matrix in the spin subspace can be written as

$$\hat{S}^{\sigma,\sigma'}_{M,M'}(\lambda) = \frac{A\delta_{\sigma,\sigma'}\delta_{M,M'} + B\delta_{-\sigma',\sigma}\delta_{M',M+2\sigma}}{\lambda - \theta - i(2S'+1)/2} , \qquad (7.120)$$

where σ (σ') and M (M') are the electron and impurity spin S' components before (after) scattering, and

$$A = \lambda - \theta - i(2S'+1)\left[\frac{1}{2} - (\sigma M|M+\sigma)(\sigma' M'|M'+\sigma')\right] ,$$
$$B = i(2S'+1)(\sigma M|M+\sigma)(\sigma' M'|M'+\sigma') , \qquad (7.121)$$

where $S = S' + \frac{1}{2}$. The impurity $\hat{S}$ matrix is generally a two-parameter function (a discrete parameter is the spin of the impurity and the coupling to the host, θ, is a continuous parameter), which differs the impurity matrix $\hat{S}$ from $\hat{X}$ ($\hat{X} = \hat{S}(\theta = 0)$ for $S = \frac{1}{2}$). Matrices $\hat{X}$ satisfy the Yang–Baxter relation

$$\hat{X}_{12}(\lambda_1 - \lambda_2)\hat{X}_{13}(\lambda_1 - \lambda_3)\hat{X}_{23}(\lambda_2 - \lambda_3)$$
$$= \hat{X}_{23}(\lambda_2 - \lambda_3)\hat{X}_{13}(\lambda_1 - \lambda_3)\hat{X}_{12}(\lambda_1 - \lambda_2) , \qquad (7.122)$$

where indices enumerate scattering host electrons. The scattering matrix $\hat{S}$ satisfies the following Yang–Baxter relation:

$$\hat{X}_{12}(\lambda_1 - \lambda_2)\hat{S}_{1,imp}(\lambda_1 - \theta)\hat{S}_{2,imp}(\lambda_2 - \theta)$$
$$= \hat{S}_{2,imp}(\lambda_2 - \theta)\hat{S}_{1,imp}(\lambda_1 - \theta)\hat{X}_{12}(\lambda_1 - \lambda_2) , \qquad (7.123)$$

where the indices for the matrices S show which particles scatter. The monodromy matrix in the spin subspace on the inhomogeneous lattice is defined as

$$L(\lambda, p_1, \ldots, p_N, \theta)$$
$$= \hat{X}_{01}(p_1 - \lambda)\hat{X}_{02}(p_2 - \lambda)\hat{X}_{0N}(p_N - \lambda)S_{0imp}(\theta - \lambda) , \qquad (7.124)$$

where the index 0 defines the auxiliary subspace, and λ is the spectral parameter. With respect to the states of the auxiliary space the monodromy matrix Eq. (7.124) forms 2×2 matrix. Monodromy matrices satisfy the Yang–Baxter relation

$$\begin{aligned}&\hat{X}_{12}(\lambda-\lambda')L(\lambda,v_1,\ldots,\theta)L(\lambda',v_1,\ldots,\theta)\\&\quad=L(\lambda',v_1,\ldots,\theta)L(\lambda,v_1,\ldots,\theta)\hat{X}_{12}(\lambda-\lambda')\ ,\end{aligned}\qquad(7.125)$$

which is the direct consequence of Eqs. (7.122) and (7.123). The exact integrability of the system follows from the fact that transfer matrices ($T(\lambda)=\mathrm{tr}_0\,L(\lambda,p_1,\ldots,\theta)$, which is the trace over the states of the auxiliary particle) commute with different spectral parameters, because any functions of $T(\lambda)$ commute mutually and with the transfer matrix (*i.e.*, we have an infinite number of conservation laws). Let us construct the matrix $T_j(p_j)$ defined as

$$\begin{aligned}T_j(p_j)=&\hat{X}^{-1}_{j,j-1}(p_j-p_{j-1})\cdots\hat{X}^{-1}_{j,1}(p_j-p_1)\times\cdots\times\hat{S}^{-1}_{j,imp}(p_j-\theta)\\&\times\hat{X}^{-1}_{j,N}(p_j-p_N)\times\cdots\times\hat{X}^{-1}_{j,j+1}(p_j-p_{j+1})\ .\end{aligned}\qquad(7.126)$$

The action of the matrix $T_j(p_j)$ implies periodic boundary conditions, *i.e.*, that one has to interchange a given electron with all other electrons in the periodic box, including the impurity. Corresponding eigenvalue of $T_j(p_j)$ is $\exp(ik_jL)$, which is related to p_j for the supersymmetric t-J chain for $V=-J/4$, $J=2$ as $[(2p_j+i)/(2p_j-i)]^L$. Substituting $\lambda=v_j$ into the monodromy matrix we see that $T(\lambda=v_j)=T_j(v_j)$. Consider components of the monodromy matrix taken in the subspace of the auxiliary particle, $L_{i,j}$ ($i,j=1,2$). The transfer matrix is, naturally, $T(\lambda)=L_{11}(\lambda)+L_{22}(\lambda)$. These operators obey commutation relations, which follow from Eq. (7.125). Let us denote the vacuum state Ω_0 as $L_{21}\Omega_0=0$. The vacuum state is the eigenstate of diagonal matrix elements with eigenvalues

$$\begin{aligned}\mathcal{L}_{22}(\lambda)&=\frac{\theta-\lambda-i(2S'-1)/2}{\theta-\lambda-i(2S'+1)/2}\prod_{j=1}^{N}\frac{v_j-\lambda}{v_j-\lambda+i}\ ,\\\mathcal{L}_{11}(\lambda)&=\frac{\theta-\lambda+i(2S'+1)/2}{\theta-\lambda-i(2S'+1)/2}\ .\end{aligned}\qquad(7.127)$$

On the other hand, the operator L_{12} has the properties like a "spin-lowering" one, such that the vector

$$\Omega(\alpha_1, \ldots, \alpha_M) = \prod_{\beta=1}^{M} L_{12}(\alpha_\beta)\Omega_0 \tag{7.128}$$

corresponds to M flipped spins. Bethe ansatz equations are the conditions on the sets of parameters p_j and α_β, under which the state Ω is the eigenstate of the transfer matrix T. The application of the operator $L_{11} + L_{22}$ produces the eigenvalue (with the reproduction of the state Ω)

$$\mathcal{L}_{11}(\lambda) \prod_{\beta=1}^{M} \frac{\lambda - \alpha_\beta + i}{\lambda - \alpha_\beta} + \mathcal{L}_{22}(\lambda) \prod_{\beta=1}^{M} \frac{\lambda - \alpha_\beta - i}{\lambda - \alpha_\beta} \tag{7.129}$$

and unwanted terms. The condition of cancellation of those unwanted terms is

$$\frac{\theta - \alpha_\gamma + i(2S'+1)/2}{\theta - \alpha_\gamma - i(2S'-1)/2} \prod_{j=1}^{N} \frac{p_j - \alpha_\gamma + i}{p_j - \alpha_\gamma} = \prod_{\beta=1}^{M} \frac{\alpha_\gamma - \alpha_\beta - i}{\alpha_\gamma - \alpha_\beta + i} . \tag{7.130}$$

Substituting $\lambda = p_j$, $\alpha_\gamma = \lambda_\gamma + i/2$ and using Eq. (7.126) we obtain Eq. (7.116).

We shall present results here for the periodic case; the ones for open boundary conditions can be obtained in a similar way. The results will be presented for the generic case N even. In the framework of the string hypothesis we can write thermodynamic Bethe ansatz equations for densities of a supersymmetric t-J chain for $V = -J/4$, $J = 2$ as (here we keep the same notations as in Chapter 4). After the Fourier transformation we have

$$\begin{aligned}
&\sigma_{m+1,h}(\omega) + \sigma_{m+1,h}(\omega) + \delta_{m,2S'} \frac{e^{i\theta\omega}}{2\pi L} \\
&\quad = 2\cosh(\omega/2)[\sigma_m(\omega) + \sigma_{m,h}(\omega)] \ , \ m \geq 1 \ , \\
&\sigma'_h(\omega) + 1 + \frac{e^{(-S'|\omega| + i\theta\omega)}}{2\pi L} = 2e^{|\omega|/2}\cosh(\omega/2)[\sigma'(\omega) + \sigma'_h(\omega)] + \rho(\omega) \ , \\
&\sigma_{1,h}(\omega) + \sigma'_h(\omega) + 1 = 2\cosh(\omega/2)[\rho(\omega) + \rho_h(\omega)] \ .
\end{aligned} \tag{7.131}$$

Thermodynamic Bethe ansatz equations for dressed energies for the chain with an impurity coincide with Eq. (4.104). The internal energy, the number of electrons and the magnetization of a supersymmetric t-J chain for $V = -J/4$, $J = 2$ with an impurity are given by

$$E = -2N + 2\pi L \int_{-\infty}^{\infty} dp a_1(p)\rho(p) + 2\pi \int_{-\infty}^{\infty} d\lambda a_2(\lambda)\sigma'(\lambda) ,$$

$$N = L \int_{-\infty}^{\infty} dp \rho(p) + 2L \int_{-\infty}^{\infty} d\lambda \sigma'(\lambda) , \tag{7.132}$$

$$M^z = S + \frac{L}{2} \int_{-\infty}^{\infty} dp \rho(p) - L \sum_{n=1}^{\infty} n \int_{-\infty}^{\infty} d\lambda \sigma_n(\lambda) .$$

The Helmholtz free energy of a supersymmetric t-J chain with an impurity for $V = -J/4$, $J = 2$ is equal to

$$\frac{F}{T} = -L \int_{-\infty}^{\infty} d\lambda a_2(\lambda) \ln[1 + \kappa^{-1}(\lambda)] - L \int_{-\infty}^{\infty} dp a_1(p) \ln[1 + \xi^{-1}(k)]$$
$$- \int_{-\infty}^{\infty} d\lambda (G_{2S'+1}(\lambda - \theta) \ln[1 + \kappa(\lambda)] + G_0(\lambda - \theta) \ln[1 + \eta_{2S'}(\lambda)]) , \tag{7.133}$$

where $a_n(x) = 2n/(4x^2 + n^2)$ and the $G_n(x)$ is the Fourier transform of $\exp(-n|\omega|/2)/2\cosh(\omega/2)$.

At high temperatures we obtain

$$f_{imp} = -T \ln(Z_S + \exp[g(\theta) - \mu] Z_{S-(1/2)}) , \tag{7.134}$$

where Z_S is the partition function of the free spin S:

$$Z_S = \frac{\sinh[(2S+1)H/2T]}{\sinh(H/2T)} , \tag{7.135}$$

the chemical potential μ is measured from the bottom of the conduction band, and the function $g(\theta)$ measures the admixture of states with the spin S and $S - \frac{1}{2}$ (it is even in θ and monotonically decreases with θ for positive θ with $g(\pm\infty) = 0$). For $\mu = 0$, *i.e.*, for the empty band the configuration with $S - \frac{1}{2}$ is favored, while for $\mu = 2\ln 2$ (half-filling) the favored configuration is rather S. For a fixed H the specific heat of an impurity displays a Schottky anomaly, while the magnetic susceptibility follows the Curie law at high temperatures.

In the ground state integral equations for densities (we here write down only equations for the part of order of L^{-1}, dropping the superscript; equations for dressed energies and the ones for densities of order of 1 coincide with the ones from Chapter 4) are:

$$\rho_h(p) + \rho(p) + \int_{|\lambda|>Q} d\lambda a_1(p-\lambda)\sigma'(\lambda) = a_{2S}(p-\theta) ,$$
$$\sigma'_h(\lambda) + \sigma'(\lambda) + \int_{|\lambda|>Q} d\lambda' a_2(\lambda-\lambda')\sigma'(\lambda') \tag{7.136}$$
$$+ \int_{|p|>B} dp a_1(p-\lambda)\rho(p) = a_{2S+1}(\lambda-\theta) .$$

In the absence of a magnetic field, $B \to \infty$, we can obtain analytical results for the valence of an impurity

$$n_{imp} = \int_{|p|>B} dp\rho(p) + 2\int_{|\lambda|>Q} d\lambda\sigma'(\lambda) , \tag{7.137}$$

which is equal to $n_{imp} = (2S+1)Q/2\pi(Q^2-\theta^2)$ for large Q (we assumed that $Q \gg |\theta|$), *i.e.*, for low electron density, and $n_{imp} = 1 - O(Q)$ for small Q, *i.e.*, for the electron density close to half-filling. Hence, as a function of the band filling the valence of an impurity smoothly varies between 0 (for $N \to 0$) and 1 (for $N \to L$). The valence is a decaying function of θ for fixed band filling. It is clear, because the larger θ pertain to weaker coupling of an impurity to the host. The valence is maximum for $\theta = 0$ which is the resonance situation (the impurity level is situated at the Fermi point for the Dirac sea of pairs). The impurity valence also decreases as a function of S close to half filling, and increases for higher values of the impurity spin for small total number of electrons in the system. The magnetization of an impurity for $H = 0$ is $S' = S - \frac{1}{2}$.

For $H \neq 0$ we can obtain the valence of an impurity for $S - \frac{1}{2} \gg |Q-\theta|$ as $n_{imp} \approx 2\sqrt{|Q-\theta|}/\pi(2S-1)$, and for the opposite case $S - \frac{1}{2} \ll |Q-\theta|$ as $n_{imp} = \frac{1}{2} + (1/\pi)[\ln 2\sqrt{|Q-\theta|} - (2S-1)/2\sqrt{|Q-\theta|}]$, where

$$Q^2 = \frac{2}{3\zeta(3)}\left(2\ln 2 - \mu + \frac{H^2}{4\pi}\right) . \tag{7.138}$$

We point out the magnetic field dependence of the impurity valence in both limits. When switching on the magnetic field the valence of an impurity becomes smaller than unity even at half filling. This is connected with the fact that the magnetic field acts twofold: it affects the magnetization of an impurity, and also creates spin excitations of the host (which carry charge too) and changes the number of charged excitations, *e.g.*, destroying pairs, which carry zero spin. This is the manifestation of correlations between electrons in the host.

The ground state magnetization is the sum of two contributions, the magnetization arising from the valence admixture and the one due to spin degrees of freedom of an impurity. Since the magnetic field is usually much smaller than the band width, the former contribution is small (and linear in H), and can be neglected. Then the Fredholm equation, which describes only the "Kondo-like" spin excitations is

$$\rho_h(p) + \rho(p) - \int_{|p'|>B} dp' G_1(p-p')\rho(p') = G_{2S-1}(p-\theta) \ . \qquad (7.139)$$

Here we assumed that $0 < 1 \ll \theta$. Then one can introduce the Kondo temperature *via* $\pi(\theta - B) = \ln(H/T_K)$, and we obtain the solution for the magnetization of an impurity

$$m_{imp}^z = \mu_i \left(1 \pm \frac{1}{2|\ln(H/T_K)|} - \frac{\ln|\ln(H/T_K)|}{4\ln^2(H/T_K)} + \cdots \right) , \qquad (7.140)$$

where we use for $H \gg T_K$ the lower sign and $\mu_i = S$, and for $H \ll T_K$ we use the upper sign and $\mu_i = S - \frac{1}{2}$ for $S > \frac{1}{2}$, and $\mu_i = H/T_K$ for $S = \frac{1}{2}$. This means that the impurity spin is underscreened at low fields to the value $S - \frac{1}{2}$ for $S > \frac{1}{2}$, while for $S = \frac{1}{2}$ it is totally screened with the finite magnetic susceptibility (inverse proportional to the Kondo temperature). For high enough values of the magnetic field the impurity spin behaves as an asymptotically free spin S. We see, that the Kondo temperature depends on the band filling *via* B. If charge fluctuations are totally suppressed, for $N = L$, the Kondo temperature is $T_K = E_F \exp(-\pi|\theta|)$. Notice that for open boundary conditions one has similar behaviour of an impurity itself (contributions from open edges also appear, see Chapter 6) with the renormalized Kondo temperature, in which $\exp(-\pi|\theta|)$ is replaced by $[2\cosh(\pi\theta)]^{-1}$. The magnetic susceptibility at $H = 0$ of the impurity spin is Curie like ($\sim T^{-1}$) in the Kondo limit with the Curie constant $S(S+1)/3$ for $T \gg T_K$ and with $[S^2 - (1/4)]/3$ at $T \ll T_K$ for $S > \frac{1}{2}$ and it is finite for low temperatures for $S = \frac{1}{2}$. We emphasize on the corrections due to the mixed valence of an impurity: they shift the value of the Kondo temperature, *e.g.*, as $T_K \to T_K(1 + 2\zeta(3)Q^3)$ for $Q \ll 1$. This is also the manifestation of correlations between electrons in the host. It is important to notice that only unbound electron excitations which carry spin can screen the spin of an impurity. Spin-singlet pairs only renormalize the valence of an impurity, but their distribution affects the distribution of unbound electron excitations. At $H \geq H_s$, in the spin-saturation phase, the magnetization of an impurity is equal to $M^z_{imp} = S' + (n_{imp}/2)$, where n_{imp} is the valence of an impurity. The specific heat of an impurity exhibits two features: a Kondo resonance at $T \sim T_K$ and a Schottky peak at $T \sim \sqrt{\pi^3/eH}$. For higher values of H both peaks can merge into one.

As it is clear from Eq. (7.115), the imaginary θ implies the non-Hermitian Hamiltonian for an impurity situated in the bulk of a correlated electron chain. The only possibility to consider imaginary θ with the Hermitian Hamiltonian is to study an impurity at the edge of an open t-J chain with zero boundary fields/potentials. Imaginary θ with $\theta^2 > (S + \frac{1}{2})^2$ pertains to the ferromagnetic coupling of an impurity to the correlated electron host (other cases correspond to antiferromagnetic impurity-host interactions). For imaginary θ we can divide it into its integer and fractional part, $2|\theta| = [2|\theta|] + \{2|\theta|\}$. In this case the fractional part can define the "Kondo temperature" of an impurity $T_K \sim [\cos(\pi\{2|\theta|\}/2)]^{-1}$. Hence, such a "crossover scale" is larger than the characteristic energy of low-lying spin excitations of the bulk, which, in turn, defines the critical field H_s of a quantum phase transition to the spin-saturated (ferromagnetic) phase. Hence, there is only one special point in the behaviour of such an impurity, H_s. Another feature of imaginary θ is that incident and reflected waves effectively scatter off different effective "impurity spins", $S - \frac{1 \pm [2|\theta|]}{2}$ and $S \pm \frac{[2|\theta|]}{2}$. Negative effective "impurity spins" signal the onset of local levels (related to bound states, caused by an impurity situated at the edge, however their appearance is not connected with a boundary potential, as in Chapter 6, but only with the ferromagnetic coupling of an impurity to the host). These levels can influence the remnant entropy of an impurity.

It is interesting to mention that there is another possibility to include a magnetic impurity into an integrable correlated electron chain. We can consider the impurity scattering matrix in the spin subspace being similar to an exchange impurity of the Kondo problem

$$\hat{S}^{\sigma,\sigma'}_{M,M'}(\lambda) = \sqrt{\frac{x^2 + c^2}{4x^2 + (2S+1)^2c^2}} \frac{2}{x + ic}[(x + i(c/2))\delta_{\sigma,\sigma'}\delta_{M,M'} + ic\vec{\sigma}_{\sigma\sigma'}\vec{S}_{MM'}] . \tag{7.141}$$

We shall not present the investigation of this case here, but rather refer the interested reader to original publications.

Now let us consider the behaviour of an impurity in the supersymmetric t-J chain with $V = 3J/4$ and $J = 2$. Here we shall consider a slightly different impurity with the impurity scattering matrix in the spin subspace

$$\hat{S}^{\sigma,\sigma'}_{M,M'}(\lambda) = \frac{A\delta_{\sigma,\sigma'}\delta_{M,M'} - B\delta_{\sigma',-\sigma}\delta_{M',M+2\sigma}}{\lambda - \theta + i(2S+1)/2} \tag{7.142}$$

where σ (σ') and M (M') are the electron and impurity spin S' components of the before (after) scattering, and

$$\begin{aligned} A &= \lambda - \theta + i(2S+1)\left[\frac{1}{2} - (\sigma M|M+\sigma)(\sigma' M'|M'+\sigma')\right], \\ B &= i(2S+1)(\sigma M|M+\sigma)(\sigma' M'|M'+\sigma') \ . \end{aligned} \tag{7.143}$$

The Clebsch–Gordan coefficient selects the way how the impurity interacts with itinerant electrons. The impurity can temporarily absorb the spin of one conduction electron and form an effective spin $S' = S - \frac{1}{2}$. The Bethe ansatz describes such a system as the solution of equations

$$\begin{aligned} e_1^L(p_j) &= (-1)^N \prod_{\substack{l=1 \\ l\neq j}}^{N} e_2(p_j - p_l) \prod_{\beta=1}^{M} e_1^{-1}(p_j - \lambda_\beta) \ , \quad j = 1, \ldots, N \ , \\ e_{2S}(\lambda_\alpha - \theta) \prod_{j=1}^{N} e_1(\lambda_\alpha - p_j) &= \prod_{\substack{\beta=1 \\ \beta\neq\alpha}}^{M} e_2(\lambda_\alpha - \lambda_\beta) \ , \quad \alpha = 1, \ldots, M \ , \\ E = 2N - 4\sum_{j=1}^{N}(4p_j^2+1)^{-1} \ , &\quad M^z = \frac{N}{2} + S - M \ . \end{aligned} \tag{7.144}$$

The most interesting properties of an impurity is in the ground state and at low temperatures. In the absence of a magnetic field the impurity valence is equal to $n_{imp} = 2QG_{2S}(\theta)$ for low electron density $Q \to 0$, where Q is the Fermi point of charged low-lying excitations. The energy of an impurity in this limit is $e_{imp} = -2n_{imp}$. On the other hand, for large electron density $Q \to \infty$ we obtain $n_{imp} = \frac{1}{3}$, independent on θ and

$$e_{imp} = -\frac{2}{3}\mathrm{Re}\left[\psi\left(\frac{S+2-i\theta}{3}\right) - \psi\left(\frac{S+1-i\theta}{3}\right)\right] , \tag{7.145}$$

where $\psi(x)$ is a digamma function. For this case $H = 0$ the magnetization of an impurity is equal to $m^z_{imp} = S - \frac{1}{2}$. The charge susceptibility of an impurity is equal to the one of the host per site. The energy is a monotonically increasing function of $|\theta|$, and the valence of an impurity is a monotonically decreasing one. The behaviour of the magnetization and magnetic susceptibility of an impurity is similar to the one of the previous case, *i.e.*, depending on the value of the impurity spin, it is either underscreened by low-lying spin-carrying excitations of the host, or totally screened for $S = \frac{1}{2}$ with the finite magnetic susceptibility. At very large fields $H \geq H_s$ the impurity magnetization becomes equal to S. The characteristic crossover scale, that defines the low-energy behaviour of a magnetic impurity is the Kondo temperature, which is proportional to $\exp(-\pi|\theta|)$. It means that θ measures the resonance shift. The smaller θ (*i.e.*, the stronger the coupling of the impurity site to the host), the larger Kondo scale.

It is also important to consider the behaviour of a magnetic impurity in the supersymmetric t-J chain with the "easy-axis" magnetic anisotropy, considered in Chapter 4, with the host Hamiltonian Eq. (4.100). This case is important, because here we can study the behaviour of a magnetic impurity in the correlated host with spin-gapped excitations. The structure of the Hamiltonian of an impurity is similar to Eq. (7.55). The energy of the system is given by (with standard notations)

$$E = -2\sum_{j=1}^{N} \frac{1 - \cos(2v_j)\cosh(\eta)}{\cosh(\eta) - \cos(2v_j)} + \text{const} . \tag{7.146}$$

The z-projection of the magnetic moment of the system is $M^z = S + \frac{(N-1)}{2} - M$. The Bethe ansatz description (here we present the case of periodic boundary conditions) of that model is based on the solution of the following equations

$$\frac{\sin(\lambda_\alpha - \theta + i(2S-1)\frac{\eta}{2})}{\sin(\lambda_\alpha - \theta - i(2S-1)\frac{\eta}{2})} \prod_{j=1}^{N} \frac{\sin(\lambda_\alpha - v_j + i\frac{\eta}{2})}{\sin(\lambda_\alpha - v_j - i\frac{\eta}{2})} = \prod_{\beta=1}^{M} \frac{\sin(\lambda_\alpha - \lambda_\beta + i\eta)}{\sin(\lambda_\alpha - \lambda_\beta - i\eta)} ,$$

$$\frac{\sin(v_j - \theta + iS\eta)}{\sin(v_j - \theta - iS\eta)} \left[\frac{\sin(v_j + i\frac{\eta}{2})}{\sin(v_j - i\frac{\eta}{2})}\right]^L = \prod_{\alpha=1}^{M} \frac{\sin(v_j - \lambda_\alpha + i\frac{\eta}{2})}{\sin(v_j - \lambda_\alpha - i\frac{\eta}{2})} , \tag{7.147}$$

where $j = 1, \ldots, N$ and $\alpha = 1, \ldots, M$. Only the first factor on the left hand sides of Eqs. (7.147) correspond to the impurity, while the energy, Eq. (7.146), depends only implicitly on the impurity. Bethe ansatz equations are again independent of the position of an impurity in the chain.

Dropping the analysis of the Bethe ansatz solution we only present results here. Energies of unbound electron states are gapped for an external magnetic field less than a critical value, H_c, see Chapter 4. H_c is one half of the minimal external magnetic field necessary to depair a singlet bound state (pair). If the value of the external magnetic field is larger than H_s, the magnetization is maximal, *i.e.*, saturated. At this saturation field the system undergoes a second order phase transition into the ferromagnetic spin-polarized state, in which there are no pairs because the dressed energy of unbound electrons is gapped. This behaviour is similar to a type-II superconductor in a magnetic field: for $H \leq H_c$ there are only Cooper-pairs, while for $H_c \leq H \leq H_s$ pairs and unbound electrons co-exist, which is reminiscent of the Meissner effect. Note, however, that in a one-dimensional electron gas there is no true superconducting order with off-diagonal long range order, but correlation functions of singlet pairs and/or unbound electrons fall off with power-laws for long times and/or distances. For $H \geq H_s$ it is straightforward to obtain the ground state energy. In the intermediate phase, $H_c \leq H \leq H_s$, however, the ground state energy depends on the filling of both Dirac seas.

We first consider the case $H < H_c$, where the ground state consists only of singlet pairs ($2M = N$). The magnetization of an impurity is exactly S' for $H \leq H_c$ for both, open or periodic, boundary conditions. This implies that the Kondo effect is absent in this model, due to the spin-gap, induced by the Ising-like ("easy-axis") magnetic anisotropy. The low-lying excitations do not carry spin, and, consequently, cannot couple to the spin S' to form a magnetic moment of spin S for $H \leq H_c$. Recall that $S' = S - 1/2$ represents the effective spin of the low temperature fixed point. The valence varies as a function of the number of electrons in the system from one for a filled band or the maximal number of conduction electrons to zero for an empty band of conduction electrons.

Next we consider the situation $H_c \leq H \leq H_s$, where both, unbound electrons and singlet pairs, have gapless low-lying excitations, *i.e.*, form Dirac seas. The valence of an impurity again depends on the density of electrons, and, interestingly, also on the external magnetic field. Due to the van Hove singularity of the empty band of unpaired electron states, the magnetization of the host is proportional to $\sqrt{H - H_c}$ for fields H slightly larger than H_c. This feature is characteristic of a Pokrovsky–Talapov level-crossing transition, which is the analog of a second order phase transitions in one-dimension. The magnetization of an impurity is driven by the host,

$$M^z_{imp} = S - \frac{1}{2} + f_S(\theta, \eta)\sqrt{H - H_c} \, , \tag{7.148}$$

where $f_S(\theta, \eta)$ is also a function of the band filling. The magnetic susceptibility of an impurity has a square root singularity as H_c is approached from above. This is also very different from the standard Kondo effect, where for spin-$\frac{1}{2}$ the magnetic susceptibility of an impurity is finite for small magnetic fields. For open boundary conditions the magnetic susceptibility of an impurity diverges as strongly as the magnetic susceptibility of open edges themselves (all inversely proportional to $\sqrt{H - H_c}$). This is also very different from the usual behaviour of a magnetic impurity in an open t-J chain with SU(2) spin symmetry and gapless excitations, where the magnetic susceptibility of edges diverges (though logarithmically), while the one for an impurity of spin $S = \frac{1}{2}$ remains finite. With increasing magnetic field the population of the Dirac sea of singlet pairs gradually decreases until H_s is reached, which is the field at which the band is empty. For fields larger than the saturation field H_s the magnetization of an impurity is equal to $M_{imp}^z = S - [(1 - n_{imp})/2]$, where n_{imp} is the valence of the impurity.

For imaginary θ, the Hamiltonian of an impurity if placed in the bulk, *i.e.*, not at the edge, is non-Hermitian (the energy eigenvalues are real, though). This is independent of the boundary conditions (open or periodic). In this case incoming and reflecting waves of electrons "see" two different effective spins of the impurity corresponding to $S' \pm \frac{[2|\theta|]}{2}$. However, the Ising magnetic anisotropy of the model again suppresses any manifestation of the Kondo effect, since only spin-singlet pairs are gapless for $H \leq H_c$, but cannot screen effective spins of an impurity. For fields slightly larger than H_c the van Hove singularity of the empty band of unbound electron states manifests itself, rather than the weaker logarithmic Kondo singularities.

At finite but low temperatures the magnetic susceptibility of an impurity (as well as the susceptibility of edges of an open chain) is exponentially small for $H < H_c$ and $H > H_s$. At $H = H_c$ or H_s the magnetic susceptibility and the Sommerfeld coefficient of the specific heat display the $\sqrt{T}$ features corresponding to the van Hove singularities of empty bands. For $H_c < H < H_s$, on the other hand, the magnetic susceptibility is finite for $S = \frac{1}{2}$ as $T \to 0$ and Curie-like for $S \geq \frac{1}{2}$. The specific heat is proportional to the temperature everywhere away from van Hove singularities.

Here it is instructive to compare the behaviour of an impurity in the supersymmetric t-J chain with the behaviour of the Anderson impurity model with the Hamiltonian

$$\mathcal{H}_A = \sum_{\mathbf{k},\sigma} \epsilon_k \psi^\dagger_{\mathbf{k},\sigma} \psi_{\mathbf{k},\sigma} + \sum_{\mathbf{k},\sigma} V_{\mathbf{k},\sigma} (\psi^\dagger_{\mathbf{k},\sigma} d_\sigma + d^\dagger_\sigma \psi_{\mathbf{k},\sigma}) + \sum_\sigma \epsilon_d n_\sigma + U n_\uparrow n_\downarrow \,, \tag{7.149}$$

where the notations are similar to Eq. (7.94), with $n_\sigma = d_\sigma^\dagger d_\sigma$, $d_\sigma^\dagger$ creates an electron on the impurity orbital, U denotes the Coulomb repulsion, and ϵ_d measures the energy of the impurity orbital from the Fermi level. The interaction of conduction electrons with localized electrons is described by the magnitude of the hybridization, $V_{\mathbf{k},\sigma}$, which is often supposed to be k- and σ-independent, $V_{\mathbf{k},\sigma} = V$. Using similar arguments as for the Kondo problem, see above, one can reduce the Hamiltonian to the one-dimensional one, in which the relativistic dispersion law for itinerant electrons (linearized about Fermi points) is considered (we put the Fermi velocity equal to 1),

$$\mathcal{H}_A = \sum_\sigma \int dx \left(-i\psi_\sigma^\dagger(x)\frac{d}{dx}\psi_\sigma(x) + V\delta(x)[\psi_\sigma^\dagger(x)d_\sigma + d_\sigma^\dagger\psi_\sigma(x)] + \epsilon_d n_\sigma \right) + U n_\uparrow n_\downarrow \ . \tag{7.150}$$

By solving the stationary Schrödinger equation with the Hamiltonian Eq. (7.150) for one conduction electron and one electron, localized on the orbital, and for two conduction electrons, then by using Yang–Baxter equations for the subsequent two-particle scattering matrices, one finds that eigenfunctions and eigenstates are parametrized by the solution of the following equations (obtained for periodic boundary conditions in a box of length L)

$$\begin{aligned} \prod_{j=1}^{N} e_1(\lambda_\alpha - g(k_j)) &= -\prod_{\beta=1}^{M} e_2(\lambda_\alpha - \lambda_\beta) \ , \qquad \alpha = 1, \ldots, M \ , \\ e_{V^2}(p_j - \epsilon_d)e^{-ik_jL} &= \prod_{\alpha=1}^{M} e_1(g(k_j) - \lambda_\alpha) \ , \qquad j = 1, \ldots, N \ , \end{aligned} \tag{7.151}$$

where $g(k) = (2k - 2\epsilon_d - U)^2/8UV^2$. The energy of the system is given by

$$E = \sum_{j=1}^{N} k_j \ . \tag{7.152}$$

Theorists often study the situation in which the Coulomb repulsion of electrons localized on orbitals is considered to be large, so that in this limit one has to replace $g(k) \to k/V^2$. One can see that the Bethe ansatz equations of the Anderson model (after the renormalization $\lambda_\alpha \to \lambda_\alpha/V^2$) for large U (which excludes a double occupation) are similar to the ones of the supersymmetric t-J chain for $V = -J/4$ and $J = 2$, for an impurity with $S' = 0$ ($S = \frac{1}{2}$) but with two differences: namely, with the presence of an additional parameter V^2, and with the linearized dispersion law. It is interesting to notice that the ground state of the Anderson impurity model also pertains to the filling of Dirac seas for unbound electron excitations and spin-singlet pairs, as for the t-J chain. This is why, the analogy in the behaviours of the Anderson impurity model and an impurity in a correlated electron chain becomes transparent. It is known that an Anderson impurity, depending on the value of ϵ_d (which works analogous to θ), reveals non-magnetic regime with $n_{imp} \sim 0$, mixed valence regime and the magnetic regime (where the valence of an impurity is close to 1). In the magnetic regime, where charge fluctuations are suppressed, the behaviour of the impurity spin is similar to the behaviour of a Kondo magnetic impurity, with the characteristic Kondo temperature $T_K \sim \exp(-\pi|\varepsilon_d|/V^2)$.

Let us now consider how an integrable impurity behaves in a Hubbard chain. Bethe ansatz equations can be obtained within the second scheme (introducing the impurity scattering matrix in the spin subspace). For the repulsive Hubbard chain let us study the behaviour of an impurity with the scattering matrix

$$\hat{S}^{\sigma,\sigma'}_{M,M'}(\lambda) = \sqrt{\frac{4x^2+U^2}{16x^2+(2S+1)^2U^2}}\frac{4}{2x+iU} \times \left[(x+i(U/4))\delta_{\sigma,\sigma'}\delta_{M,M'} + i(U/2)\vec{\sigma}_{\sigma\sigma'}\vec{S}_{MM'}\right] . \quad (7.153)$$

This two-particle scattering matrix satisfies Yang–Baxter relations with two-particle scattering matrices of electrons for the Hubbard chain, *cf.* Chapter 4. By using the method, described above, Bethe ansatz equations for this system with periodic boundary conditions can be written as

$$\begin{aligned} \frac{\lambda_\alpha - \theta + iUS/2}{\lambda_\alpha - \theta - iUS'/2}\prod_{j=1}^{N}\frac{\lambda_\alpha - \sin k_j + iU/4}{\lambda_\alpha - \sin k_j - iU/4} &= \prod_{\substack{\beta=1 \\ \beta\neq\alpha}}^{M}\frac{\lambda_\alpha - \lambda_\beta + iU/2}{\lambda_\alpha - \lambda - iU/2} , \\ e^{ik_jL+i\phi_j} &= \prod_{\beta=1}^{M}\frac{\sin k_j - \lambda_\beta + iU/4}{\sin k_j - \lambda_\beta - iU/4} , \end{aligned} \quad (7.154)$$

where $j = 1,\ldots,N$ and $\alpha = 1,\ldots,M$ and $\phi_j = \tan^{-1}[2(\sin k_j - \theta)/U] - \tan^{-1}[4(\sin k_j - \theta)/U(2S+1)]$. The magnetization of the model is $M^z = (N/2) + S - M$, and the energy is $E = -2\sum_{j=1}^{N}\cos k_j + \text{const}$. The reader is already aware that one can divide everything into the part, describing the host, and the one, describing the impurity. We shall write only equations and results for an impurity, because the host part behaves in the same way as it is presented in Chapter 4.

The most interesting behaviour is in the ground state, where Bethe ansatz integral equations for densities (we keep the notations of Chapter 4 and drop the superscript, which denotes an impurity) are:

$$\begin{aligned}\rho(k) + \rho_h(k) &= \cos k \int_{-B}^{B} d\lambda a_{U/4}(\lambda - \sin k)\sigma(\lambda) \\ &- \frac{\cos k}{2}[a_{(2S+1)U/4}(\sin k - \theta) - a_{U/2}(\sin k - \theta)] \ , \\ \sigma(\lambda) + \sigma_h(\lambda) + \int_{-B}^{B} d\lambda' a_{U/2}(\lambda - \lambda')\sigma(\lambda') \\ &= \int_{-Q}^{Q} dk a_{U/4}(\lambda - \sin k)\rho(k) + a_{SU/2}(\lambda - \theta) \ .\end{aligned} \quad (7.155)$$

The energy, magnetization and the valence of the impurity are given by

$$\begin{aligned} e_{imp} &= -2\int_{-Q}^{Q} dk \cos k\rho(k) \ , \qquad n_{imp} = \int_{-Q}^{Q} dk\rho(k) \ , \\ m_{imp}^z &= S + \frac{1}{2}\int_{-Q}^{Q} dk\rho(k) - \int_{-B}^{B} d\lambda\sigma(\lambda) \ . \end{aligned} \quad (7.156)$$

For $H = 0$ we have $B = \infty$. For the half-filled host (*i.e.*, for the insulator) the valence of an impurity is zero independent on θ and S. In the limit $U \to 0$ the valence is $n_{imp} = 1$ for $|\theta| < \sin Q$ and zero otherwise, while for $U \to \infty$ it is always zero. The valence monotonically decreases with increasing U and has its maximum at $Q = \pi/2$. The energy of an impurity monotonically increases with $|\theta|$, while the valence monotonically increases for the metallic case. The effect is largest when the impurity parameter θ lies in the Dirac sea for charged excitations, because it is in resonance with itinerant electron states. The magnetization of an impurity $m_{imp}^z = S - \frac{1}{2}$ for $H = 0$. For the weak magnetic field it is totally screened by host excitations for $S = \frac{1}{2}$ with the finite magnetic susceptibility. The relation of the magnetic susceptibilities of the impurity and the host for large $|\theta|$ and sufficiently large U is

$$\frac{\chi_{imp}}{\chi_{host}} = \frac{e^{2\pi|\theta|U} + n_{imp}}{N/L} \ . \quad (7.157)$$

For $S > \frac{1}{2}$ in the metallic case in the Kondo limit, where charge fluctuations can be neglected we obtain

$$m^z_{imp} = \mu_i \left(1 \pm \frac{1}{2|\ln(H/T_K)|} - \frac{\ln|\ln(H/T_K)|}{4\ln^2(H/T_K)} + \cdots \right) , \qquad (7.158)$$

where $T_K \sim \exp(-2\pi|\theta|U)$. Here we use for $H \gg T_K$ the lower sign and $\mu_i = S$, and for $H \ll T_K$ we use the upper sign and $\mu_i = S - \frac{1}{2}$. This implies that the impurity spin is underscreened at low fields to the value $S - \frac{1}{2}$, and for high enough values of the magnetic field the impurity spin behaves as the asymptotically free spin S. Naturally, at $H = H_s$ the impurity spin is S, and the impurity susceptibility reveals a square root divergence, as the consequence of the van Hove singularity of the empty band.

For the attractive Hubbard chain $U < 0$ we study the impurity scattering matrix of the form

$$\hat{S}^{\sigma,\sigma'}_{M,M'}(x) = \frac{A\delta_{\sigma,\sigma'}\delta_{M,M'} + B\delta_{-\sigma',\sigma}\delta_{M',M+2\sigma}}{x - \theta - iU(2S'+1)/4} , \qquad (7.159)$$

where

$$\begin{aligned} A &= x - \theta - i\frac{U(2S'+1)}{4}[1 - 2(\sigma M|M+\sigma)(\sigma' M'|M'+\sigma')] , \\ B &= iU(2S'+1)(\sigma M|M+\sigma)(\sigma' M'|M'+\sigma')/2 , \end{aligned} \qquad (7.160)$$

with $S = S' + \frac{1}{2}$. This two-particle scattering matrix also satisfies Yang–Baxter relations with the two-particle scattering matrices of electrons for the Hubbard chain. Bethe ansatz equations for such a system with periodic boundary conditions can be written as

$$\begin{aligned} \frac{\lambda_\alpha - \theta + iUS'/2}{\lambda_\alpha - \theta - iUS'/2} \prod_{j=1}^{N} \frac{\lambda_\alpha - \sin k_j + iU/4}{\lambda_\alpha - \sin k_j - iU/4} &= \prod_{\substack{\beta=1 \\ \beta\neq\alpha}}^{M} \frac{\lambda_\alpha - \lambda_\beta + iU/2}{\lambda_\alpha - \lambda - iU/2} , \\ e^{ik_j L} \frac{\sin k_j - \theta + iU(2S'+1)/4}{\sin k_j - \theta - iU(2S'+1)/4} &= \prod_{\beta=1}^{M} \frac{\sin k_j - \lambda_\beta + iU/4}{\sin k_j - \lambda_\beta - iU/4} , \end{aligned} \qquad (7.161)$$

where $j = 1, \ldots, N$ and $\alpha = 1, \ldots, M$. In the ground state Bethe ansatz integral equations for densities of an impurity are:

$$\rho(k) + \rho_h(k) + \cos k \int_{-Q}^{Q} d\lambda a_{U/4}(\lambda - \sin k)\sigma'(\lambda)$$
$$= \cos k \; a_{U(2S'+1)/4}(\sin k - \theta) \ ,$$
$$\sigma'(\lambda) + \sigma'_h(\lambda) + \int_{-Q}^{Q} d\lambda' a_{U/2}(\lambda - \lambda')\sigma'(\lambda') \tag{7.162}$$
$$= - \int_{-B}^{B} dk a_{U/4}(\lambda - \sin k)\rho(k) + a_{(S'+1)U/2}(\lambda - \theta) \ .$$

The valence of an impurity, magnetization and energy are given by

$$n_{imp} = \int_{-B}^{B} dk\rho(k) + 2\int_{-Q}^{Q} d\lambda\sigma'(\lambda) \ , \quad m_{imp}^z = S' + \frac{1}{2}\int_{-B}^{B} dk\rho(k) \ ,$$
$$e_{imp} = -2\int_{-B}^{B} dk \cos k\rho(k) \tag{7.163}$$
$$- 4\mathrm{Re}\int_{-Q}^{Q} d\lambda\sqrt{1 - [\lambda - i(U/4)]^2}\sigma'(\lambda) \ .$$

For $H < H_c$ the valence of an impurity varies between 0 and 1. For $U \to 0$ we have $n_{imp} = 1$ for $|\theta| < Q$ and zero otherwise. For large $U \to \infty$ the impurity valence tends to zero. It is a monotonically decreasing function of θ, while the energy of an impurity is a monotonically increasing function. For $H < H_c$, where only spin-singlet pairs have their Dirac sea, the impurity magnetization is equal to $S - \frac{1}{2}$. When H_c is approached from above the impurity magnetic susceptibility has the square root singularity, similar to the case of an impurity in the anisotropic t-J chain. At H_s, where the Dirac sea for pairs is empty, the magnetization of an impurity is equal to $m_{imp}^z = S - \frac{1}{2} + \frac{n_{imp}}{2}$. We see, that such a behaviour of an impurity, where there is no characteristic Kondo logarithmic dependencies, is typical for a magnetic impurity embedded into a one-dimensional correlated electron host with spin-gapped low-lying excitations. The strong van Hove singularity of the empty Dirac sea (the singularity due to the quantum phase transition in the host) is more stronger than weak Kondo logarithms in these cases. On the other hand, the valence of an impurity depends on the band filling and on the parameter of an impurity θ, but also depends on the value of the external magnetic field. The last feature is the manifestation of strong correlations in the chain.

Some other possibility of integrable impurities in correlated electron hosts are known. It is worthwhile mentioning the model, in which an impurity site differs from other host sites of the supersymmetric t-J chain by different representation. For example, this site permits four states, unlike other sites of the chain, where two states with electron either with spin up or down and the empty state are possible. For other models one replaces an empty state of an impurity by the state with two electrons. Hamiltonians of such models are very complicated, too, but the characteristic feature of Bethe ansatz solvable Hamiltonians with an impurity survives. An impurity is coupled to two neighboring sites of the host, it renormalizes the interaction and hopping between these sites of the host, and causes three-site terms, which violate T and P symmetry, but preserve PT symmetry. Again, choosing open boundary conditions with zero boundary fields and potentials one can avoid unwanted terms in the impurity Hamiltonian. For instance, the valence of an impurity with four possible states varies with the band filling from 0 to 2, as expected. The magnetic susceptibility of such an impurity is finite. There exists some critical electron density, which depends on the external magnetic field H, at which the charge susceptibility of an impurity (charge stiffness) has a feature. We again refer the interested reader to original publications.

To summarize, in this chapter we presented to the reader with exact results for behaviours of single magnetic impurities in quantum spin chains and correlated electron chains. We started with the description of magnetic impurities in simple XY chains, then followed by the description of single magnetic impurities in Heisenberg quantum spin chains. Finally, the Bethe ansatz description of the behaviour of single magnetic and hybridization impurities in correlated electron chains was presented. We compared the behaviour of an impurity with the behaviour of host sites and with the behaviour of edges of open chains and with the behaviours of the Anderson impurity model and the Kondo model.

Studies of the behaviour of a magnetic impurity in an isotropic XY chain can be found in [Tjon (1970); Kleiner and Tsukernik (1975); Kleiner and Tsukernik (1980)]. The behaviour of an impurity in a dimerized spin-$\frac{1}{2}$ XY chain is presented in [Zvyagin and Segal (1995)]. The first Bethe ansatz study of a spin-S impurity in a spin-$\frac{1}{2}$ periodic chain is [Andrei and Johannesson (1984)], for higher-spin host see [Schlottmann (1991)]. The reader can find the study of a two-parametric magnetic impurity (which can have the same spin as the host ones) in an open and periodic spin chain in [Frahm and Zvyagin (1997b)], for thermodynamics see also [Zvyagin (2002)]. The behaviour of a magnetic impurity in a uniaxial spin chain is described in [Schlottmann (1999);

Schlottmann (2000)]. The comparison of characteristics of a magnetic impurity in a quantum spin chain and in a metal (for Bethe ansatz equations of a Kondo impurity, see, *e.g.*, [Wiegmann (1981)] for a single-channel case and [Tsvelick and Wiegmann (1984)] for a multi-channel situation) can be found in [Zvyagin (2002)]. The first co-ordinate Bethe ansatz description of a magnetic impurity in a correlated electron gas with a δ-function attraction was in [Schulz (1987)]. The two-parametric magnetic impurity in a supersymmetric t-J periodic chain was introduced and solved in [Schlottmann and Zvyagin (1997a)]. For a comparison use the exact solution of the Anderson impurity model [Wiegmann and Tsvelick (1983)]. For an impurity in an open t-J chain, and for the algebraic Bethe ansatz description of the magnetic impurity in an integrable chain consult [Zvyagin (1997)], see also [Zvyagin and Johannesson (1998)], where the ferromagnetic coupling of an impurity to the host was considered, and [Zvyagin (2003)]. The reader can find the Bethe ansatz description of an impurity in a Hubbard chain in [Zvyagin and Schlottmann (1997)]. Other (nonmagnetic) integrable impurities in correlated electron chains are presented, *e.g.*, in [Bedürftig, Eßler and Frahm (1996); Foerster, Links and Tonel (1999)]. An interesting approach for integrable impurities, which can be simultaneously scatterers and reflectors was recently proposed in [Mintchev, Ragoucy and Sorba (2002)].

Chapter 8

Correlated Quantum Chains with a Finite Concentration of Impurities

In this chapter we shall present Bethe ansatz results for thermodynamic characteristics of a finite concentration of impurities in quantum spin and correlated electron chains. We shall mainly consider two effects: the onset of impurities-induced bands and the behaviour of disordered impurities. Both these effects qualitatively differ the situation with the finite concentration of impurities from the behaviour of a single impurity.

8.1 Impurities' Bands

The problem of the behaviour of many impurities in quantum chains is a special problem of great interest. From the previous chapter the reader already knows how the behaviour of an isolated (single) impurity differs from the behaviour of host particles in quantum correlated chains. If the number of impurities, N_i, is much less then the number of host sites, L, (or, in other words, the concentration of impurities $c \ll 1$) one can consider those impurities as independent and their contribution to characteristics of the studied system can be considered as additive. On the other hand, the reader is aware that, *e.g.*, in metals with a finite concentration of magnetic impurities, a reflection of conduction electrons off magnetic impurities produces an impurity-impurity long-range interaction (known as the Ruderman–Kittel–Kasuya–Yosida, or RKKY, coupling). Naturally, properties of interacting magnetic impurities in metals are then very different from those of isolated impurities. This, actually, determines the aim of this chapter: we want to inform the reader about exact results, which are known for the situation with many impurities in correlated electron hosts (in the insulator case, where only spin degrees of freedom possess the dynamics, and in the metallic case).

Let us start from the description of the behaviour of impurities of a finite concentration in quantum spin chains. Generally speaking, one can solve the stationary Schrödinger equation for the XY spin-$\frac{1}{2}$ chain with many embedded impurities, because by using the Jordan–Wigner transformation one can exactly transform such a Hamiltonian to a quadratic form of Fermi operators. However, the explicit diagonalization of that form is a very difficult problem (one needs to diagonalize $L \times L$ matrix of the tridiagonal, in the case of the nearest-neighbor interactions, different from each other matrix elements).

One of the great advantages of Bethe ansatz integrable impurities, studied in the previous chapter, is the possibility to find thermodynamic characteristics of quantum chains with many impurities. It is based on the fact that, by construction, one can introduce any number of L-operators of integrable impurities into the monodromy operator of a quantum exactly solvable chain. All monodromy operators, obtained that way, satisfy intertwining (Yang–Baxter) relations with R-matrices of the host, and, hence, transfer matrices of those monodromies with different spectral parameters commute. This constitutes the exact integrability of such systems. Naturally, all integrable impurities, studied in the previous chapter, which can be introduced to a monodromy, preserving the Bethe ansatz solvability, have the same property: they do not produce any reflection. This, naturally, limits the applicability of exact results, which can be obtained. For example, it is impossible to have the RKKY-like interaction between impurities. Nevertheless, in the framework of the Bethe ansatz it is possible to introduce an interaction between the nearest-neighboring impurities, which can produce features of thermodynamic characteristics of integrable impurities, reminiscent of real magnetic impurities in metals with the RKKY interaction.

Actually, the Hamiltonian of a spin chain with the finite concentration of impurities, c, is the same as Eqs. (7.48), (7.55), or (7.78), with many impurities embedded between sites of the host chain. Suppose the concentration of impurities is such that many of them are situated at the nearest-neighboring links of the original host chain. Then, it is obvious from the construction of the Hamiltonian of impurities and its' algebraic Bethe ansatz, that one can consider the other Hamiltonian, for which impurities and host sites are interchanged. Naturally, this corresponds to the situation, in which "impurities" define "host" chain, and original host sites play the role of impurities situated at links between the sites with impurities and coupled to the neighboring sites with impurities. The situation is reminiscent of zig-zag spin chains, in which there is a nearest-neighbor and the next to nearest neighbors interaction. Additional Hamiltonian has the same structure as the original one, by definition. It is easy to check that transfer matrices of these two Hamiltonians commute for any spectral parameters in the case if shifts, θ, are the same for both systems. This fact depends neither on how many neighboring impurities are connected with each other (or, in other words, how long these effective spin clusters are, connected to each other *via* zig-zag-like interaction, which is determined by the parameter θ), nor on overall multipliers (J) in front of each Hamiltonian. Then, the commutation of transfer matrices implies that these Hamiltonians have the same set of eigenfunctions. Summarizing, one can introduce a direct coupling between integrable impurities situated at neighboring links of the host Bethe ansatz-integrable chain, for the illustration, see Fig. 8.1, and, in the case if interactions between the host and impurities and impurity-impurity interaction are related to the same parameter θ, such an impurity-impurity interaction does not violate the exact integrability of the problem. Naturally, this fact pertains not only to quantum spin chains, but, also, to correlated electron systems with impurities, constructed this way.

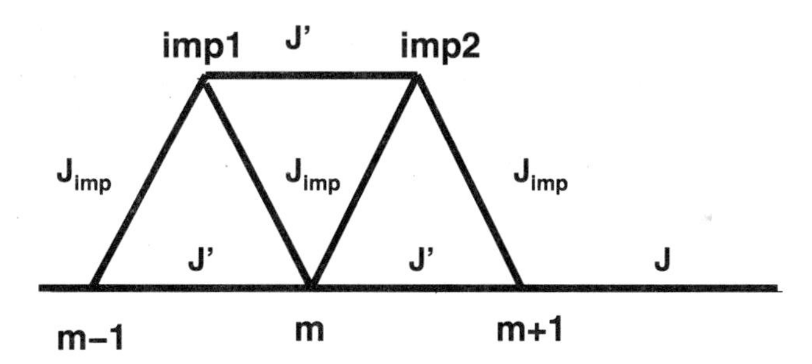

Fig. 8.1 Illustration of impurity-host and impurity-impurity interactions. In the simplest case of the isotropic Heisenberg spin-$\frac{1}{2}$ chain the local impurity-host exchange constant is $J_{imp} = 4J/[4\theta_j^2 + (2S' + 1)^2]$, and $J' = \theta_j^2 J_{imp}$.

Bethe ansatz equations for a spin S Heisenberg (Takhtajan–Babujian) chain with $N_i = cL$ (please, do not confuse this with the coupling constant of the algebraic Bethe ansatz description, set to unity here) embedded impurities with spin S' can be written as:

$$e_{2S}^{L}(\lambda_j)e_{2S'}^{N_i}(\lambda_j - \theta) = \prod_{\substack{l=1 \\ l \neq j}}^{M} e_2(\lambda_j - \lambda_l) , \qquad j = 1, \ldots, M , \tag{8.1}$$

where $e_n(x) = (2x + in)/(2x - in)$, the total magnetic moment is $M^z = LS + N_i S' - M$, and the energy is

$$E = E_0 - \sum_{j=1}^{M} \left(\frac{J(1-c)S}{\lambda_j^2 + S^2} + \frac{JcS'}{(\lambda_j - \theta)^2 + (S')^2} - H \right) , \tag{8.2}$$

where E_0 is the energy of the ferromagnetic state (of the mathematical vacuum with all spins directed upward). For $c = 0$ or $c = 1$ we recover the standard Bethe ansatz equations for a Takhtajan–Babujian homogeneous spin-S or spin-S' chain, *cf.* Chapter 5. Notice, that for $c = \frac{1}{2}$ the system reduces to an alternating spin-$S - S'$ zig-zag-like chain. It is clear also that for $N_i = 1$ the situation reduces to the case of a single impurity, studied in the previous chapter. The reader can, obviously, see that Bethe ansatz equations do not depend on whether θ is real or complex. Moreover, we see from the expression for the energy that the energy is ever real, does not matter whether θ is real or complex. However, as the reader knows from the previous chapter, the Hamiltonian of an integrable system is non-Hermitian for imaginary θ (it is senseless to consider only two impurities at edges of an open chain to look for a concentration dependence of thermodynamic characteristics of a system with a finite concentration of impurities). This is why, from now on in this chapter we limit ourselves by real θ. Also, we shall consider here the case of $L + N_i$ even.

In the framework of the string hypothesis integral Bethe ansatz equations for for dressed energies are (with the notations, similar to previous chapters)

$$\begin{aligned} T \ln[1 + \eta_n(\lambda)] - T \sum_{n=1}^{\infty} \int_{-\infty}^{\infty} d\lambda' A_{nm}(\lambda - \lambda') \ln[1 + \eta_m^{-1}(\lambda)] \\ = nH - \pi J(1-c) \sum_{l=1}^{\min(n,2S)} a_{n+2S+1-2l}(\lambda) \\ - \pi Jc \sum_{l=1}^{\min(n,2S')} a_{n+2S'+1-2l}(\lambda - \theta) , \end{aligned} \tag{8.3}$$

where ρ_n, ρ_{nh}, $\varepsilon_n = T \ln(\rho_{nh}/\rho_n) = T \ln \eta_n$ are the density, density of holes, and the dressed energy of the spin string of length n. The set of integral equations for densities of spin strings of length n has the form

$$\rho_{nh}(\lambda) + \sum_{n=1}^{\infty} \int_{-\infty}^{\infty} d\lambda' A_{nm}(\lambda - \lambda')\rho_m(\lambda') = (1-c)$$
$$\times \sum_{l=1}^{\min(n,2S)} a_{n+2S+1-2l}(\lambda) + c \sum_{l=1}^{\min(n,2S')} a_{n+2S'+1-2l}(\lambda - \theta) \ . \quad (8.4)$$

The internal energy per site is

$$e = e_0 + \sum_{n=1}^{\infty} \int_{-\infty}^{\infty} d\lambda \rho_n(\lambda) \Bigg[nH - \pi J(1-c) \sum_{l=1}^{\min(n,2S)} a_{n+2S+1-2l}(\lambda)$$
$$- \pi Jc \sum_{l=1}^{\min(n,2S')} a_{n+2S'+1-2l}(\lambda - \theta) \Bigg] , \quad (8.5)$$

the Helmholtz free energy per site and the magnetization per site are equal to

$$f = f_0 - T \int_{-\infty}^{\infty} \frac{d\lambda}{2\cosh(\pi\lambda)} [(1-c)\ln[1+\eta_{2S}(\lambda)] + c\ln[1+\eta_{2S'}(\lambda-\theta)]] ,$$
$$m^z = cS' + (1-c)S - \sum_{n=1}^{\infty} n \int_{-\infty}^{\infty} d\lambda \rho_n(\lambda) , \quad (8.6)$$

where

$$f_0 = J(1-c)S\left(\psi[(1/4) + (S/2)] - \psi[(3/4) + S/2)]\right)$$
$$+ JcS'\left(\psi[(1/4) + (S'/2) - i\theta] - \psi[(3/4) + S/2) - i\theta]\right) , \quad (8.7)$$

and $\psi(x)$ are digamma functions.

Let us consider the actions of parameters of impurities, S' and θ, separately. First, let us investigate which contribution $S \neq S'$ yields for $\theta = 0$. In such a case in the ground state only two kinds of strings, the ones of length $2S$, and the ones of length $2S'$ can have negative energies, *i.e.*, their Dirac seas. Energies of all other states (solutions to Bethe ansatz equations) are non-negative. At $H = 0$ the solution of the ground state equations for dressed energies is a singlet with

$$\frac{\varepsilon_{2S}}{(1-c)} = \frac{\varepsilon_{2S'}}{c} = -\frac{\pi J}{2\cosh(\pi\lambda)} ,$$
$$\frac{\rho_{2S}}{(1-c)} = \frac{\rho_{2S'}}{c} = \frac{1}{2\cosh(\pi\lambda)} . \quad (8.8)$$

It is important to point out here that the above results are valid only for finite c. On the other hand, if $c \to 0$ ($N_i = 1$), the situation is different since only spin strings of length $2S$ (but not of $2S'$) have their Dirac sea. In this limit $\varepsilon_{2S'}$ vanishes, and one has $\rho_{2S'} = [2L\cosh(\pi\lambda)]^{-1}$. Consequently, the limit $N_i \to 1$ ($c \to 0$) is singular. For isolated impurities at $H = T = 0$ one gets the remnant magnetization $S' - S$ for $S' > S$ and critical non Fermi liquid behaviour for $S' < S$, *i.e.*, it is not a singlet. As the reader knows from the previous chapter, there exists a remnant entropy of a single impurity for $S \neq S'$. Hence, the situation with a single impurity can be qualitatively different from the case of a finite concentration of such impurities. It is namely because the onset of the new band of impurities' states (the new Dirac sea), $\varepsilon_{2S'}$, which appears only for finite concentrations of similar impurities, due to the interaction between nearest impurities.

To be concrete, let us write down integral equations for dressed energies and densities for the case $S = \frac{1}{2}$ and $S' = 1$ (other cases can be studied in a similar way). Those ground state integral equations have the form

$$\begin{aligned}
\varepsilon_n(\lambda) &= nH - \pi J g_n(\lambda) - \int_{-B_1}^{B_1} d\lambda' K_{n1}(\lambda - \lambda')\varepsilon_1(\lambda') \\
&\quad - \int_{-B_2}^{B_2} d\lambda' K_{n2}(\lambda - \lambda')\varepsilon_2(\lambda') , \\
\rho_n(\lambda) + \rho_{nh}(\lambda) &= g_n(\lambda) - \int_{-B_1}^{B_1} d\lambda' K_{n1}(\lambda - \lambda')\rho_1(\lambda') \\
&\quad - \int_{-B_2}^{B_2} d\lambda' K_{n2}(\lambda - \lambda')\rho_2(\lambda') ,
\end{aligned} \tag{8.9}$$

where $n = 1, 2$ and

$$\begin{aligned}
&K_{11}(x) = a_2(x) , \qquad K_{22}(x) = a_4(x) + 2a_2(x) , \\
&K_{12}(x) = K_{21}(x) = a_3(x) + a_1(x) , \\
&g_1(x) = (1 - c)a_1(x) + ca_2(x) , \\
&g_2(x) = (1 - c)a_2(x) + ca_1(x) + ca_3(x) .
\end{aligned} \tag{8.10}$$

Here $B_{1,2}$ are the Fermi points of Dirac seas, determined from the conditions $\varepsilon_{1,2}(\pm B_{1,2}) = 0$. Naturally, for $H = 0$ both $B_{1,2} = \infty$. Both these limits decrease with increasing the value of the magnetic field H. However, for c small B_2 decreases faster than B_1. This situation holds for any $S' > S$.

The monotonic decrease of the values of the Fermi points implies that the Dirac seas for ε_{2S} and $\varepsilon_{2S'}$ are gradually depleted with the field. The magnetization is zero at $H = 0$ and is monotonically increased with H until the Dirac sea for bound spin states with higher spin (for $c < (1-c)$) becomes empty. For example, for $S = \frac{1}{2}$ the Dirac sea for $\varepsilon_{2S'}$ ($S' > \frac{1}{2}$) is depleted first at the critical field $H_c(c)$, defined as

$$H_c = \frac{J}{2S'}\left[\frac{1-c}{S'} + c\sum_{l=1}^{2S'}\frac{2}{4S'+1-2l}\right] + \frac{1}{2S'}\int_{-B_1}^{B_1} d\lambda \varepsilon_1(\lambda)[a_{2S'+1}(\lambda) + a_{2S'-1}(\lambda)] \ . \tag{8.11}$$

As the consequence of the van Hove singularity of the empty band (Dirac sea) for ε_{2S} the magnetization per site has a cusp with the infinite slope when $H_c(c)$ is approached from below. At $H_c(c)$ the ground state magnetic susceptibility diverges. This value of the magnetic field pertains to the second order quantum phase transition. Naturally, for $c \to 0$, the critical field $H_c(c)$ tends to zero, giving rise to the singular behaviour of a single impurity at $H = T = 0$. For $H = H_c$ only the Dirac sea ε_{2S} ($S < S'$) is partially filled. The magnetization increases monotonically with increasing field until the Dirac sea ε_{2S} (for $S < S'$ and $c < (1-c)$) becomes empty. For example, for $S = \frac{1}{2}$ this happens at

$$H_s(c) = J\frac{2S'(1-c)+c}{S'} \ . \tag{8.12}$$

At this quantum critical line the second order quantum phase transition to the spin-saturated (ferromagnetic) phase takes place. Close to $H = H_s(c)$ the ground state magnetic susceptibility diverges as $1/\sqrt{H_s - H}$ (and the magnetization has a cusp). For $H > H_s$ the magnetic susceptibility is zero and the ground state magnetization per site is $(1-c)S + cS'$. $H_c(c)$ decreases linearly with c, while $H_s(c)$ increases linearly with c. For $c > 1-c$ the Dirac seas for ε_{2S} and $\varepsilon_{2S'}$ interchange their roles in the ground state behaviour in the external magnetic field.

The low temperature specific heat of the chain is proportional to T (*i.e.*, the Sommerfeld coefficient is constant), except of critical lines $H_{c,s}(c)$, where it is proportional to $\sqrt{T}$.

Now let us study the effect of a nonzero coupling constant θ. Here we limit ourselves, *e.g.*, with the case $S = S' = \frac{1}{2}$ (other cases can be studied analogously). Consider the ground state behaviour of a Heisenberg spin chain with a finite concentration of impurities. The ground state pertains to solutions of Bethe ansatz equations with negative energies. The set of integral equations for densities of rapidities and dressed energies of low-lying excitations (spinons for $S = S' = \frac{1}{2}$) are:

$$\rho_1(\lambda) + \rho_{1h}(\lambda) + \int_{(B)} d\nu a_2(\lambda - \nu)\rho_1(\nu) = (1-c)a_1(\lambda) + ca_1(\lambda - \theta) \quad (8.13)$$

and

$$\varepsilon_1(\lambda) + \int_{(B)} d\nu a_2(\lambda - \nu)\varepsilon_1(\nu) = H - \pi J(1-c)a_1(\lambda) - \pi J c a_1(\lambda - \theta) \ . \quad (8.14)$$

Integrations are performed over the domain (B), determined in such a way that dressed energies inside these intervals are negative. The limits of integrations are determined by zeros of dressed energies; they are Fermi points for each Dirac sea. The ground state for $H = 0$ is known, see above for $S = S' = \frac{1}{2}$. There Fermi points for the Dirac sea of spinons are $\pm\infty$. In the nonzero external magnetic field $H \neq 0$ the situation appears to be very different for different values of the coupling constant θ and c. Let us for simplicity shift $\lambda \to \lambda - (\theta/2)$ (nothing, naturally, depends on such a shift for a periodic system).

For small $\theta < \theta_c(c)$ (the critical value of θ_c is dependent of the concentration of impurities c) there is only one Fermi sea for spinons (*i.e.*, there is only one minimum in the distribution of dressed energies), *cf.* Fig. 8.2. For $H \geq H_s$ and $\theta < \theta_c$ the system is in the spin-saturated, ferromagnetic phase at zero temperature and spinons become gapped. The magnetic susceptibility manifests a square root singularity at H_s: $\chi(H) \sim 1/\sqrt{(H_s - H)}$. It is related to the one-dimensional van Hove singularity of an empty Dirac sea of spinons. In a weak magnetic field the magnetic susceptibility is proportional to $\chi \propto (\pi^2 J)^{-1}(1 + (2|\ln AH|)^{-1} - (\ln|\ln AH|/4\ln^2 AH) + \cdots)$ where A is a non-universal constant (logarithmic corrections appear due to SU(2)-symmetry of the model).

At $\theta = \theta_c$ there is also only one minimum in the distribution of dressed energies. However, this extremum is more flat than for $\theta < \theta_c$. Instead of the behaviour of the dressed energy near the minimum $\sim \lambda^2$ for $\theta < \theta_c$, for $\theta = \theta_c$ it is proportional to λ^4, *cf.* Fig. 8.3. This is why, the magnetic susceptibility manifests the different singularity $\chi(H) \sim 1/(H_s - H)^{3/4}$ at H_s for $\theta = \theta_c$. For $H \geq H_s$ elementary excitations are gapped and the ground state magnetic susceptibility is zero.

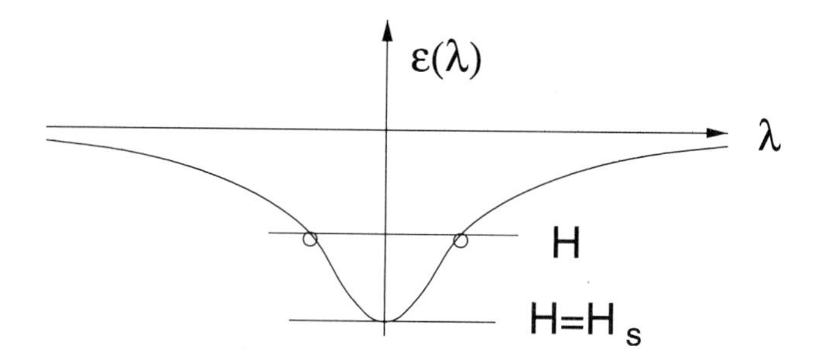

Fig. 8.2 Dressed energy of spinons as a function of a spectral parameter of a Heisenberg spin chain with a finite concentration of impurities for $\theta < \theta_c$. For illustrative purposes we shifted the spectral parameter by $-\theta/2$ and consider the case $c = \frac{1}{2}$.

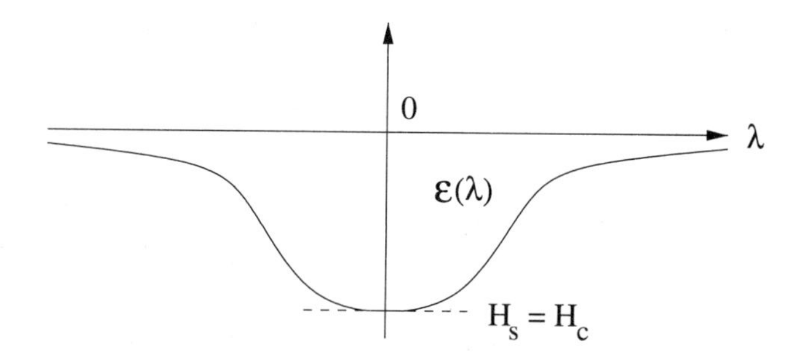

Fig. 8.3 The same as in Fig. 8.2 but for $\theta = \theta_c$.

For $\theta > \theta_c$ two minima for the dressed energy of spinons appear at $\lambda = \pm\theta/2$, *i.e.*, there are two minima and one maximum in the distribution of dressed energies, *cf.* Fig. 8.4.

Hence, the ground state behaviour strongly depends on the value of an external magnetic field and the concentration of impurities. There are two critical values of the field for $\theta > \theta_c$. For $H \geq H_s$ the system is in the spin-saturated phase with gapped spinon excitations. The magnetic susceptibility also manifests square-root singularity at H_s ($\chi \sim 1/\sqrt{H_s - H}$). For $H < H_s$ spinons are gapless. There exists an additional critical value of an external magnetic field, H_c, which also depends on c and θ (it is, unfortunately, impossible to find the explicit analytic expression for H_c, because one needs a numerical solution of the Fredholm integral equation with finite limits of integration). For $H < H_c$ and $\theta > \theta_c$ there is only one Dirac sea for spinons. However, for $H > H_c$ the behaviour of a spin chain is drastically changed: there are two Dirac seas for spinons (with four Fermi

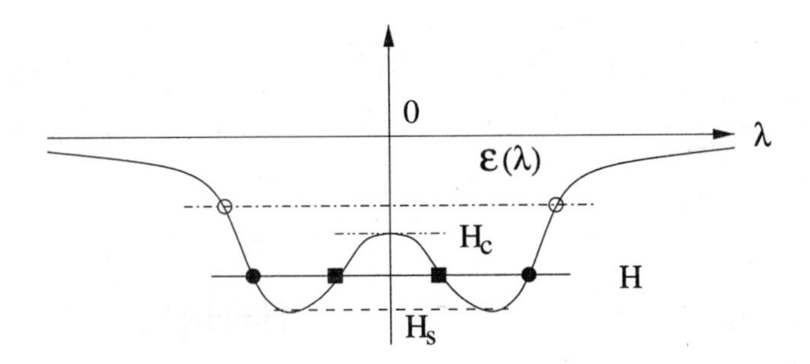

Fig. 8.4 The same as in Fig. 8.2 but for $\theta > \theta_c$. Filled circles and filled squares denote Fermi points for "particles" and "holes" for $H \neq 0$. Notice that for $H < H_c$ there are only two Fermi points (open circles).

points). One can also speak about the onset of the Dirac sea of "holes" of spinons (related to the onset of the maximum in the dependence of the dressed energy on λ) for $H > H_c$. According to this picture the critical field H_c pertains to the van Hove singularity of the empty band of these "holes" of spinons. Notice, that at $\theta = \theta_c$, $H_c = H_s$, *i.e.*, this point is tricritical. Fillings of these two Dirac seas for spinons for $H > H_c$, $\theta > \theta_c$ are not independent. This is the direct consequence of the fact that the same magnetic field determines fillings of the Dirac seas for "particles" and "holes", or, in other words, fillings of two Dirac seas for spinons centered at $\lambda = \pm\theta/2$. The Dirac sea for "holes" disappears, naturally, for $H \to H_c$, $\theta \to \theta_c$. We have to point out here that there is a crucial difference between behaviours of a quantum spin chain with a finite concentration of impurities and correlated electron models in the metallic phase. In the later case two Dirac seas of the ground states are connected with different kinds of excitations, *e.g.*, unbound electrons and spinons for a repulsive Hubbard chain, or Cooper-like singlet pairs and unbound electrons for a supersymmetric t-J chain for $V = -J/4$ or an attractive Hubbard chain. They pertain to two different kinds of Lagrange multipliers: the chemical potential and magnetic field. Thus, low-lying excitations are practically independent of each other (spin-charge separation). [Note that spin and charge sectors are connected though *via* integral equations. This is the consequence of the fact that unbound electrons carry both charge and spin.] On the other hand, two Dirac seas appear for the same kinds of excitations for a spin chain with a finite concentration of impurities. Their fillings are governed by the same Lagrange multiplier, the magnetic field. Two Dirac seas appear for nonzero c due to two minima in the bare energy distribution and correspond to nonzero θ in Bethe ansatz equations. The ground state phase diagram of a spin chain with a finite concentration of impurities $H-\theta$ is presented in Fig. 8.5.

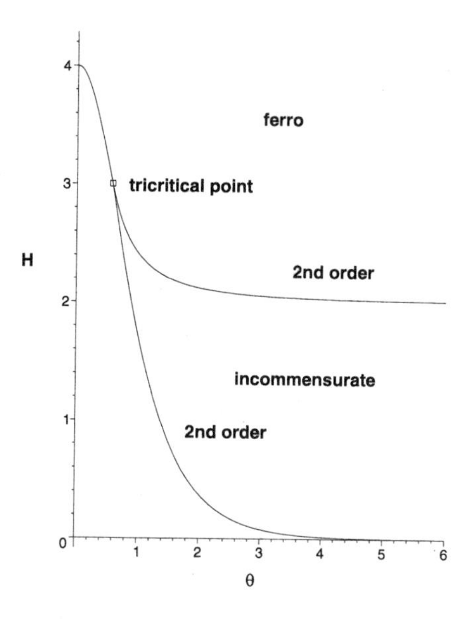

Fig. 8.5 The ground state phase diagram of a spin-$\frac{1}{2}$ Heisenberg antiferromagnetic chain with a finite concentration of impurities as a function of the magnetic field H and coupling constant of impurities θ.

The limit $c \to 0$ ($N_i = 1$) is again singular. For $c \to 0$ the critical field H_c goes to zero. The magnetic susceptibility of the impurity spin is, though, finite for $S = S'$, at low temperatures and it is determined by the parameter θ (*i.e.*, related to its' Kondo temperature). On the other hand, for $c \neq 0$ there is a second order quantum phase transition at $H = H_c$ with the square root singularity in the behaviour of the magnetic susceptibility at $T = 0$ for $\theta > \theta_c(c)$. At $H_c = H_s$ for $\theta = \theta_c(c)$ the singularity is more weak.

The low temperature behaviour of the magnetic susceptibility and the low-temperature Sommerfeld coefficient of the specific heat have square root features at $H = H_c$ for $\theta > \theta_c$, and in the tricritical point they are proportional to $T^{-3/4}$. The quantum phase transition at H_c is a transition between a commensurate (for $H < H_c$) and an incommensurate phase. Out of the lines of phase transitions the magnetic susceptibility is finite at low temperatures and Curie-like at high temperatures, while the Sommerfeld coefficient is finite, reminiscent of the Fermi liquid behaviour.

Summarizing, a finite concentration of impurities in quantum spin chains causes the onset of quantum phase transitions (related either to the different coupling of impurities to the host, or to different spins of impurities, or to both reasons together). These quantum phase transitions are consequences of the appearance of additional Dirac seas (bands) caused by the finite concentration of impurities. The limiting procedure to the behaviour of a single impurity is singular, because of van Hove singularities of those empty one-dimensional bands.

Now, let us consider the behaviour of correlated electron chains with a finite concentration of magnetic impurities.

Let us start with a supersymmetric t-J chain with a finite concentration of impurities. For $V=-J/4$, $J=2$ situation Bethe ansatz equations are

$$\begin{aligned} e_{2S-1}^{N_i}(\lambda_\alpha-\theta)\prod_{j=1}^{N}e_1(\lambda_\alpha-p_j) &= \prod_{\beta=1}^{M}e_2(\lambda_\alpha-\lambda_\beta)\ , \quad \alpha=1,\ldots,M\ , \\ e_{2S}^{N_i}(p_j-\theta)e_1^L(p_j) &= \prod_{\alpha=1}^{M}e_1(p_j-\lambda_\alpha)\ , \quad j=1,\ldots,N\ . \end{aligned} \tag{8.15}$$

The total magnetization is $M^z=(1-c)(N/2)+cS-M$, where $c=N_i/L$ is the concentration of impurities and the energy of the system is equal to

$$\begin{aligned} E = E_0 &+ 2\pi\sum_{j=1}^{N}[(1-c)a_1(p_j)+ca_{2S}(p_j-\theta)] \\ &- 2\pi c\sum_{\alpha=1}^{M}a_{2S-1}(\lambda_\alpha-\theta)\ , \end{aligned} \tag{8.16}$$

where the last two terms are related to impurities. If the energy is measured from the bottom of the conduction band, then $E_0=0$. Naturally, for $N_i=1$ we obtain Bethe ansatz equations for a single impurity. For $c=0$ (and $N_i=0$) we obtain standard Bethe ansatz equations for a supersymmetric t-J chain. At half filling, on the other hand, the model reduces to the spin-$\frac{1}{2}$ Heisenberg chain with a finite concentration of spin-S impurities (this can be seen more directly in the Sutherland's version of the Bethe ansatz equations, *cf.* Chapter 6).

As the reader already knows, the manifestation of differences in the behaviour of a quantum correlated chain with a finite concentration of impurities is most pronounced in the ground state. Let us write down the set of integral Bethe ansatz equations (we keep the notations of previous chapters) for dressed energies at $T = 0$

$$\varepsilon(p) = 2\pi(1-c)a_1(p) + 2\pi c a_{2S}(p-\theta) - \mu - \frac{H}{2}$$
$$- a_1(p-\lambda) * \psi(\lambda) + a_{2S-1}(p-\lambda) * \phi_{2S-1}(\lambda) \, ,$$
$$\psi(\lambda) = 2\pi(1-c)a_2(\lambda) + 2\pi c a_{2S+1}(\lambda-\theta) - 2\mu$$
$$- a_1(p-\lambda) * \varepsilon(p) - a_2(\lambda-\lambda') * \psi(\lambda) \, , \tag{8.17}$$
$$\phi_{2S-1}(\lambda) = (2S-1)H - 2\pi c \sum_{l=1}^{2S-1} a_{2l-1}(\lambda-\theta) - a_{2S-1}(\lambda-p) * \varepsilon(p)$$
$$- \left(a_{4S-2}(\lambda-\lambda') + 2\sum_{l=1}^{2S-2} a_{2l}(\lambda-\lambda') \right) * \phi_{2S-1}(\lambda') \, ,$$

where $*$ denotes the convolution over the intervals for which dressed energies are negative. One can see that a finite concentration of magnetic impurities results in the onset of an additional Dirac sea for spin strings of length $2S-1$. The reader can check, that for $c \to 0$ the second term in the third equation becomes zero and, hence, the dressed energy of spin strings of length $2S-1$ becomes positive for any H, like dressed energies of other spin excitations. Densities satisfy the equations

$$\rho(p) + \rho_h(p) = (1-c)a_1(p) + c a_{2S}(p-\theta) - a_1(p-\lambda) * \sigma'(\lambda)$$
$$- a_{2S-1}(p-\lambda) * \sigma_{2S-1}(\lambda) \, ,$$
$$\sigma'(\lambda) + \sigma'_h(\lambda) = (1-c)a_2(\lambda) + c a_{2S+1}(\lambda-\theta)$$
$$- a_1(p-\lambda) * \rho(p) - a_2(\lambda-\lambda') * \sigma'(\lambda) \, ,$$
$$\sigma_{2S-1}(\lambda) + \sigma_{2S-1,h}(\lambda) \tag{8.18}$$
$$= c\sum_{l=1}^{2S-1} a_{2l-1}(\lambda-\theta) + a_{2S-1}(\lambda-p) * \rho(p)$$
$$- \left(a_{4S-2}(\lambda-\lambda') + 2\sum_{l=1}^{2S-2} a_{2l}(\lambda-\lambda') \right) * \sigma_{2S-1}(\lambda') \, .$$

The number of electrons and the magnetization are given by

$$N = \int dp\rho(p) + 2\int d\lambda\sigma'(\lambda) \, , \; M^z = \frac{1}{2}\int d\lambda\sigma_{2S-1,h}(\lambda) \, , \tag{8.19}$$

and the ground state energy is

$$E = 2\pi \int dp[(1-c)a_1(p) + ca_{2S}(p-\theta)]\rho(p)$$
$$+ 4\pi \int d\lambda[(1-c)a_2(\lambda) + ca_{2S+1}(\lambda-\theta)]\sigma'(\lambda)$$
$$- 2\pi c(2S-1)\sum_{l=1}^{2S-1} \int d\lambda a_{2l-1}(\lambda-\theta)\sigma_{2S-1}(\lambda) . \qquad (8.20)$$

Naturally, the results for $N_i = 1$ coincides with the ones from the previous chapter. We see that again, the transition $c \to 0$ (with $N_i = 1$) is singular (for $S \neq \frac{1}{2}$, because of the additional Dirac sea for spin strings of length $2S-1$ and for $S = \frac{1}{2}$ because of the additional Dirac sea for $\theta > \theta_c$ in a nonzero magnetic field).

For $H = 0$ and $S \neq \frac{1}{2}$ we have $\phi_{2S-1}(\lambda) = -\pi c/\cosh[\pi(\lambda-\theta)]$. Hence, the Dirac sea of spin strings has no holes and the magnetization is zero. This differs from the case of a single impurity with $S > \frac{1}{2}$, for which the remnant magnetization, and, hence, the remnant entropy appears, *cf.* the previous chapter. The integral equation for dressed energies of spin-singlet pairs is decoupled for $H = 0$ because of the spin gap for unbound electron excitations

$$\psi(\lambda) = 2\pi(1-c)a_2(\lambda) + 2\pi ca_{2S+1}(\lambda-\theta) - 2\mu - a_2(\lambda-\lambda') * \psi(\lambda'). \quad (8.21)$$

The integration limits depend on the band filling. For $\theta < \theta_{c1}(c, S)$ there is only one Dirac sea for pairs. However, for $\theta > \theta_{c1}$ an additional Dirac sea can appear depending on the band filling. Hence, in this case we can observe a quantum phase transition between commensurate and incommensurate phases, which is governed by the number of electrons in a correlated electron chain. This transition is of the second order for $\theta > \theta_{c1}$ and there is a tricritical point at $\theta = \theta_{c1}$. The charge stiffness and the Sommerfeld coefficient of the specific heat are divergent at the critical line $\mu = \mu_{c1}$, with the low-temperature square root singularities. On the other hand, at the tricritical point their singularities are weaker, proportional to $T^{-3/4}$.

On the other hand, for $H > H_s$ the Dirac sea for spin-singlet pairs is depopulated and the Dirac sea for spin strings is also empty. The ground state behaviour is determined by unbound electron excitations with dressed energies

$$\varepsilon(p) = 2\pi(1-c)a_1(p) + 2\pi ca_{2S}(p-\theta) - \mu - \frac{H}{2} . \qquad (8.22)$$

The number of electrons is determined by the integral over p at which the energy of unbound electrons is negative. The saturation field H_s is given by the lowest magnetic field so that the Dirac sea for pairs is empty, *i.e.*, when $\mu = \frac{H}{2}$. In general, the saturation field has to be determined numerically. The magnetization in the spin-saturated phase is $M^z = (1-c)\frac{N}{2L} + cS$.

At nonzero magnetic field for small θ only one minimum exists for spin strings at $\lambda = 0$. Hence, in this region only one critical value of the magnetic field, H_s exists, at which the system undergoes a transition to the spin-saturated phase. For θ larger than some, generally speaking, other critical value $\theta_{c2}(c, S)$, there appears an additional Dirac sea for spin strings of length $2S-1$. This additional Dirac sea also affects the behaviour of the ground state characteristics of a correlated electron chain with a finite concentration of impurities. It reveals itself in the onset of an additional quantum critical point at H_c, which describes the quantum phase transition from a commensurate phase at $H < H_c$ to an incommensurate phase for $H > H_c$ (both for $\theta > \theta_{c2}$). This phase transition is related to the van Hove singularity of an additional Dirac sea of spin strings of length $2S-1$, which becomes empty at $H = H_c$. For larger values of the magnetic field a phase transition to the spin-saturated, ferromagnetic phase at $H = H_s$ takes place. At the lines $H = H_c$ and $H = H_s$ the magnetic susceptibility has square root singularities for the values of the magnetic field larger (smaller) than H_c (H_s), and at low temperatures the magnetic susceptibility and the Sommerfeld coefficient are proportional to $T^{-1/2}$. At the tricritical point $\theta = \theta_{2c}$, $H = H_c = H_s$ the quantum phase transition is weaker, with divergences, proportional to $T^{-3/4}$. Out of the lines of phase transitions the magnetic susceptibility is finite at low temperatures and Curie-like at high temperatures, while the Sommerfeld coefficient is finite, reminiscent of the Fermi liquid behaviour.

The situation with $S = \frac{1}{2}$ differs from the considered above, because there are no only spin-carrying low-lying excitations. In this case it is more convenient to use the expression for the magnetization

$$M^z = \frac{1}{2}\int dp\rho(p) \ . \tag{8.23}$$

However, dressed energies of spin-singlet pairs and unbound electron excitations can have additional Dirac seas due to nonzero θ for large θ. Hence, additional quantum phase transitions due to a finite concentration of impurities can also take place. These phase transitions again manifest that the limit $N_i = 1$ ($c \to 0$) is singular.

In the general case, to know the quantative behaviour of a supersymmetric t-J chain with a finite concentration of impurities one has to solve integral equations for dressed energies and densities numerically.

Now let us briefly consider the behaviour of a supersymmetric t-J chain with $V = 3J/4$ and $J = 2$ with a finite concentration of impurities. The Bethe ansatz describes such a system as the solution of equations

$$
\begin{aligned}
&e_1^L(p_j) = (-1)^N \prod_{\substack{l=1\\ l\neq j}}^{N} e_2(p_j - p_l) \prod_{\beta=1}^{M} e_1^{-1}(p_j - \lambda_\beta) \ , \quad j = 1, \ldots, N \ , \\
&e_{2S}^{N_i}(\lambda_\alpha - \theta) \prod_{j=1}^{N} e_1(\lambda_\alpha - p_j) = \prod_{\substack{\beta=1\\ \beta\neq\alpha}}^{M} e_2(\lambda_\alpha - \lambda_\beta) \ , \quad \alpha = 1, \ldots, M \ , \\
&E = -2\pi(1-c) \sum_{j=1}^{N} a_1(p_j) - 2\pi c \sum_{\alpha=1}^{M} a_{2S}(\lambda_\alpha - \theta) \ , \\
&M^z = \frac{(1-c)N}{2} + Sc - M \ ,
\end{aligned}
\tag{8.24}
$$

where the energy is measured from the bottom of the conduction band. The model reveals the most interesting properties in the ground state and at low temperatures.

The analysis is similar to the cases, considered above. The ground state set of Bethe ansatz equations for dressed energies is

$$
\begin{aligned}
&\varepsilon_1 + a_2 * \varepsilon_1 + (a_{2S-1} + a_{2S+1}) * \varepsilon_{2S} - a_1 * \phi_1 - a_{2S} * \phi_{2S} \\
&\qquad = -2\pi(1-c)a_1(p) - \mu - \frac{H}{2} \ , \\
&\varepsilon_{2S} + K_{2S} * \varepsilon_{2S} + (a_{2S-1} + a_{2S+1}) * \varepsilon_1 - K_1 * \phi_{2S} - a_{2S} * \phi_1 \\
&\qquad = -2\pi(1-c)a_{2S}(p) - 2S\mu - SH \ , \\
&\phi_1 + a_2 * \phi_1 + (a_{2S-1} + a_{2S+1}) * \phi_{2S} - a_1 * \varepsilon_1 - a_{2S} * \varepsilon_{2S} \\
&\qquad = H - 2\pi c a_{2S}(\lambda - \theta) \ , \\
&\phi_{2S} + K_{2S} * \phi_{2S} + (a_{2S-1} + a_{2S+1}) * \phi_1 - K_1 * \varepsilon_{2S} - a_{2S} * \varepsilon_1 \\
&\qquad = 2SH - 2\pi c K_1(\lambda - \theta) \ ,
\end{aligned}
\tag{8.25}
$$

where

$$
K_1(x) = \sum_{l=1}^{2S} a_{2l-1}(x) \ , \ K_{2S}(x) = a_{4S}(x) + 2 \sum_{l=1}^{2S-1} a_{2l}(x) \ . \tag{8.26}
$$

This set implies that the finite concentration of impurities causes the onset of additional Dirac seas for charged bound states of the length $2S$ and spin strings of the length $2S$, comparing to the case without impurities, *cf.* Chapter 4. The set of Bethe ansatz equations for densities is

$$\begin{aligned}
&\rho_1 + \rho_{1,h} + a_2 * \rho_1 + (a_{2S-1} + a_{2S+1}) * \rho_{2S} \\
&\qquad - a_1 * \sigma_1 - a_{2S} * \sigma_{2S} = (1-c)a_1(p) \ , \\
&\rho_{2S} + \rho_{2S,h} + K_{2S} * \rho_{2S} + (a_{2S-1} + a_{2S+1}) * \rho_1 \\
&\qquad - K_1 * \sigma_{2S} - a_{2S} * \sigma_1 = (1-c)a_{2S}(p) \ , \\
&\sigma_1 + \sigma_{1,h} + a_2 * \sigma_1 + (a_{2S-1} + a_{2S+1}) * \sigma_{2S} - a_1 * \rho_1 \\
&\qquad - a_{2S} * \rho_{2S} = c a_{2S}(\lambda - \theta) \ , \\
&\sigma_{2S} + \sigma_{2S,h} + K_{2S} * \sigma_{2S} + (a_{2S-1} + a_{2S+1}) * \sigma_1 - K_1 * \rho_{2S} \\
&\qquad - a_{2S} * \rho_1 = cK_1(\lambda - \theta) \ .
\end{aligned} \tag{8.27}$$

The analysis of the ground state and low temperature behaviour of this model is very similar to the above one. The case $N_i = 1$ ($c \to 0$) is singular, because of van Hove singularities of additional Dirac seas. These additional Dirac seas for a finite concentration of impurities are related either to $S \neq \frac{1}{2}$ or to large enough θ. Then additional quantum phase transitions as a function of an applied magnetic field or governed by the filling of the system with electrons take place with special divergent behaviours of spin or charge susceptibilities and the Sommerfeld coefficient of the low-temperature specific heat. The interesting feature of the behaviour of a finite concentration of spin $S = \frac{1}{2}$ impurities is the onset of two additional (together with the quantum critical point of a transition to the spin-saturated phase H_s, which is equal to $H_s = H_s(c=0) + 2c/(\theta^2+1)$) quantum critical points, at which the magnetic susceptibility diverges.

Finally, let us investigate the behaviour of a finite concentration of magnetic impurities in an attractive Hubbard chain. This case is of interest physically, because it manifests how a finite concentration of magnetic impurities changes the ground state properties of a system with gapped spin-carrying low-lying excitations. Bethe ansatz equations can be written as

$$\begin{aligned}
&\left(\frac{\lambda_\alpha - \theta + iU(S-\frac{1}{2})}{\lambda_\alpha - \theta - iU(S-\frac{1}{2})}\right)^{N_i} \prod_{j=1}^{N} \frac{\lambda_\alpha - \sin k_j + iU/4}{\lambda_\alpha - \sin k_j - iU/4} = \prod_{\substack{\beta=1 \\ \beta\neq\alpha}}^{M} \frac{\lambda_\alpha - \lambda_\beta + iU/2}{\lambda_\alpha - \lambda - iU/2} \ , \\
&e^{ik_jL}\left(\frac{\sin k_j - \theta + iUS/2}{\sin k_j - \theta - iUS/2}\right)^{N_i} = \prod_{\beta=1}^{M} \frac{\sin k_j - \lambda_\beta + iU/4}{\sin k_j - \lambda_\beta - iU/4} \ ,
\end{aligned} \tag{8.28}$$

where $j = 1, \ldots, N$ and $\alpha = 1, \ldots, M$. In the ground state Bethe ansatz integral equations for the densities are:

$$\begin{aligned}
&\rho(k) + \rho_h(k) + \cos k \int_{-Q}^{Q} d\lambda a_{U/4}(\lambda - \sin k)\sigma'(\lambda) \\
&\quad = \frac{1-c}{2\pi} + c \cos k \; a_{SU/2}(\sin k - \theta) \,, \\
&\sigma'(\lambda) + \sigma'_h(\lambda) - \frac{1-c}{\pi} \mathrm{Re} \frac{1}{\sqrt{1 - [\lambda - i(U/4)]^2}} - c a_{(2S+1)U/4}(\lambda - \theta) \\
&\quad = - \int_{-Q}^{Q} d\lambda' a_{U/2}(\lambda - \lambda')\sigma'(\lambda') - \int_{-B}^{B} dk a_{U/4}(\lambda - \sin k)\rho(k) \,.
\end{aligned} \tag{8.29}$$

The energy is given by

$$\begin{aligned}
e = c e_{imp} &- 2(1-c) \int_{-B}^{B} dk \cos k \rho(k) \\
&- 4(1-c) \mathrm{Re} \int_{-Q}^{Q} d\lambda \sqrt{1 - [\lambda - i(U/4)]^2} \sigma'(\lambda) \,,
\end{aligned} \tag{8.30}$$

where e_{imp} is the energy of impurities per site, and the number of electrons and magnetization per site are equal to

$$\begin{aligned}
\tfrac{N}{L} &= \int_{-B}^{B} dk \rho(k) + 2 \int_{-Q}^{Q} d\lambda \sigma'(\lambda) \,, \\
\frac{M^z}{L} &= cS + \frac{1-c}{2} \int_{-B}^{B} dk \rho(k) \,.
\end{aligned} \tag{8.31}$$

The expression for the energy of impurities can be derived from the algebraic Bethe ansatz. It is very cumbersome. The total energy can be approximated for $c \ll 1$ as

$$\begin{aligned}
e = &-2 \int_{-B}^{B} dk \cos k [1 + c R_\rho] \rho(k) \\
&- 4 \mathrm{Re} \int_{-Q}^{Q} d\lambda \sqrt{1 - [\lambda - i(U/4)]^2} [1 + c R_{\sigma'}] \sigma'(\lambda) \,,
\end{aligned} \tag{8.32}$$

where $R_\rho = \rho_{imp}/\rho_{host}$ and $R_{\sigma'} = \sigma'_{imp}/\sigma'_{host}$, $\rho = \rho_{host} + (1/L)\rho_{imp}$, $\sigma' = \sigma'_{host} + (1/L)\sigma'_{imp}$. Then the ground state equations for an attractive Hubbard chain with a finite concentration of magnetic impurities can be approximated as

$$\varepsilon(k) + \int_{-Q}^{Q} d\lambda a_{U/4}(\lambda - \sin k)\psi(\lambda) = -2\cos k[1 + cR_\rho] - \mu - \frac{H}{2} ,$$

$$\psi(\lambda) + \int_{-Q}^{Q} d\lambda' a_{U/2}(\lambda - \lambda')\psi(\lambda') = -\int_{-B}^{B} dk a_{U/2}(\lambda - \sin k)\varepsilon(k) \tag{8.33}$$

$$- 4\mathrm{Re}\sqrt{1 - [\lambda - i(U/4)]^2}[1 + cR_{\sigma'}] - 2\mu .$$

Naturally, the Fermi points are determined by the conditions $\varepsilon(\pm B) = 0$ and $\psi(\pm Q) = 0$. Since the total number of electrons is conserved, the limits of integration are renormalized due to the nonzero impurity concentration c.

One can check that the answers for a single impurity $c \to 0$ ($N_i = 1$) coincide with the ones given in Chapter 7.

Let us calculate the spin gap G, which is the smallest energy required to depair a singlet bound state. In zero magnetic field it is given by

$$G = -2[1 + cR_\rho] - \mu - \int_{-Q}^{Q} d\lambda a_{U/4}(\lambda)\psi(\lambda) . \tag{8.34}$$

The analysis of the numerical solution of the above presented integral equations shows that the spin gap decreases with the concentration of impurities as the consequence of the term, proportional to R_ρ, which is in general larger than the renormalization of the chemical potential as a function of c (observe that $(d\mu/dc) < 0$). The decrease is linear with the concentration. At some critical concentration c_{cr} the gap can be closed. The value dG/dc is negative and monotonically increases with θ. $|\theta| < Q$ is in-resonance case (here an isolated impurity lies in the continuum of charge rapidities of Cooper pairs), while the large $|\theta|$ describes the off-resonance situation. The mechanism of the reducing of a spin gap due to magnetic impurities is then of the pair weakening type rather than pair breaking as for standard magnetic impurities in a Bardeen–Cooper–Schrieffer superconductor. No unpaired electrons are generated as long as there is a spin gap. The pair weakening decreases with increasing spin for small $|\theta|$, while it is reversed for large absolute values of θ. For $S = \frac{1}{2}$ and very large $|\theta|$ the spin gap slightly increases with the impurity concentration. This is related to the fact that a finite concentration of magnetic impurities works twofold: it increases the density of states of pairs, supporting pairs, but, on the other hand, as usual for magnetic impurities, it reduces "superconducting" properties.

For the impurity concentration larger than c_{cr} a fraction of itinerant electrons is depaired and spontaneously magnetized. It follows from the fact that $\varepsilon(k)$ is negative, and $M^z \neq 0$ for $H = 0$. This differs the behaviour of a finite concentration of magnetic impurities introduced into a correlated electron chain with the spin gap. As the reader saw, magnetic impurities in correlated electron chains without spin gaps, see above studied cases of supersymmetric t-J chains, are antiferromagnetically correlated, and the ground state is a magnetic singlet.

8.2 Disodered Ensembles of Impurities in Correlated Chains. "Quantum Transfer Matrix" Approach

In the previous section we studied the behaviour of many similar impurities in quantum correlated chains. However, sometimes impurities possess characteristics different from each other. This is why, it is important to investigate how the disorder in the distribution of characteristics of impurities can affect thermodynamic behaviour of quantum chains.

It is known that the theoretical description of disordered systems is more complicated than the study of homogeneous ones. So far, only a few exact results are known about the behaviour of disordered systems. Bethe ansatz solvable models give a rare opportunity to find exactly (and in many cases analytically) thermodynamic characteristics of ensembles of impurities with randomly distributed parameters. Actually, it is, probably, clear to the reader, who already knows the structure of Bethe ansatz integrable chains with embedded impurities. The fact that impurity L-operators (introduced into monodromy matrices) satisfy intertwining (Yang–Baxter) relations with R-matrices of the host, and, the exact integrability, as the consequence of these algebraic constructions, does not depend on the values of characteristics of an impurity (*e.g.*, on the values of θ and S'). Also, this fact does not depend on how many impurities are introduced. Hence, this crucial feature gives the possibility to consider as many different impurities of the structure, studied in Chapter 7, as it is necessary, and all models, constructed this way will be integrable. Naturally, this does not pertain to boundary impurities, which introduce reflections. This is why, the main feature of disordered impurities, which permit exact solutions, is the absence of reflection (*i.e.*, these impurities are only elastic scatterers and do not cause relaxation). Naturally, this condition prohibits the presence of local levels caused by impurities (bound states caused by an interaction are possible to study though). To summarize, the Bethe ansatz gives the unique possibility to study the common effect of disorder and interactions, which is impossible, to the best of our knowledge, with the help of any other method.

To start, let us consider quantum spin chains with embedded impurities, studied in the previous chapter, each of which is characterized by its own spin S'_j and own coupling to the host chain, *i.e.*, own θ_j. According to the previous section, we can also study neighboring impurities, coupled to each other (in the case if they have the same θ_j). However, in what follows we shall study the situation, in which lengths of such clusters are small compared with the length of a chain L.

From the previous chapter the reader, in fact, is informed that due to the linearity of integral thermodynamic Bethe ansatz equations for densities the contribution from each impurity can be considered as additive. This is, naturally, based on the use of the string hypothesis. Now we shall study a different version of Bethe ansatz thermodynamics, which is not based on the string hypothesis in what follows. For the consideration of one-dimensional inhomogeneous quantum spin chain at any temperature we choose a suitable lattice path integral representation by some mapping, preserving integrability. Following several authors, (here it is worthwhile to mention M. Suzuki, M. Inoue, T. Koma, A. Klümper, P. A. Pearce and others, who contributed to the development of this method), we propose to study an associated two-dimensional classical vertex model instead of the direct treatment of a one-dimensional quantum system, as we used in previous chapters. The connection of exactly Bethe ansatz solvable quantum one-dimensional problem with the one of classical models of statistical mechanics is well known and it has been studied in several monographs, to which we refer the reader.

Let us start with the R-matrix Eq. (7.76) which describes the behaviour of a quantum spin-S chain with the uniaxial magnetic anisotropy. Let us for definiteness consider the case of the "easy plane" anisotropy. One can introduce R-matrices of different type, related to the initial one by an anti-clockwise rotation $\bar{R}^{\mu\nu}_{\alpha\beta}(u) = R^{\alpha\beta}_{\nu\mu}(u)$ and $\tilde{R}^{\mu\nu}_{\alpha\beta}(u) = R^{\beta\alpha}_{\mu\nu}(u)$ by a clockwise rotation. Then the transfer matrix $\bar{\tau}(u, \{\theta\}^{L}_{i=1})$ can be constructed in a way similar to the case of a standard transfer matrix constructed with the help of usual R-matrices. Here θ_j characterizes the coupling of the j-th impurity to the host. Then we substitute $u = -J\sin\eta/NT$, where N is the Trotter number. We observe that

$$[\tau(u)\bar{\tau}(u)]^{N/2} = e^{-\mathcal{H}/T} + \mathcal{O}(1/N) \ . \tag{8.35}$$

Hence, the partition function of the quantum one-dimensional system is identical to the partition function of an inhomogeneous classical vertex model with alternating rows on a square lattice of size $L \times N$

$$Z = \lim_{N\to\infty} \mathrm{tr}[\tau(u)\bar{\tau}(u)]^{N/2} . \tag{8.36}$$

Interactions on the two-dimensional lattice are four-spin interactions with coupling parameters depending on $(NT)^{-1}$ and interaction parameters θ_i, where i is the number of the column to which the considered vertex of the lattice belongs. Note that interactions are homogeneous in each column, but vary from column to column. This is similar to the *McCoy–Wu model*, which is the two-dimensional Ising model with a columnar disorder. [However in its' one-dimensional realization the Hamiltonian of the McCoy–Wu model can be mapped on a quadratic fermion form by means of the Jordan–Wigner transformation, *i.e.*, there are no interactions in that model; models, considered in this section, definitely reveal an essential coupling between particles.] We study this system in the thermodynamic limit $N, L \to \infty$ using an approach, which is based on a transfer matrix describing transfers in horizontal direction. Corresponding column-to-column transfer matrices are referred to as *quantum transfer matrices* (an external magnetic field H is included by means of twisted boundary conditions)

$$\tau_{QTM}(\theta_j, u) = \sum_{\mu} e^{\mu_1 H/T} \prod_{i=1}^{N/2} R^{\mu_{2i-1}\mu_{2i}}_{\alpha_{2i-1}\beta_{2i-1}}(u + i\theta_j) \times \tilde{R}^{\mu_{2i}\mu_{2i+1}}_{\alpha_{2i}\beta_{2i}}(u - i\theta_j) . \tag{8.37}$$

The illustration of a quantum transfer matrix is presented in Fig. 8.6. In general all quantum transfer matrices corresponding to L columns are different. However, all these operators commute pairwise. Therefore, the Helmholtz free energy per lattice site of the considered one-dimensional quantum chain can be calculated from the largest eigenvalue of a quantum transfer matrix (corresponding to only one eigenstate). The Helmholtz free energy per site f of a one-dimensional inhomogeneous quantum spin chain is given by only the largest eigenvalue of the quantum transfer matrix Λ_{QTM} as

$$f = -\lim_{L\to\infty} \frac{T}{L} \sum_{i=1}^{L} \lim_{N\to\infty} \ln \Lambda_{QTM}(\theta_i, u) , \tag{8.38}$$

where $u = -\frac{J\sin\eta}{TN}$ and the dependence on N is understood implicitly.

Let us consider the hierarchy of quantum transfer matrices acting on the subspace $\otimes^N V_{2S}$ (subscripts determine spins of scatterers) with T_n being a

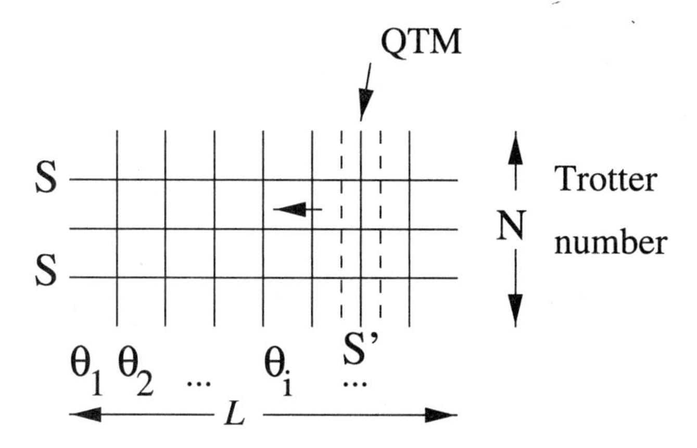

Fig. 8.6 The classical two-dimensional model with four-spin interaction around vertices and alternating coupling parameters from column to column, related to a quantum one-dimensional chain.

member of such hierarchy with the auxiliary subspace V_n (here the index n determines the spin of the auxiliary particle, *i.e.*, the auxiliary particle with spin S' scatters off N spins S). By means of a Bethe ansatz procedure we find the eigenvalue of the quantum transfer matrix to be given by

$$\Lambda_{QTM}(\theta_i) = \frac{\Lambda_{2S'}(\frac{2\theta_i}{\eta})}{\prod_{p=1}^{2S'}(\sinh(ip\eta))^{N/2}} \tag{8.39}$$

and

$$\Lambda_p(x) = \sum_{l=1}^{p+1} \lambda_l^{(p)}(x) , \tag{8.40}$$

where

$$\begin{aligned} \lambda_l^{(p)}(x) &= \psi_l^{(p)}(x) e^{H(p+2-2l)/T} \\ &\quad \times \frac{Q[x+i(p+1)]Q[x-i(p+1)]}{Q[x+i(2l-p-1)]Q[x+i(2l-p-3)]} , \\ \psi_l^{(p)}(x) &= \prod_{z=1}^{p-l+1} \phi_-[x-i(p-2S-2z)]\phi_+[x+i(p-2S+2-2z)] \\ &\quad \times \prod_{z=1}^{l-1} \phi_-[x-i(p-2S+2-2z)]\phi_+[x+i(p-2S-2z)] \end{aligned} \tag{8.41}$$

with $p \geq 2$, $\Lambda_0 = 1$ and

$$\Lambda_1(x) = \phi_+[x - i(2S-1)]\phi_-[x - i(2S+1)]e^{H/T}\frac{Q(x+2i)}{Q(x)} + \phi_-[x + i(2S-1)]\phi_+[x + i(2S+1)]e^{-H/T}\frac{Q(x-2i)}{Q(x)} . \quad (8.42)$$

Here we have dropped the dependence on u and θ_i, which are fixed, and consider the dependence on the spectral parameter x explicitly. We have used

$$\phi_\pm(x) = \sinh^{N/2}[\eta\frac{x \pm iu'}{2}] , \qquad Q(x) = \prod_{j=1}^{m} \sinh[\eta\frac{x - x_j}{2}] \quad (8.43)$$

with the renormalized $u' = 2u/\eta$. Here $\{x_j\}_{j=1}^{m}$ is the set of Bethe ansatz rapidities which are subject to the "local" Bethe ansatz equations

$$\frac{\phi_-[x_j + i(2S-1)]\phi_+[x_j + i(2S+1)]}{\phi_+[x_j - i(2S-1)]\phi_-[x_j - i(2S+1)]} = -e^{2H/T}\frac{Q(x_j+2i)}{Q(x_j-2i)} \quad (8.44)$$

where m is the number of the roots of the "local" Bethe ansatz equations, being different for different eigenstates of the quantum transfer matrix. For the largest eigenvalue we have to take $m = NS$. However, we shall not solve Eq. (8.44) directly, but rather shall be interested in the functional properties of the eigenvalue of a quantum transfer matrix. Note that $\Lambda_0 = 1$ and

$$\Lambda_p(x+i)\Lambda_p(x-i) = f_p(x) + \Lambda_{p-1}(x)\Lambda_{p+1}(x) , \quad (8.45)$$

where $p \geq 1$ and

$$f_n(x) = \prod_{j=1}^{n}\prod_{\pm} \phi_\pm[x \pm i(n-2S-2j+1)]\phi_\pm[x \pm i(2S-n+2j+1)] . \quad (8.46)$$

For this purpose we introduce auxiliary functions $y_n(x)$, $Y_n(x) = 1 + y_n(x)$, $b(x)$, $\bar{b}(x)$, $B(x) = 1 + b(x)$ and $\bar{B}(x) = 1 + \bar{b}(x)$ by

$$\begin{aligned}
y_n(x) &= \Lambda_{n-1}(x)\Lambda_{n+1}(x)/f_n(x) \ , \ n \geq 1 \ , \\
b(x) &= \frac{\lambda_1^{(2S')}(x+i) + \cdots + \lambda_{2S'}^{(2S')}(x+i)}{\lambda_{2S'+1}^{(2S')}(x+i)} \ , \\
\bar{b}(x) &= \frac{\lambda_2^{(2S')}(x-i) + \cdots + \lambda_{2S'+1}^{(2S')}(x-i)}{\lambda_1^{(2S')}(x-i)} \ ,
\end{aligned} \tag{8.47}$$

where $n \geq 1$. Then one can straightforwardly check that $(y_0 = 0)$

$$\begin{aligned}
& y_n(x+i)y_n(x-i) = Y_{n-1}(x)Y_{n+1}(x) \ , \\
& \Lambda_{2S'}(x+i) = B(x)\lambda_{2S'+1}^{(2S')}(x+i) \\
& = e^{-2S'H/T} \prod_{\pm}\prod_{j=1}^{2S'} \phi_\pm[x + i(2j + 2S - 2S' - \pm 1)] \frac{Q(x - 2iS')}{Q(x + 2iS')} \ , \\
& \Lambda_{2S'}(x-i) = \bar{B}(x)\lambda_1^{(2S')}(x-i) \\
& = e^{2S'H/T} \prod_{\pm}\prod_{j=1}^{2S'} \phi_\pm[x - i(2j + 2S - 2S' \pm 1)] \frac{Q(x + 2iS')}{Q(x - 2iS')} \ .
\end{aligned} \tag{8.48}$$

The first set of equations is known as the *fusion hierarchy* (so-called Y-system). Let us use the first $2S' - 2$ equations of the Y-system as they are. In the equation for $y_{2S'-1}$ we replace $Y_{2S'}(x)$ by $B(x)\bar{B}(x)$, due to

$$Y_p(x) = B(x)\bar{B}(x) \ , \tag{8.49}$$

i.e., we have

$$y_{2S'-1}(x-i)y_{2S'-1}(x+i) = Y_{2S'-2}(x)B(x)\bar{B}(x) \ . \tag{8.50}$$

Then it obviously yields

$$\begin{aligned}
b(x) &= e^{(2S'+1)H/T} \prod_{\pm} \frac{\phi_\pm[x + i(2S - 2S' \pm 1)]\Lambda_{2S'-1}(x)}{\prod_{j=1}^{2S'} \phi_\pm[x + i(2j + 2S - 2S' \pm 1)]} \\
& \quad \times \frac{Q[x + i(2S' + 2)]}{Q(x - 2iS')} \ , \\
\bar{b}(x) &= e^{-(2S'+1)H/T} \prod_{\pm} \frac{\phi_\pm[x + i(2S - 2S' \pm 1)]\Lambda_{2S'-1}(x)}{\prod_{j=1}^{2S'} \phi_\pm[x - i(2j + 2S - 2S' \pm 1)]} \\
& \quad \times \frac{Q[x - i(2S' + 2)]}{Q(x + 2iS')}
\end{aligned} \tag{8.51}$$

and

$$\Lambda_{k-1}(x-i)\Lambda_{k-1}(x+i) = Y_{k-1}(x) f_{k-1}(x) , \tag{8.52}$$

which are the consequences of definitions.

The reader can see that these auxiliary functions are analytic, non-zero and have constant asymptotic behaviour for the strip $-1 < \operatorname{Im} x \leq 0$ for $b(x)$ and $B(x)$, for the strip $0 \leq \operatorname{Im} x < 1$ for $\bar{b}(x)$ and $\bar{B}(x)$ and for the strip $-1 \geq \operatorname{Im} x \geq 1$ for y_n and Y_n. Introducing $a(x) = b(\frac{2}{\pi}(x+i\epsilon))$ and $\bar{a}(x) = \bar{b}(\frac{2}{\pi}(x-i\epsilon))$ (infinitesimal $\epsilon > 0$), taking the logarithmic derivative of these functions, then Fourier transforming the equations, eliminating the functions $Q(x)$ and finally inverse-Fourier transforming, we obtain the final set of nonlinear integral equations. Eventually, we take the limit $N \to \infty$. Proceeding this way we find for our system the following set of nonlinear integral equations for the "energy density" functions a, $\bar{a}$, $A = 1 + a$, $\bar{A} = 1 + \bar{a}$, y_n and Y_n, dependent of the spectral parameter x:

$$\begin{aligned}
&\ln y_1(x) = \int k'(x-y) \ln Y_2(y) dy , \\
&\ln y_j(x) = \int k'(x-y) \ln[Y_{j-1}(y) Y_{j+1}(y)] dy , \ 2 \leq j \leq 2S' - 1 , \\
&\int [k'(x-y) \ln Y_{2S'-2}(y) + k'(x-y+i\epsilon) \ln A(y) \\
&\qquad + k(x-y-i\epsilon) \ln \bar{A}(y)] dy = \ln y_{2S'-1}(x) ,
\end{aligned} \tag{8.53}$$

and

$$\begin{aligned}
&\int [k(x-y) \ln A(y) - k(x-y-i\pi+i\epsilon) \ln \bar{A}(y) \\
&\qquad + k'(x-y+i\epsilon) \ln Y_{2S'-1}(y)] dy = \ln a(x) + \frac{v^F}{T \cosh x} - \frac{\pi H}{2(\pi - 2S\eta)T} \\
&\int [k(x-y) \ln \bar{A}(y) - k(x-y+i\pi-i\epsilon) \ln A(y) \\
&\qquad + k'(x-y-i\epsilon) \ln Y_{2S'-1}(y)] dy = \ln \bar{a}(x) + \frac{v^F}{T \cosh x} + \frac{\pi H}{2(\pi - 2S\eta)T} ,
\end{aligned} \tag{8.54}$$

where $v^F = \pi J \frac{\sin(2\eta S)}{4\eta S}$ is the Fermi velocity of low-lying excitations, with kernel functions

$$k(x) = \frac{1}{2\pi}\int d\omega \frac{\sinh[(\frac{\pi^2}{2\eta} - \frac{2S+1}{2}\pi)\omega]\cos(x\omega)}{2\cosh(\frac{\pi\omega}{2})\sinh(\frac{\pi-2S\eta}{2\eta}\pi\omega)} \tag{8.55}$$

and

$$k'(x) = \frac{1}{2\pi}\int d\omega \frac{\cos(x\omega)}{2\cosh(\frac{\pi\omega}{2})} . \tag{8.56}$$

It is important to emphasize that this set of nonlinear equations does not depend on θ_j. The Helmholtz free energy per site f is given by

$$f(x) = e_0(x) - \frac{T}{2\pi}\int \frac{\ln A(y)dy}{\cosh(x-y+i\epsilon)} - \frac{T}{2\pi}\int \frac{\ln \bar{A}(y)dy}{\cosh(x-y-i\epsilon)} , \tag{8.57}$$

where e_0 is the ground state energy. The Helmholtz free energy of the total quantum spin chain with impurities is

$$F = \sum_j f[\frac{\pi}{\eta}\theta_j + i\pi(S'-S)] , \tag{8.58}$$

where the sum is taken over all the sites (for sites without impurities we get $f(0)$). Notice that for $S' < S$ one has to put $\ln Y_{2S'}$ into Eqs. (8.53) and (8.54) instead of $\ln A\bar{A}$, as follows from Eq. (8.49).

These equations can be easily solved numerically for arbitrary values of the magnetic field and temperature. The random distribution of the values θ_j can be described by a distribution function $P(\theta_j)$. It is worthwhile to emphasize here the simplicity of the derived equations: For each impurity there are only two parameters, the real and imaginary shifts of the spectral parameter in the formula for the free energy per site Eq. (8.57). Then the exact solvability of the problem for any number of impurities permits to introduce the distribution of these shifts (related to strengths of impurity-host couplings pertained to *local Kondo temperatures*, which one can introduce to describe the local behaviour of each magnetic impurity in a quantum correlated chain, *cf.* the previous chapter, and spins of impurities). One has only $2S'+1$ non-linear integral equations, Eqs. (8.53) and (8.54), to solve, and the answer can in principle be obtained for arbitrary temperature and magnetic field ranges.

It turns out that for the most important case $S = S' = \frac{1}{2}$ the set of integral equations is considerably simplified. We have $y_n = 0$ ($Y_n = 1$), and only two integral equations totally describe thermodynamics of a disordered ensemble of spin-$\frac{1}{2}$ integrable impurities in a spin-$\frac{1}{2}$ chain.

The next-largest eigenvalue of the quantum transfer matrix is related to the correlation length ξ of the longitudinal spin-spin correlation function as

$$\xi^{-1} + iP^F = -\lim_{N\to\infty} \ln \frac{\Lambda^{(2)}_{QTM}}{\Lambda_{QTM}} + \cdots , \tag{8.59}$$

where P^F is the Fermi momentum of low-lying excitations. For the simplest case $S = S' = \frac{1}{2}$ the nonlinear equations, which describe thermodynamics of such an excitation are two last equations from Eqs. (8.54) with $y_n = 0$ ($Y_n = 1$) and with the addition of driving terms

$$\begin{aligned} &\pm i\pi \mp 2i\pi k[x - \lambda_1 \mp i(\pi/2)] \mp 2i\pi k[x - \lambda_2 \mp i(\pi/2)] \\ &+ \ln \frac{\sinh \frac{\eta}{\pi-\eta}[x - \lambda_0 \mp i(\pi/2)]}{\sinh \frac{\eta}{\pi-\eta}[x - \lambda_0 \pm i(\pi/2)]} , \end{aligned} \tag{8.60}$$

for the equations for $a(x)$ and $\bar{a}(x)$, respectively, where $\lambda_{0,1,2}$ define the positions of holes, related to elementary excitations (with $\lambda_{1,2}$ on the real axis, or forming a conjugate pair and complex rapidity λ_0 with $\text{Im}\lambda_0 = \eta/2$). These parameters are not arbitrary, but satisfy coupled equations

$$\begin{aligned} a[\lambda_1 + i(\pi/2)] = a[\lambda_2 + i(\pi/2)] = a[\lambda_0 + i(\pi/2)] = -1 , \\ \bar{a}[\lambda_1 - i(\pi/2)] = \bar{a}[\lambda_2 - i(\pi/2)] = \bar{a}[\lambda_0 - i(\pi/2)] = -1 . \end{aligned} \tag{8.61}$$

The next-largest eigenvalue of the quantum transfer matrix is then

$$\ln \frac{\Lambda^{(2)}_{QTM}(x)}{\Lambda_{QTM}(x)} = \ln \left(\tanh \frac{1}{2}(x - \lambda_1) \tanh \frac{1}{2}(x - \lambda_2) \right) . \tag{8.62}$$

Naturally, the method, described above, is valid for the study of homogeneous systems, too. We would like to emphasize that in the low temperature regime lattice effects are non-essential and couplings of impurities to the host can be considered as contact ones in the thermodynamic limit.

The coupling of an impurity to the host (J^j_{imp}) is determined by the constant θ_j. The reader already knows that precisely this constant determines the effective Kondo temperature of the impurity, *e.g.*, in a Heisenberg spin chain *via* $T_{jK} \propto \exp(-\pi|\theta_j|)$. For energies higher than this crossover Kondo scale one has an asymptotically free impurity spin S', while for lower energies the impurity spin is underscreened for $S' > S$ (with the Curie-like behaviour of a remnant effective spin $S' - S$), totally screened for $S' = S$ (with the usual marginal Fermi liquid-like behaviour persisting with the finite susceptibility and linear temperature dependence of the specific heat at low temperature, and, hence, a finite Wilson ratio in the ground state) and overscreened for $S' < S$ with the critical non Fermi liquid behaviour of a single impurity spin. It is similar to the findings in the theory of a Kondo impurity in a free electron matrix. In other words, θ_j measures the shift of the Kondo resonance (higher values of $|\theta_j|$ correspond to lower values on the Kondo scale) of the impurity level with host spin excitations, similar to the standard picture of the Kondo effect in a free electron host.

One can see from Eqs. (8.53)–(8.58) that for low T the temperature behaviour of the magnetic susceptibility and specific heat of isolated impurities strongly depends on relative values of host spins S and impurity spin S'. This fact, naturally, agrees with the description of thermodynamics of quantum spin chains with the help of the string hypothesis, presented in previous chapters.

For $S < S'$ an impurity is underscreened by low-lying excitations of a chain. Naturally, the total low-temperature magnetic susceptibility of any disordered ensemble of such impurities is also divergent at low temperatures. On the other hand, for $S' < S$ the spins of low-lying excitations of an antiferromagnetic "easy-plane" chain overscreen the spin of a single magnetic impurity. This yields the critical behaviour, which reveals itself in divergences of the $T \to 0$ magnetic susceptibility of a single magnetic impurity and of the low-temperature Sommerfeld coefficient of the specific heat. The total low-temperature magnetic susceptibility and the Sommerfeld coefficient of any disordered ensemble of such impurities are also divergent at low temperatures. Here the disorder of distributions of impurity-host couplings does not yield any qualitative changes, introducing only specific additional features of the non Fermi liquid behaviour of the total system, which is already present for a single magnetic impurity.

A more interesting situation is for the case $S' = S$. Here the solution of Eqs. (8.53)–(8.57) can be obtained analytically at low temperatures. We know that at sufficiently low temperatures the functions a and $\ln A$ manifest a sharp crossover behaviour, reminiscent of a step function: $|a| \ll 1$ and $|\ln A| \ll 1$ for $x < \ln \alpha T_{jK}/T$ and $|a|, |\ln A| \sim O(1)$ for $x > \ln \alpha T_{jK}/T$, where α is some constant and for small anisotropy $T_{jK} = v^F \exp(-\pi|\theta_j|/\eta)$, *cf.* Chapter 6. We can introduce the scaling functions $\ln a^{\pm} = \ln a(\pm[x + \ln(\alpha T_{jK}/T)])$, $\ln \bar{a}^{\pm} = \ln \bar{a}(\pm[x + \ln(\alpha T_{jK}/T)])$, $\ln A^{\pm} = \ln A(\pm[x + \ln(\alpha T_{jK}/T)])$, $\ln \bar{A}^{\pm} = \ln \bar{A}(\pm[x + \ln(\alpha T_{jK}/T)])$, $\ln y_p^{\pm} = \ln y_p(\pm[x + \ln(\alpha T_{jK}/T)])$ and $\ln Y_p^{\pm} = \ln Y_p(\pm[x + \ln(\alpha T_{jK}/T)])$, where $p = 1, ..., 2S' - 1$. In terms of those scaling functions Eqs. (8.53) and (8.54) are renormalized in such a way that driving terms in the last two equations for $H = 0$ become proportional to $v \exp(-x \pm i\epsilon)$ (where small corrections of order of $O(T)$ were neglected). Hence, the only asymptotic behaviour of A and $\bar{A}$ at large spectral parameter is essential. Then it is not difficult to obtain the low temperature behaviour of the Helmholtz free energy per site (for $H = 0$, α is a constant)

$$f(\theta_j) = e_0(\theta_j) - \frac{\pi S T^2}{2(S+1)T_{jK}}\left[1 + \frac{3S^3}{[\ln(\alpha T_{jK}/T)]^3}\right] + \cdots . \tag{8.63}$$

In the presence of a weak magnetic field $H \ll T$ we can calculate temperature corrections to the Helmholtz free energy per site

$$f(\theta_j) = e_0^j(\theta_j, H) - \frac{\pi S T^2}{2(S+1)T_{jK}} - \frac{SH^2}{2\pi T_{jK}}\left[1 + \frac{S}{\ln(\alpha T_{jK}/T)} + \frac{S^2 \ln|\ln(\alpha T_{jK}/T)|}{\ln^2(\alpha T_{jK}/T)}\right] + O(T^2) . \tag{8.64}$$

Proceeding as above for the correlation length (for $S = S' = \frac{1}{2}$) we obtain for the low temperature dependence of the correlation length

$$\ln \frac{\Lambda_{QTM}^{(2)}}{\Lambda_{QTM}} = \pi\Delta \frac{T}{T_{jK}} + iP^F , \tag{8.65}$$

where $P^F = (N - M)\pi$ and $2\Delta = (\Delta N)^2 + (\Delta M)^2 + 2n$, where ΔN, ΔM and n are integers or half-integers determining the spin projection, momentum and the number of particle-hole excitations related to $\lambda_{1,2,0}$, Δ is known as the *conformal dimension* (see the next chapter).

For a single impurity $P(\theta_j) = \delta(\theta_j - \theta)$ we immediately recover the Kondo behaviour of an asymptotically free spin (characteristic for a Kondo impurity in a free electron host and for a single impurity in a Heisenberg antiferromagnetic chain, *cf.* the previous chapter). For the homogeneous case we put $\theta_j = 0$ (it means that $T_{jK} \to v^F$ where v^F is the Fermi velocity of low-lying excitations, *e.g.*, spinons for $S = \frac{1}{2}$). The reader can see that the only one parameter gets renormalized in the disordered case — the Fermi velocity of U(1)-symmetric low-lying excitations. The Kondo scale plays the role of a "local Fermi velocity" for an impurity.

Considered model permits to average over a distribution of θ_j (or local Kondo temperatures), because of the factorization of the Helmholtz free energy of the total system. This is the consequence of the Bethe ansatz integrability of a model (*i.e.*, of the only elastic scattering off impurities). Note that θ_j-dependence, present in low energy characteristics, results only in universal scales, T_{jK}, (it is not so for higher energies, but the latters are irrelevant for low temperature disorder-driven divergences). Hence for low energies we can use distributions of T_{jK}, which are also more appropriate in the connection to experiments on quantum spin chains with disordered impurities. This is why, the main features of low energy characteristics of a disordered integrable spin chain are determined by distributions of local Kondo temperatures for impurities. Let us consider the strong disorder distribution, which starts with the term $P(T_{jK}) \propto G^{-\lambda}(T_{jK})^{\lambda-1}$ ($\lambda < 1$) valid till some energy scale G, do not confuse with the spin gap of previous sections, for the lowest values of T_{jK} (that distribution was shown to pertain to real disordered quantum spin chains and some heavy fermion alloys). Now we can calculate the low temperature behaviour of the average magnetic susceptibility, Sommerfeld coefficient of the specific heat (for the most interesting case $S = S' = \frac{1}{2}$) of the form (the lower limit of the integral over the distribution of T_{jK} gives a regular contribution)

$$\langle \chi \rangle \propto \langle \gamma \rangle \sim G^{-\lambda} T^{\lambda-1} \,. \tag{8.66}$$

These formulas definitely manifest low-temperature divergences of $\langle \chi \rangle$ and $\langle \gamma \rangle$ and the strong renormalization in a disordered spin chain as compared to the homogeneous situation. The ground state average magnetization reveals $\langle M^z \rangle \sim (H/G)^{\lambda}$ behaviour, also different from the homogeneous case. It is interesting to notice that the average correlation length for $S = S' = \frac{1}{2}$ is $\langle \xi \rangle \sim (\pi\Delta)^{-1}(G/T)^{\lambda}$.

In the important marginal case $\lambda = 1$ logarithmic temperature divergences appear. Here one has the distribution $P(T_{jK} = 0) = P_0 \neq 0$ valid till G. Then averaging the low temperature part of the susceptibility and Sommerfeld coefficient we obtain

$$\langle \chi \rangle \propto \langle \gamma \rangle \sim -\frac{P_0}{2\pi}[\ln(G/T) + \ln\sqrt{\ln(\alpha G/T)} + \cdots] \,. \tag{8.67}$$

Here we again see low-temperature divergences of $\langle \chi \rangle$ and $\langle \gamma \rangle$ (more weak, though, comparing to the previous case). We can also calculate the low field ground state magnetization: $\langle M^z \rangle \sim HP_0[-\ln(H/G) - \ln(\ln(H/C'G)) + \cdots]$. The average correlation length is $\langle \xi \rangle \sim -(2/P_0\Delta T)[\ln(G/T) + \ln\sqrt{\ln(\alpha G/T)}]^{-1} + \cdots$.

The weak power law or logarithmic dependence pertains to the *Griffiths singularities* (due to R. B. Griffiths) in the proximity of a critical point $T = 0$. For these distributions of T_{jK} the Wilson ratio at $T = 0$ is equal to $2\pi^2/3$, characteristic for a Fermi liquid like situation. It turns out that above mentioned results for low temperatures are valid also for random ensembles of $S' = n/2$ (where n is the number of channels) multi-channel Kondo impurities with local anisotropic, generally speaking, interactions of latters with conduction electrons, because at low temperatures the difference between the energy of a quantum spin-$\frac{1}{2}$ Heisenberg chain and the spin subsystem of a Kondo model is negligible.

We can illustrate analytic results by numerical calculations for solutions of Eqs. (8.53)–(8.58). In Figs. 8.7 and 8.8 temperature dependencies for the magnetic susceptibility and the Sommerfeld coefficient for the most interesting magnetically isotropic Heisenberg antiferromagnetic spin $S = \frac{1}{2}$ chain are depicted. Solid lines show the finite values of χ and γ in this case without impurities. However, dashed and dotted lines present answers for distributions of θ_j with a *strong disorder*. The latter means that wings of distributions are large enough, comparing to maxima of distributions. The dotted line corresponds to the Lorentzian distribution with $P(\theta_j) = [(2\theta_j/\eta)^2 + \pi^2]^{-1}$. The dashed line pertains to the so-called logarithmically normal distribution with $P(\theta_j) = \exp(-[\ln(|2\theta_j/\eta| + 10^{-6}) + \frac{1}{4}]^2)/\sqrt{\pi}(|2\theta_j/\eta| + 10^{-6})$, which is also characteristic for a strong disorder. The reader can see the qualitative difference between the behaviour of $S' = S$ magnetic impurities with a strong disorder of the distribution of their couplings to the host comparing to the homogeneous spin chain. The magnetic susceptibility and the Sommerfeld coefficient strongly diverge at $T \to 0$ for strongly randomly distributed parameters of impurity-host couplings. It is in a drastic contrast with the homogeneous case. It turns out that low-temperature asymptotics of the log-normal case of the disorder are

$$\begin{aligned} c &\sim [\ln(1/T)\exp([\ln\ln(1/T)]^2)]^{-1} , \\ \chi &\sim [T\ln\ln(1/T)\exp([\ln\ln(1/T)]^2)]^{-1} , \end{aligned} \tag{8.68}$$

while for the Lorentzian distribution one has

$$\begin{aligned} c &\sim [\ln(1/T)]^{-2} , \\ \chi &\sim [T\ln(1/T)]^{-1} . \end{aligned} \tag{8.69}$$

In Figs. 8.9–8.10 similar behaviours for magnetic susceptibilities and Sommerfeld coefficients of the homogeneous case, and cases with log-normal and Lorentzian distributions (strong disorder) and the Gaussian distribution (weak disorder, see below) for the mostly anisotropic "easy-plane" case $\eta = \pi/2$ (for $S = \frac{1}{2}$ this corresponds to the isotropic XY model). The reader can see, that changes due to the nonzero magnetic anisotropy

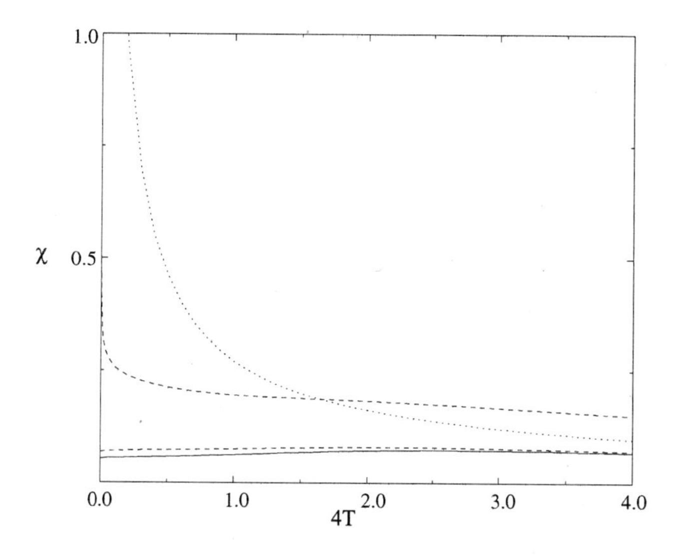

Fig. 8.7 The magnetic susceptibility ($H = 0$) of a Heisenberg spin-$\frac{1}{2}$ antiferromagnetic chain with $\frac{1}{2}$ magnetic impurities. The exchange constant of the host is $J = 2$. Solid line shows the homogeneous chain; the long-dashed line — the Gaussian distribution; the dashed line — the log-normal distribution; the dotted line — the Lorentzian distribution of θ_j.

of the "easy-plane" type are only qualitative. This is clear, because such an "easy-plane" magnetic anisotropy does not produce gaps for low-energy excitations (*i.e.*, it is marginally irrelevant perturbation from the renormalization group viewpoint), and, hence, the system remains in a critical regime.

On the other hand, a weak disorder does not produce such qualitative changes in the behaviour of random ensembles of disordered magnetic impurities. By a weak disorder we mean a narrow distribution of θ_j. Long-dashed lines of Figs. 8.7–8.10 depict the temperature behaviour of the ensemble of magnetic impurities with the weak Gaussian distribution of θ_j (which is close to a single impurity distribution $P(\theta_j) = \delta(\theta_j)$). The reader can see that such a narrow distribution (weak disorder) does not yield divergences of the low-temperature magnetic susceptibility and Sommerfeld coefficient of the specific heat. The reason for such a different behaviour of wide and narrow distributions of the parameters, which define impurity-host couplings (or strong–weak disorder, respectively), is clear. At low energies a local Kondo temperature defines the crossover scale for the behaviour of a magnetic impurity. For the case $S' = S$ a single magnetic impurity is screened by low-lying excitations of the host for $T < T_{jK}$, and

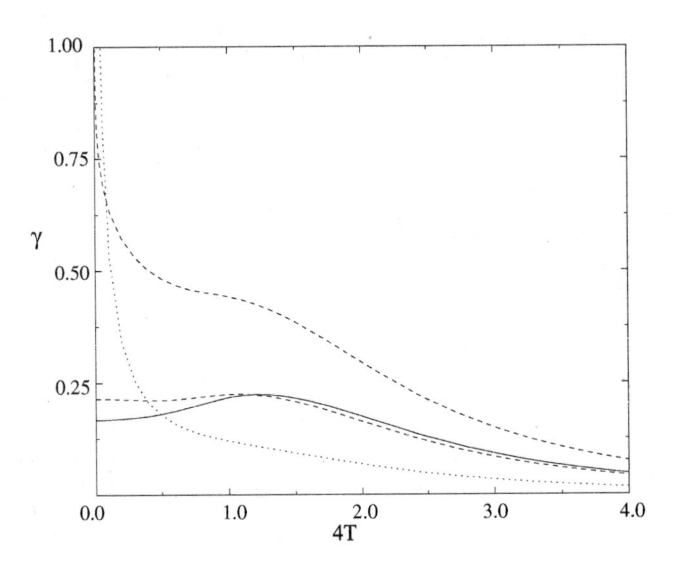

Fig. 8.8 The Sommerfeld coefficient γ ($H = 0$) of the same chain. The solid line shows the homogeneous chain; the long-dashed line — the Gaussian distribution; the dashed line — the log-normal distribution; the dotted line — the Lorentzian distribution of θ_j.

is not screened for $T > T_{jK}$ (with the Curie-like behaviour of the unscreened remnant spin). For ensembles of magnetic impurities with a weak disorder the temperature is larger than the average Kondo temperature of the ensemble of impurities, and, hence, the total magnetic susceptibility and the Sommerfeld coefficient are finite for $T \to 0$. Contrary, for a strong disorder, many local Kondo temperatures are less than the temperature. Those impurities remain unscreened by low-lying excitations of the host, and, hence, the total magnetic susceptibility and the Sommerfeld coefficient become divergent for $T \to 0$.

Finally we would like to attract the attention of the reader, to emphasize how the magnetic field lifts the degeneracy. In Figs. 8.11–8.12 the temperature behaviour of magnetic susceptibilities and Sommerfeld coefficients for isotropic cases for the log-normal and Lorentzian distributions (*cf.* Figs. 8.7-8.8), but for the nonzero magnetic field $H = 0.2$ are depicted. The reader can clearly see that such a field removes divergences in the low-T susceptibilities and Sommerfeld coefficients for models with a strong disorder. As an example, the temperature dependences of the same values for $H = 0.2$ are shown for a homogeneous chain. It turns our that the weak magnetic field does not yield any qualitative changes in the temperature behaviour, as expected.

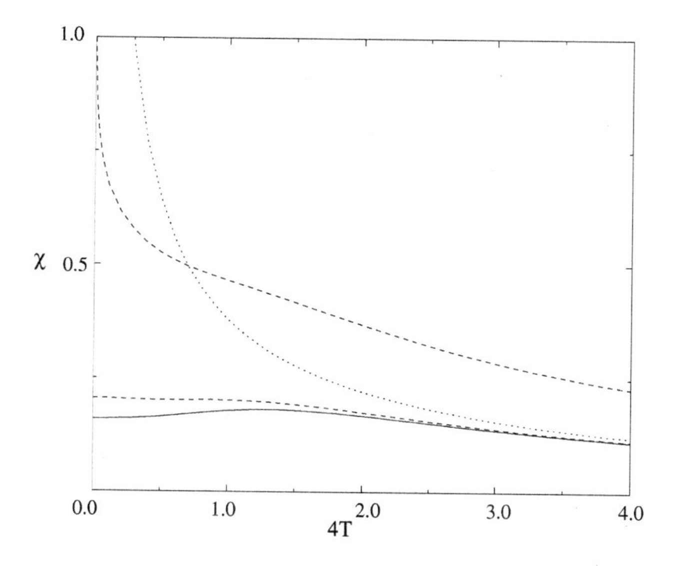

Fig. 8.9 The magnetic susceptibility of an isotropic XY chain ($\eta = \pi/2$) with $\frac{1}{2}$ magnetic impurities for $H = 0$. The solid line shows the homogeneous chain; the long-dashed line — the Gaussian distribution; the dashed line — the log-normal distribution; the dotted line — the Lorentzian distribution of θ_j.

For higher values of spins the changes, as compared to the case $S' = S = \frac{1}{2}$, are only quantative. For example, the values of χ and c become larger for larger spin values. However, there are no drastic changes in the behaviour of disordered ensembles of impurities, in comparison with the case, discussed above. This seems to be natural, because only low-lying excitations (those, which have Dirac seas in the ground state) are responsible for the Kondo-like screening of spins of impurities, while other excitations (which quasi-energies are described by y_p and Y_p), are more higher-energetic.

Now let us turn to the behaviour of ensembles of disordered impurities in correlated electron chains, where not only spin dynamics, but also charge dynamics is permitted. Here the ideology of the Bethe ansatz consideration is very similar to the one for quantum spin chains, because we, actually, used the concrete structure of R-matrices, *etc.* only to have the concrete realization of Yang–Baxter algebras. This is why, we can apply the powerful machinery of the Bethe ansatz to correlated electron chains with ensembles of integrable impurities with randomly distributed parameters, too.

As an example, let us consider the characteristics of a supersymmetric t-J chain with integrable impurities. We limit ourselves to the case of $S = \frac{1}{2}$

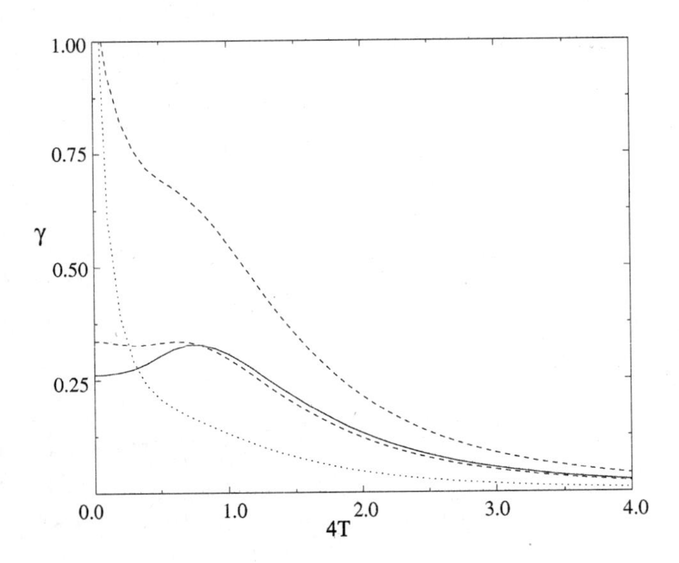

Fig. 8.10 The Sommerfeld coefficient γ of the same XY chain. The solid line shows the homogeneous chain for $H = 0$; the long-dashed line — the Gaussian distribution; the dashed line — the log-normal distribution; the dotted line — the Lorentzian distribution of θ_j.

impurities, because, as the reader already saw, namely in this case the renormalization of thermodynamic characteristics is the most dramatic.

In the ground state we obtain the valence of a single jth impurity $n_j(\theta_j) = \frac{1}{2} + (1/\pi)[\ln 2\sqrt{|Q-\theta_j|}]$, where $Q^2 = [2/3\zeta(3)][2\ln 2 - \mu + (H^2/4\pi)]$. Actually, we see that the result for the average ground state valence of disordered impurities per site is not dependent on the universal energy scale T_{jK}, but it also gets renormalized for disordered impurities, depending on the distribution of θ_j. The energy scale for the renormalization of the average valence is much larger for charge degrees of freedom than for spin degrees of freedom of correlated chains, since the magnetic field is usually much smaller than the band width. The magnetization per site for small magnetic fields is $m_j^z(\theta_j) = (H/T_{jK})[1 + (1/2|\ln(H/T_{jK})|) - (\ln|\ln(H/T_{jK})|/4\ln^2(H/T_{jK})) + \cdots]$, which does show the universal energy scale T_{jK}. This local Kondo scale depends on the band filling. If charge fluctuations are totally suppressed, for $N = L$, the local Kondo temperature is $T_{jK} = v_s^F \exp(-\pi|\theta_j|)$, where v_s^F is the Fermi velocity of a spin-carrying low-lying excitation. Corrections due to the mixed valence of each impurity shift the value of the local Kondo temperature, *e.g.*, as $T_{jK} \to T_{jK}(1 + 2\zeta(3)Q^3)$ for $Q \ll 1$. This is again the manifestation of correlations between electrons.

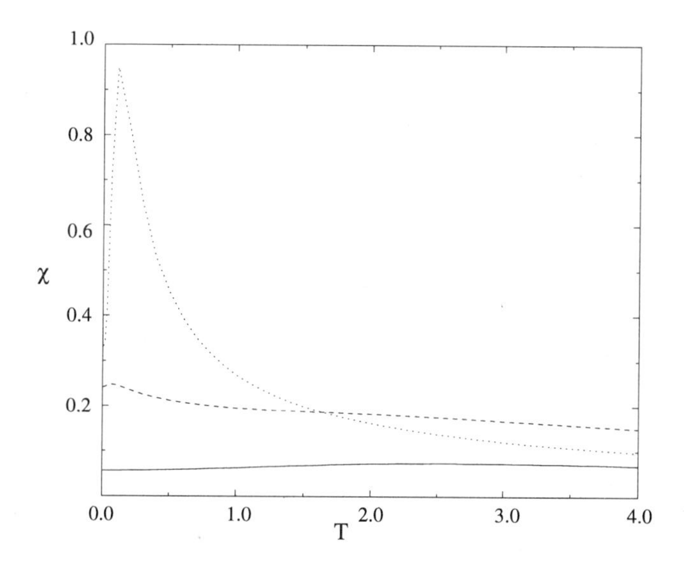

Fig. 8.11 The magnetic susceptibility of a Heisenberg spin-$\frac{1}{2}$ antiferromagnetic chain with $\frac{1}{2}$ magnetic impurities for $H = 0.2$. The solid line shows the homogeneous chain; the dashed line — the log-normal distribution; the dotted line — the Lorentzian distribution.

For nonzero temperatures we can apply the quantum transfer matrix method. The nonzero elements of the R-matrix of the related to the supersymmetric t-J chain two-dimensional Perk–Schultz model can be written as

$$\begin{aligned} & R^{\alpha\alpha}_{\alpha\alpha}(x) = \frac{ic + \epsilon_\alpha x}{ic} \ , \ R^{\mu\mu}_{\alpha\alpha}(x) = \frac{\epsilon_\alpha \epsilon_\mu x}{ic} \ , \\ & R^{\mu\alpha}_{\alpha\mu}(x) = 1 + \text{sign}(\alpha - \mu) x \ , \end{aligned} \tag{8.70}$$

where x is the spectral parameter, $\alpha, \mu = 1, 2, 3$ and $\epsilon_{1,2,3}$ are the Grassmann parities, see Chapter 5. Introducing, as above, clockwise and anticlockwise rotated matrices, we find again that the Helmholtz free energy per site is obtained from the largest eigenvalue of the quantum transfer matrix.

Using the analytical properties of the quantum transfer matrix we can derive the following set of nonlinear integral equations for the "energy density" functions, which at low T are closely related to Gibbs' exponents of "dressed energies" of spin, $a(x)$ and $\bar{a}(x)$, and charge, $c(x)$, excitations of the supersymmetric t-J chain, the solution of which describes thermodynamics of the supersymmetric t-J model for any values of temperature,

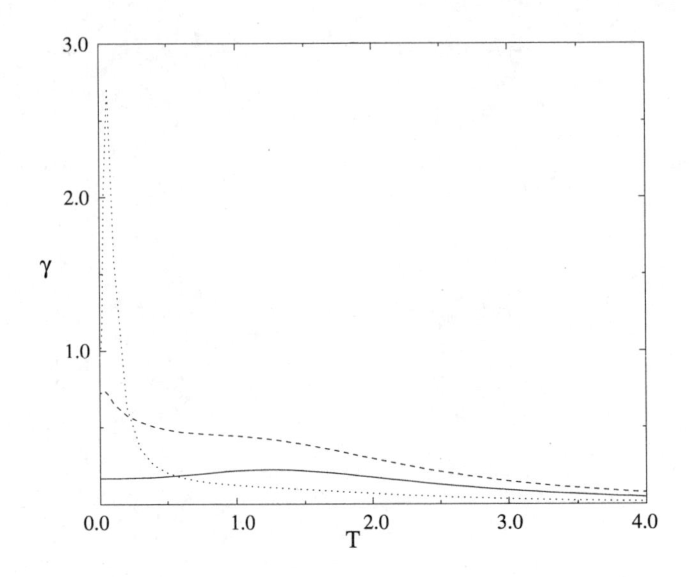

Fig. 8.12 The Sommerfeld coefficient γ of the same chain. The solid line shows the homogeneous chain for $H = 0.2$; the dashed line — the log-normal distribution; the dotted line — the Lorentzian distribution.

magnetic field and chemical potential (related to the total number of electrons in the system)

$$
\begin{aligned}
T \ln a(x) &= -2\pi \Psi_a(x + i\epsilon) + \mu + (H/2) \\
&\quad - T\Psi_a * \ln(1 + \bar{a})|_{x+2i\epsilon} - T\Psi_a * \ln(1 + c)|_{x+i\epsilon} , \\
T \ln c(x) &= -2\pi \Psi_c(x) + 2\mu - T\Psi_c * \ln(1 + c) \\
&\quad - T\Psi_a * \ln(1 + \bar{a})|_{x+i\epsilon} - T\Psi_{\bar{a}} * \ln(1 + a)|_{x-i\epsilon} ,
\end{aligned}
\tag{8.71}
$$

where $*$ means convolution,

$$
\begin{aligned}
\Psi_a(x) &= \frac{1}{2\pi x(x - i)} , \qquad \Psi_{\bar{a}}(x) = \frac{1}{2\pi x(x + i)} , \\
\Psi_c(x) &= \frac{1}{\pi(x^2 + 1)} ,
\end{aligned}
\tag{8.72}
$$

and $0 < \epsilon < 1$. The equation for $\bar{a}(x)$ is obtained from the one for a by the replacements $i \to -i$, $H \to -H$ and $a \leftrightarrow \bar{a}$. The derivation of this set of integral equations is reminiscent of the above derivation of nonlinear integral equations for a spin chain and we refer the interested reader to the original publication. The Helmholtz free energy per site of a supersymmetric t-J

chain with impurities does depend on θ_j and is given by

$$f(\theta_j) = 2\mu - T \ln c(\theta_j) \tag{8.73}$$

and the total Helmholtz free energy is $F = \sum_{j=1}^{L} f(\theta_j)$, where for the host sites we put $\theta_j = 0$. For a single impurity $P(\theta_j) = \delta(\theta_j - \theta)$ and we recover in the Kondo limit the logarithmic Kondo behaviour of an asymptotically free spin (which is characteristic both to a Kondo impurity in a free electron host and to a single magnetic impurity in a t-J chain). For the case of the homogeneous chain we put $\theta_j = 0$.

The numerical solution of Eq. (8.71) shows that for narrow distributions (weak disorder) a disordered supersymmetric t-J chain is in a singlet state, *i.e.*, the Kondo screening persists. For broad distributions (strong disorder) the non Fermi liquid behaviour is manifested. Here low temperature magnetic susceptibility diverges, *i.e.*, there is no Kondo quenching. The low temperature divergences disappear upon applying a finite magnetic field which restores the screening of impurities.

We can analytically solve Eq. (8.71) in several important limiting cases. First, for low T the Helmholtz free energy per site is given by

$$f(\theta_j) \approx e_0(\theta_j) - \frac{\pi T^2}{6}[v_c^{-1}(\theta_j) + v_s^{-1}(\theta_j)] + \dots , \tag{8.74}$$

where $e_0(\theta_j) \equiv e_0^j$ is the ground state energy per site, and $v_{c,s}(\theta_j)$ are Fermi velocities of charge and spin low-lying excitations of the supersymmetric t-J chain taken at the associated Fermi points shifted by θ_j. For $\theta_j = 0$ it is the known low temperature limit of the homogeneous host. The only low energy parameters which get renormalized by the disorder are the effective "local velocities" of low-lying charge and spin excitations. For low densities of electrons (where $\mu \ll T$) for $H = 0$ we obtain the Helmholtz free energy per site (we put $\epsilon = 1/2$)

$$f(\theta_j) \approx e_0^j - T \ln(1 + 2e^{-1/T(\theta_j^2+1)}) . \tag{8.75}$$

For the high density regime $\mu \gg T$ one can use the approximation $\ln c \approx \ln(1 + c)$. This yields

$$T \ln a = \frac{\pi}{\cosh \pi x} + \frac{H}{2} + Tk(x) * \ln(1 + a) - Tk(x + i) * \ln(1 + \bar{a}) , \tag{8.76}$$

and similar for $\bar{a}$, with the kernel

$$k(x) = \frac{1}{2\pi} \int d\omega \frac{e^{i\omega x}}{1 + e^{|\omega|}} . \tag{8.77}$$

The free energy per site becomes

$$f(x) = e_0(x) + \mu - T \int dy \frac{\ln(1+a)(1+\bar{a})}{\cosh \pi(x-y)} . \tag{8.78}$$

The reader can recognize in these nonlinear equations the ones of Eq. (8.54) in the limit $\eta \to 0$ and $S = S' = \frac{1}{2}$, *i.e.*, the equations for a disordered Heisenberg spin $\frac{1}{2}$ chain. It is clear, because in the limit of large μ the electron density per site is equal to 1 (the largest possible value for a t-J model).

In the low temperature regime lattice effects are non-essential and couplings of impurities to the host can be considered as contact ones. Typically corrections to low temperature asymptotics of thermodynamic characteristics of quantum chains possessing SU(2) spin symmetry manifest logarithmic behaviour (singularities), see below, Chapter 10. Their origin can be traced back to marginal operators existing for models with SU(2) spin symmetry (present in a supersymmetric t-J chain). To know how logarithmic singularities in the low temperature susceptibility and specific heat get renormalized for a disordered supersymmetric t-J chain in the high density regime (which is the most important because it pertains to the Kondo, magnetic, behaviour of impurities) we perform an analytic low-temperature study of Eqs. (8.71). We again introduce scaling functions $a_\pm(x) \equiv a(\pm x \pm Ln)$, where $Ln = \ln(\alpha T_{jK}/T)$ (α is a constant) *etc.* Eq. (8.71) are transformed so that for the new set of scaling functions the only known asymptotic behaviour of "energy density" functions $1 + a_\pm$ and $1 + \bar{a}_\pm$ at large spectral parameter enters. Then we obtain (at $H = 0$) the Helmholtz free energy of the dense limit of a supersymmetric t-J chain per site

$$f(T_K^j) = e_0^j + \mu - \frac{\pi T^2}{6T_{jK}}[1 + 3(2Ln)^{-3}] + \cdots . \tag{8.79}$$

For a weak nonzero magnetic field $H \ll T$ we calculate logarithmic temperature corrections for the Helmholtz free energy per site as

$$\begin{aligned} f(T_K^j) = e_0^j(H) + \mu - \frac{\pi T^2}{6T_{jK}} \\ - \frac{H^2}{4\pi T_{jK}}[1 + (2Ln)^{-1} - (2Ln)^{-2}\ln(2Ln)] + O(T^2) . \end{aligned} \tag{8.80}$$

Notice that for the dense limit of the low temperature behaviour of a t-J chain with disordered impurities the dependence on θ_j enters only as T_{jK}, *i.e.*, as distributions of "characteristic velocities" of spin excitations (or crossover scales, which pertain to each impurity). It is not the case for higher energies and for lower densities, but those are not important for low temperature disorder-driven divergences. Hence, for low energies we can use distributions of T_{jK}, which are also more appropriate in connection to experiments. For the case of a homogeneous chain $T_{jK} \to v_s^F$, where v_s^F is the Fermi velocity of low-lying spin-carrying excitations of a supersymmetric t-J chain. We can apply the results obtained above for a Heisenberg spin chain for the dense limit of a supersymmetric t-J chain with disordered impurities. Notice that the average compressibility for the high density limit also reveals the low temperature divergence.

Summarizing, in this chapter we considered exact Bethe ansatz solutions for spin and correlated electron chains with finite concentrations of magnetic impurities. For similar impurities we showed how the finite concentration of them and interactions between them yield impurity bands, which drastically change the behaviour of the system. We also presented exact results for thermodynamic characteristics of correlated electron and spin chains with disordered ensembles of magnetic impurities. For this purpose we used the "quantum transfer matrix" approach, which description is presented to the reader.

Quantum spin chains with a finite concentration of impurities were introduced in [Schlottmann (1994)], see also [de Vega and Woynarovich (1992)]. The special case of zig-zag spin and correlated electron chains was reviewed in [Zvyagin (2001b)]. The reader can find the description of impurities' bands in integrable correlated electron chains with a finite concentration of impurities in [Schlottmann and Zvyagin (1997b); Schlottmann (1998a); Schlottmann (1998b)]. The description of the McCoy–Wu model can be found in [McCoy and Wu (1973)]. The reader can find an information about the Griffiths phase in [Griffiths (1969)]. For the quantum transfer matrix approach we refer to [Suzuki (1985); Suzuki and Inoue (1987); Inoue and Suzuki (1988); Koma (1990); Pearce and Klümper (1991); Klümper, Batchelor and Pearce (1991); Klümper (1993); Klümper (1998)]. In particular, thermodynamic quantum transfer matrix equations for a supersymmetric t-J chain and a repulsive Hubbard chain are derived and studied in [Jüttner, Klümper and Suzuki (1997); Jüttner, Klümper and Suzuki (1998)]. The description of disordered ensembles of spin-$\frac{1}{2}$ impurities in Heisenberg and Heisenberg–Ising spin chains can be found in [Klümper and Zvyagin (1998); Zvyagin (2000); Klümper and Zvyagin (2000)]. It was generalized in [Zvyagin (2002)] for any values of spins of the host and impurities and for ensembles of disordered Kondo impurities in metals for single- and multi-channel situations. The exact solution for a correlated electron t-J chain with randomly distributed impurities was obtained in [Zvyagin (2001a)].

Chapter 9

Finite Size Corrections in Quantum Correlated Chains

In this chapter we shall study the next order corrections in L^{-1} to thermodynamic characteristics of correlated quantum chains. These corrections are related to quantum topological effects in these chains (like persistent currents) and to the asymptotic behaviour of correlation functions in the conformal limit.

9.1 Finite Size Corrections for Quantum Spin Chains

Let us start to consider finite size corrections with the simplest model of a spin-$\frac{1}{2}$ Heisenberg-Ising chain with periodic boundary conditions. Bethe ansatz equations Eq. (3.19) for the state with the z-projection of the total spin $S^z = (L/2) - M$ can be re-written for $J_z/J = \cos\eta$ as

$$z_L(\lambda_j) = \frac{J_j}{L} , \tag{9.1}$$

where we introduced the so-called *counting function*

$$z_L(x) = \frac{1}{2\pi}\left(p^0(x) - \frac{1}{L}\sum_{l=1}^{M}\phi^0(x-\lambda_l)\right) , \tag{9.2}$$

with

$$\begin{aligned} p^0(x) &= 2\tan^{-1}[\cot(\eta/2)\tanh(x/2)] , \\ \phi^0(x) &= 2\tan^{-1}[\cot\eta\tanh(x/2)] , \end{aligned} \tag{9.3}$$

and quantum numbers $J_j = (M+1)/2$ (mod 1). The energy and the total momentum of this state are defined as

$$E = E_0 + \sum_{j=1}^{M} \left(H - \frac{J \sin^2 \eta}{\cosh \lambda_j - \cos \eta} \right) ,$$
$$P = -\frac{2\pi}{L} \sum_{j=1}^{M} J_j , \tag{9.4}$$

where E_0 is the energy of the ferromagnetic, spin-polarized state (with $M = 0$). Let us specify the set of quantum numbers J_j. Let us choose two numbers $J^{\pm} = M/2$ (mod 1), so that

$$J^+ - J^- = M , \quad -\frac{1}{2}(J^+ + J^-) = D . \tag{9.5}$$

For J_j we take all the numbers equal to $(M+1)/2$ (mod 1) between J^+ and J^-. This pertains to a Dirac sea of M particles with D particles moved from the left Fermi point to the right one. We can also introduce the function $\sigma_L(x) = dz_L(x)/dx$. By using the Euler–Maclaurin formula we can re-write the finite sum as the series

$$\frac{1}{L} \sum_j f(\lambda_j) = \int_{\Lambda^-}^{\Lambda^+} d\lambda f(\lambda) \sigma_L(\lambda)$$
$$- \frac{1}{24L^2} \left(\frac{f'(\Lambda^+)}{\sigma_L(\Lambda^+)} - \frac{f'(\Lambda^-)}{\sigma_L(\Lambda^-)} \right) + O(L^{-3}) , \tag{9.6}$$

where the limits of integration are defined as $z_L(\Lambda^{\pm}) = J^{\pm}/L$. Then it follows that

$$\sigma_L(x) = \frac{1}{2\pi} \Bigg[\frac{dp^0(x)}{dx} - \int_{\Lambda^-}^{\Lambda^+} dy K(x-y) \sigma_L(y)$$
$$- \frac{1}{24L^2} \left(\frac{1}{\sigma_L(\Lambda^+)} \frac{dK(x-\Lambda^+)}{dx} - \frac{1}{\sigma_L(\Lambda^-)} \frac{dK(x-\Lambda^-)}{dx} \right) \Bigg] , \tag{9.7}$$

where $K(x) = d\phi^0(x)/dx$. This linear integral equation is completed by the equations, determining $\Lambda^{\pm}$:

$$\frac{1}{2} \left(\int_{\Lambda^+}^{\infty} d\lambda \sigma_L(\lambda) - \int_{-\infty}^{\Lambda^-} d\lambda \sigma_L(\lambda) \right) = \frac{D}{L} ,$$
$$\int_{\Lambda^-}^{\Lambda^+} d\lambda \sigma_L(\lambda) = \frac{M}{L} . \tag{9.8}$$

Equation (9.7) can be written in the form

$$\sigma_L(x) = \sigma(x|\Lambda^+,\Lambda^-) - \frac{1}{24L^2}\left(\frac{\rho(x|\Lambda^+,\Lambda^-)}{\sigma_L(\Lambda^+)} + \frac{\rho(-x|-\Lambda^-,-\Lambda^+)}{\sigma_L(\Lambda^-)}\right), \tag{9.9}$$

where $\sigma(x|\Lambda^+,\Lambda^-)$ and $\rho(x|\Lambda^+,\Lambda^-)$ are the solutions of the following linear integral equations:

$$\begin{aligned}\rho(x|\Lambda^+,\Lambda^-) &= \frac{1}{2\pi}\left(\frac{dK(x-\Lambda^+)}{dx} - \int_{\Lambda^-}^{\Lambda^+} dyK(x-y)\rho(y|\Lambda^+,\Lambda^-)\right),\\ \sigma(x|\Lambda^+,\Lambda^-) &= \frac{1}{2\pi}\left(\frac{dp^0(x)}{dx} - \int_{\Lambda^-}^{\Lambda^+} dyK(x-y)\sigma(y|\Lambda^+,\Lambda^-)\right),\end{aligned} \tag{9.10}$$

with the accuracy which is necessary for our purposes it is enough to define $\Lambda^{\pm}$ of order of L^{-1}. This is why we may replace $\sigma_L(x)$ with $\sigma(x|\Lambda^+,\Lambda^-)$ in Eqs. (9.8), which yields

$$\begin{aligned}\left(\int_{\Lambda^+}^{\infty} d\lambda\sigma(\lambda|\Lambda^+,\Lambda^-) - \int_{-\infty}^{\Lambda^-} d\lambda\sigma(\lambda|\Lambda^+,\Lambda^-)\right) &= \frac{D}{N},\\ \int_{\Lambda^-}^{\Lambda^+} d\lambda\sigma(\lambda|\Lambda^+,\Lambda^-) &= \frac{M}{L}.\end{aligned} \tag{9.11}$$

The energy of the state can be written as

$$E = E_0 + L\varepsilon\left(\frac{M}{L},\frac{D}{L}\right) - \frac{1}{24L}\left(\frac{e(\Lambda^+,\Lambda^-)}{\sigma_L(\Lambda^+)} + \frac{e(-\Lambda^-,-\Lambda^+)}{\sigma_L(\Lambda^-)}\right), \tag{9.12}$$

where

$$\begin{aligned}\varepsilon\left(\frac{M}{L},\frac{D}{L}\right) &= -\int_{\Lambda^-}^{\Lambda^+} d\lambda\varepsilon^0(\lambda)\sigma(\lambda|\Lambda^+,\Lambda^-),\\ \varepsilon^0(x) &= H - \frac{J\sin^2\eta}{\cosh x - \cos\eta},\end{aligned} \tag{9.13}$$

and

$$e(\Lambda^+,\Lambda^-) = \frac{d\varepsilon_0(\Lambda^+)}{d\Lambda^+} - \int_{\Lambda^-}^{\Lambda^+} d\lambda\varepsilon_0(\lambda)\rho(\lambda|\Lambda^+,\Lambda^-). \tag{9.14}$$

When one takes the thermodynamic limit $L, M \to \infty$ keeping $(M/L) = \nu(H)$, $(D/L) = \delta$ finite, then $\varepsilon[\nu(H),\delta]$ is the internal energy per site of the infinite system. In the ground state this internal energy ε_∞ must be minimal with respect to $\nu(H)$ and δ. We perform such a minimization, expanding then $\varepsilon[(M/L),(D/L)]$ about that minimum. If $\delta = 0$, then $\Lambda^+ = -\Lambda^- = \Lambda$. We get

$$\varepsilon\left(\frac{M}{L},\frac{D}{L}\right) = \varepsilon_\infty + \frac{e(\Lambda,-\Lambda)}{\sigma(\Lambda|\Lambda,-\Lambda)} \times \left[\frac{1}{4Z^2}\left(\frac{M}{L}-\nu(H)\right)^2 + Z^2\frac{D^2}{L^2}\right] , \tag{9.15}$$

where we introduced the *dressed charge* $Z = \xi(\Lambda)$ as the solution of the following equation

$$\xi(x) = 1 - \frac{1}{2\pi}\int_{-\Lambda}^{\Lambda} dyK(x-y)\xi(y) , \tag{9.16}$$

taken at the Fermi point. Actually, it is not difficult to recognise that $\xi(x) = (\partial\varepsilon(x)/\partial H)$. The dressed charge of the excitation shows how the interaction "dresses" the "bare charge" of the "bare" energy of a low-lying excitation. The reader can see that in the considered model η measures the strength of interaction (for $\eta = \pi/2$, *i.e.*, for the isotropic XY chain, spinons are non-interacting fermions, and for them $Z = 1$). For $H = 0$ we have

$$Z = \sqrt{\frac{\pi}{2(\pi-\eta)}} . \tag{9.17}$$

For the isotropic Heisenberg chain case with $H = 0$ one obtains $Z = 1/\sqrt{2}$. On the other hand, for $H = H_s$ we get $Z = 1$. Actually, it is easy to show that the dressed charge is related to the magnetic susceptibility *via*

$$\chi = \frac{Z^2}{\pi v^F} , \tag{9.18}$$

which follows from the definition of the dressed charge.

Then we calculate

$$E = E_0 + L\varepsilon_\infty - \frac{1}{L}\frac{e(\Lambda,-\Lambda)}{\sigma(\Lambda|\Lambda,-\Lambda)}\left(\frac{1}{12} - \frac{1}{4Z^2}[M-\nu(H)L] - Z^2D^2\right) . \tag{9.19}$$

One can denote

$$v^F = \frac{e(\Lambda, -\Lambda)}{2\pi\sigma(\Lambda|\Lambda, -\Lambda)} , \tag{9.20}$$

where v^F is the Fermi velocity of a spinon (for $H = 0$ it is equal to $v^F = \pi J \sin\eta/2\eta$). It is easy to see that the equation for $\sigma(x|\Lambda, -\Lambda)$ from Eqs. (9.9) coincides with the ground state Bethe ansatz equation for the density of spinons $\rho_1(\lambda) = \sigma(\lambda|\Lambda, -\Lambda)$, with $\Lambda = B$. The reader can also check that Eq. (9.14) is the derivative of the ground state Bethe ansatz equation for the dressed energy of spinons $\varepsilon_1(\lambda)$ taken at $\lambda = B$, from which it follows that

$$e(\Lambda, -\Lambda) = \frac{\partial \varepsilon_1(\lambda)}{\partial \lambda}|_{\lambda=\Lambda} , \tag{9.21}$$

so that the above definition of the Fermi velocity coincides with the definitions of Fermi velocities of low-lying excitations (for this concrete model, spinons) which were used in previous chapters. Then, let us introduce particle-hole excitations by removing J_j from the Dirac sea and introducing J_j outside the sea. Notice that in order not to change M and D, *i.e.*, the total number of quasiparticles and the number of quasiparticles moved from the left Fermi point to the right one, the number of the particles and holes for particle-hole excitations should be equal in the vicinity of both the left and right Fermi points. We characterize the holes and particles in the vicinity of $J^{\pm}$ as

$$J_p^{\pm} = J^{\pm} \pm n_p^{\pm} , \ J_h^{\pm} = J^{\pm} \mp n_h^{\pm} , \tag{9.22}$$

where the numbers $n_{p,h}^{\pm} > 0$ are half integers. We can introduce their total numbers as

$$n^{\pm} = \sum (n_p^{\pm} + n_h^{\pm}) , \tag{9.23}$$

where $n^{\pm}$ are integers since $J_p^{\pm} = J_h^{\pm}$. The expression for the total momentum is then

$$P(M, D, n^{\pm}) = \frac{2\pi}{L}(MD - n^+ + n^-) + 2DP^F , \tag{9.24}$$

where $P^F = \frac{\pi}{2}(1 - 2m^z)$ is the Fermi momentum, and

$$E(M, D, n^{\pm}) = E_0 + L\varepsilon_{\infty} - \frac{\pi v^F}{6L} + \frac{2\pi v^F}{L}(\Delta + n^+ + n^-) , \tag{9.25}$$

where

$$\Delta = [2Z]^{-2}[M - L\nu(H)]^2 + Z^2 D^2 . \tag{9.26}$$

These equations mean that the low-energy state of a quantum spin chain with periodic boundary conditions is characterized by a set of quantum numbers, M, D and $n^{\pm}$, which defines particle excitations, excitations which manifest transfers from one Fermi point to the other one, and particle-hole excitations above the ground state.

The reader can check the above results for the case of the isotropic XY chain, for which the results for finite L are presented in Chapter 2. Defining $m_0 = M_0/L$, where M_0 relates to the total magnetization $M^z = (L/2) - M_0$ in the minimum, we get for the internal energy per site

$$\begin{aligned} e = e_0 &- \frac{J}{\pi}\sin(\pi m_0) + H m_0 \\ &- \frac{\pi J}{6L^2}\sin(\pi m_0)[1 - 3(M - m_0 L)^2] + O(L^{-3}) , \end{aligned} \tag{9.27}$$

where $\cos \pi m_0 = H/|J|$.

Equations (9.24) and (9.25) are rather universal. They can be applied to any Bethe ansatz solvable model, which can be described by only one set of rapidities, which states form the Dirac sea with the Fermi energy of low-lying excitations v^F. The only condition for the application of these equations is the metallic character of the low-lying spectrum, *i.e.*, that energies of these low-lying excitations are of order of L^{-1} (gapless). The reader can see that only definitions of $p^0(x)$ and $\phi^0(x)$ are model-dependent. Each excited state is determined by a set of quantum numbers M, D and $n^{\pm}$.

Let us now see how this description is modified when one considers finite size corrections for models with a single impurity. Calculations similar to the above yield

$$\begin{aligned} E(M, D, n^{\pm}) &= E_0 + L\varepsilon_{\infty} + e_{imp} - \frac{\pi v^F}{6L} + \frac{2\pi v^F}{L}(\Delta_{imp} + n^+ + n^-) , \\ \Delta_{imp} &= [2Z]^{-2}[M - n_{imp} - L\nu(H)]^2 + Z^2(\lambda)[D - d_{imp}]^2 , \end{aligned} \tag{9.28}$$

where n_{imp} is the valence of an impurity for models of particles with interaction (or it is related to the magnetization of the impurity for spin models)

and

$$d_{imp} = \frac{1}{2}\left(\int_{-\infty}^{-B} d\lambda \rho_1^{(1)}(\lambda) - \int_{B}^{\infty} d\lambda \rho_1^{(1)}(\lambda)\right) , \tag{9.29}$$

where B denotes the Fermi point ($B = \Lambda$ in the notations of the above) and $\rho_1^{(1)}$ satisfies an equation for density of an impurity of order of L^{-1}, *cf.* Chapter 7. Naturally, the values n_{imp} and d_{imp} are defined mod 1. They determine shifts of the values $\Delta M = M - L\nu(H)$ and $\Delta D = D - L\delta$ due to a single impurity. It is important to emphasize that a dressed charge of a quantum spin chain with a single impurity does not depend on the parameters of the impurity.

Now let us see how a finite concentration of impurities can modify the answer for finite size corrections. The reader knows from the previous chapter that a finite concentration of similar impurities is responsible for the onset of additional Dirac seas, connected with impurities. Then, each Dirac sea contributes to finite size corrections in the form, equivalent to Eqs. (9.24) and (9.25) with its own Fermi velocity and sets of quantum numbers M, D and $n^{\pm}$. Here it is important to emphasize that those quantum numbers are not all independent, because the filling of Dirac seas, caused by a finite concentration of impurities, is related to the same generalized chemical potential (for spin systems — to the magnetic field H), which governs the filling of all Dirac seas, *cf.* Chapter 8.

The interesting case, which deserves special consideration, is the case of a finite concentration of impurities with $\theta_j = \theta$, *e.g.* a finite concentration of spin-$\frac{1}{2}$ impurities in a Heisenberg spin-$\frac{1}{2}$ antiferromagnetic chain with the direct interaction between impurities situated between the neighbouring sites of the host chain for $\theta > \theta_c$ and $H > H_c$, see the previous chapter. Let us shift all rapidities by $\lambda_j \to \lambda_j + \theta/2$, to symmetrize the situation (nothing depends on that shift, naturally). The Dirac seas (*i.e.*, spinons with negative energies) are in the intervals $[-B^+, -B^-]$ and $[B^-, B^+]$ (minima in the distributions of rapidities at $\mp\theta/2$). This can be also interpreted as the symmetrically distributed (around zero) Dirac seas of "particles" for $[-B^+, B^+]$ and the Dirac sea of "holes" for $[-B^-, B^-]$. Naturally, Fermi velocities of "particles" are positive, $v^+ = (2\pi\rho(B^+))^{-1}\varepsilon'(\lambda)|_{\lambda=B^+}$, while Fermi velocities of "holes" are negative, $v^- = -(2\pi\rho(B^-))^{-1}\varepsilon'(\lambda)|_{\lambda=B^-}$. Finite size corrections to the energy for this case are

$$E = L\varepsilon_\infty - \frac{\pi}{12L}(v^+ + v^-) + \frac{\pi}{L}\left(v^+(\Delta_l^+ + \Delta_r^+) + v^-(\Delta_l^- + \Delta_r^-)\right) , \tag{9.30}$$

where the dispersion laws of "particles" and "holes" are linearized about Fermi points for each of Dirac sea. Here $\Delta^{\pm}_{l,r}$ are (superscripts denote Dirac seas; subscripts denote right and left Fermi points of each of these two Dirac seas):

$$2\Delta^{\mp}_{l,r} = \left[\frac{(Z_{-\pm}\Delta M^{+} - Z_{+\pm}\Delta M^{-})}{2\det \hat{Z}} \mp \frac{(X_{-\pm}\Delta D^{+} - X_{+\pm}\Delta D^{-})}{2\det \hat{X}}\right]^2 + 2n^{\mp}_{l,r} , \tag{9.31}$$

where the "−" sign between the terms in square brackets corresponds to particles about the right Fermi point and "+" to the ones about the left Fermi point. Here $\Delta M^{\pm}$ denote differences between the numbers of particles excited in the Dirac seas of "particles" and "holes", labelled by upper indices. $\Delta D^{\pm}$ denote the numbers of excitations transfered from the right to the left Fermi point, respectively, and $n^{\pm}_{l,r}$ are the numbers of the particle-hole excitations for each of Dirac seas (for "particles" and "holes"). $\Delta M^{\pm}$ and $\Delta D^{\pm}$ are not independent. Their values are restricted by the following relations: $\Delta M^{+} - \Delta M^{-} = \Delta M$, and $\Delta D^{+} - \Delta D^{-} = \Delta D$, where ΔM and ΔD determine in a standard way the changes of the total magnetization and the total momentum of the system, respectively, due to excitations. There are only two independent of four such possible excitations. This is the direct consequence of the fact that only one magnetic field determines the filling of Dirac seas for "particles" and "holes", or, in other words, the filling of two Dirac seas for spinons centred at $\pm\theta/2$.

The dressed charges $Z_{ik}(Q^k)$ and $X_{ik}(Q^k)$ $(i,k = +,-)$ are matrices in this phase. They can be expressed by using the solution of the integral equation

$$f(u|B^{\pm}) = \left(\int_{-B^{+}}^{B^{+}} - \int_{-B^{-}}^{B^{-}}\right) K(u-v) f(v|B^{\pm}) = K(u - B^{\pm}) , \tag{9.32}$$

with

$$\begin{aligned} X_{ik}(B^k) &= \delta_{i,k} + (-1)^k \frac{1}{2}\left(\int_{B^i}^{\infty} - \int_{-\infty}^{-B^i}\right) dv f(v|B^k) , \\ Z_{ik}(B^k) &= \delta_{i,k} - (-1)^k \int_{-B^i}^{B^i} dv f(v|B^k) . \end{aligned} \tag{9.33}$$

Notice that dressed charges depend on the value of the coupling constant θ indirectly, only *via* limits of integrations. The Dirac sea for "holes" disappears, naturally, for $H \to H_c$, $\theta \to \theta_c$. The slopes of dressed energies of "particles" and "holes" at Fermi points of Dirac seas (Fermi velocities) differ in general from each other. Hence, in this region dressed charges are 2×2 matrices. At the critical line H_c the Dirac sea of "holes" disappears (as well as components of the dressed charge matrix $\hat{X}$) with square root singularities. Note that the dressed charge Z becomes $Z = (2X)^{-1}$ at the line of the quantum phase transition H_c.

So far we have studied systems with periodic boundary conditions. Let us investigate how open boundary conditions change the answers. Proceeding as for the periodic case we obtain

$$\begin{aligned} E(M,n) &= E_0 + L\varepsilon_\infty + e_b - \frac{\pi v^F}{24L} + \frac{\pi v^F}{L}(\Delta^b + n) , \\ \Delta^b &= [2Z^2]^{-1}[M - \Theta(h_{1,L}) - L\nu(H)]^2 , \end{aligned} \tag{9.34}$$

where e_b is the energy of open boundaries (both, the contributions from open edges, and from boundary fields), *cf.* Chapter 6, and $\Theta(h_{1,L})$ is the contribution from open edges themselves and boundary fields $h_{1,L}$ of the finite size

$$\Theta(h_{1,L}) = -\frac{1}{2}\int_{-B}^{B} d\lambda \rho_1^{(1)}(\lambda) . \tag{9.35}$$

At $H = 0$ it is

$$\Theta(h_{1,L}) = \frac{\eta}{2(\pi - \eta)} - \frac{1}{2(\pi - \eta)}\sum_{1,L} \tan^{-1}\left(\cot(\eta/2)\frac{2h_{1,L} - \cos\eta + 1}{2h_{1,L} - \cos\eta - 1}\right) , \tag{9.36}$$

defined mod 1. It determines the shift of $\Delta M = M - L\nu(H)$ due to open edges and boundary fields. The reader can see the differences by comparing this with the case of periodic boundary conditions. First, there appears the contribution of order of 1, which describes the "surface energy" (boundary fields and edges of an open chain themselves), *cf.* Chapter 6. This contribution is, obviously, similar to the contribution of a single impurity. Second, similar to the contribution from a single impurity, open edges and boundary fields renormalize the shift ΔM. Third, in the contribution for ground state one has to replace $L \to 2L$ (as expected). Finally, there is only one Fermi point for open boundary conditions, and, hence, $D = 0$ (there is no transfer from one Fermi point to the other) and we can introduce particle-hole excitations about only one Fermi point. It is important to emphasize that a dressed charge of a quantum chain does not depend on the parameters of open edges and boundary fields, *i.e.*, it is *universal.*

The next order corrections in the series for spin systems, which respect SU(2) symmetry are logarithmic, due to the presence of marginal operators in the renormalization group sense. For a Heisenberg spin-$\frac{1}{2}$ antiferromagnetic chain it follows that the series with the next (logarithmic) correction is

$$E = E_0 + L\varepsilon_\infty - \frac{\pi v^F}{6L}\left[1 + \frac{3}{8\ln^3 AL} + O\left(\frac{\ln\ln AL}{4\ln^2 AL}\right)\right] , \tag{9.37}$$

where A is a constant.

9.2 Finite Size Corrections for Correlated Electron Chains

Now let us find finite size corrections to the energy and the momentum of correlated electron chains, like a Hubbard chain and a supersymmetric t-J chain, first with periodic boundary conditions. We can introduce counting functions

$$z_{i,L}(x) = \frac{1}{2\pi}\left(p_i^0(x) - \frac{1}{L}\sum_{j=1}^{2}\sum_{l=1}^{N_j}\phi_{ij}^0(x, u_{j,l})\right) , \qquad i = 1, 2 , \tag{9.38}$$

and $\sigma_{i,L} = (\partial z_{i,L}(x)/\partial x)$ with $\phi_{ij}^0(x, y) = -\phi_{ji}^0(y, x)$, which satisfy Bethe ansatz equations, *cf.* Chapter 4,

$$z_{i,L}(\lambda_j) = \frac{J_{i,j}}{L} , \tag{9.39}$$

where $J_{1,j} = N_2/2 \pmod 1$ and $J_{2,j} = (N_1 + N_2 + 1)/2 \pmod 1$. Here we denoted for a repulsive Hubbard model

$$\begin{gathered} u_{1,j} = \sin k_j , \quad u_{2,j} = \lambda_j , \quad N_1 = N , \quad N_2 = M , \\ p_1^0(x) = \sin^{-1} x , \quad p_2^0(x) = 0 , \quad \phi_{11}^0 = 0 , \\ \phi_{12}^0(x, y) = 2\tan^{-1}[4(x - y)/U] , \quad \phi_{22}^0(x, y) = 2\tan^{-1}[2(x - y)/U] . \end{gathered} \tag{9.40}$$

For an attractive Hubbard chain we have

$$\begin{gathered} u_{1,j} = \sin k_j , \quad u_{2,j} = \lambda_j , \quad N_1 = N - 2M , \quad N_2 = M , \\ p_1^0(x) = \sin^{-1} x , \quad p_2^0(x) = 2\mathrm{Re}(\sin^{-1}[x + i(U/4)]) , \end{gathered} \tag{9.41}$$

with the same $\phi^0_{ij}(x,y)$ as for a repulsive Hubbard chain. For a supersymmetric t-J chain with $V=-J/4$ and $J=2$ these functions are

$$
\begin{gathered}
u_{1,j}=p_j\ ,\quad u_{2,j}=\lambda_j\ ,\quad N_1=N\ ,\quad N_2=N-2M\ ,\\
p^0_1(x)=2\tan^{-1}2x\ ,\quad p^0_2(x)=2\tan^{-1}x\ ,\quad \phi^0_{11}=0\ ,\\
\phi^0_{12}(x,y)=2\tan^{-1}[2(x-y)]\ ,\quad \phi^0_{22}(x,y)=2\tan^{-1}(x-y)\ .
\end{gathered}
\tag{9.42}
$$

The momentum and energy of the state with N electrons, M of which having their spin down are

$$
P=\frac{2\pi}{L}\sum_{i=1}^{2}\sum_{j=1}^{N_i}J_{ij}\ ,\ E=E_0+\sum_{i=1}^{2}\sum_{j=1}^{N_i}\varepsilon^0_i(u_{i,j})
\tag{9.43}
$$

where for a repulsive Hubbard ring we have

$$
\varepsilon^0_1(x)=-\mu-\frac{H}{2}-2\cos[\sin^{-1}(x)]\ ,\quad \varepsilon^0_2(x)=H\ ,
\tag{9.44}
$$

for an attractive Hubbard model

$$
\begin{aligned}
\varepsilon^0_1(x)&=-\mu-\frac{H}{2}-2\cos[\sin^{-1}(x)]\ ,\\
\varepsilon^0_2(x)&=-2\mu-4\mathrm{Re}\sqrt{1-[x+i(U/4)]^2}\ ,
\end{aligned}
\tag{9.45}
$$

and for a supersymmetric t-J model one uses

$$
\varepsilon^0_1(x)=-\mu-\frac{H}{2}-2+2\pi a_1(x)\ ,\quad \varepsilon^0_2(x)=-2+2\pi a_2(x)-2\mu\ ,
\tag{9.46}
$$

with the same notations as in Chapter 4. Let us choose two sets of numbers $J^\pm_1=(N_2+1)/2$ (mod 1) and $J^\pm_2=(N_1+N_2)/2$ (mod 1), so that

$$
\frac{1}{2}(J^+_i+J^-_i)=D_i\ ,\quad J^+_i-J^-_i=M_i\ ,\quad i=1,2\ ,
\tag{9.47}
$$

for a Hubbard chain, and $J^+_1-J^-_1=L-N+M$ and $J^+_2-J^-_2=L-N$ for a t-J chain, which determine numbers of particles in each Dirac sea for low-lying excitations and numbers of particles which are transfered from the left Fermi point of excitations of each kind to the right Fermi point. With this choice $J_{i,j}$ are all numbers between J^+_i and J^-_i. By using the

Euler–Maclaurin formula, which we can re-write as

$$\frac{1}{L}\sum_{j=n_1}^{n_2} f(n/L) = \int_{(2n_1-1)/2L}^{(2n_2+1)/2L} dx f(x) \\ - \frac{1}{24L^2}\left[f'\left(\frac{2n_2+1}{2N}\right) - f'\left(\frac{2n_1-1}{2N}\right)\right] + O(L^{-3}) , \tag{9.48}$$

we can derive the following equations

$$\sigma_{i,L}(x) = \frac{1}{2\pi}\left[\frac{dp_i^0(x)}{dx} - \sum_{j=1}^{2}\left(\int_{\Lambda_j^-}^{\Lambda_j^+} dy K_{ij}(x,y)\sigma_{j,L}(y) \\ - \frac{1}{24L^2}\left(\frac{1}{\sigma_{j,L}(\Lambda_j^+)}\frac{\partial K_{ij}(x,\Lambda_j^+)}{\partial x} - \frac{1}{\sigma_{j,L}(\Lambda_j^-)}\frac{\partial K_{ij}(x,\Lambda_j^-)}{\partial x}\right)\right)\right] , \tag{9.49}$$

where $K_{ij}(x,y) = (\partial\phi_{ij}^0(x,y)/\partial x)$. Here $\Lambda_i^\pm$ satisfy the equations

$$z_{i,L}(\Lambda_i^\pm) = \frac{J_i^\pm}{L} . \tag{9.50}$$

Notice that for the case of a supersymmetric t-J model integrations are performed not from Λ_j^- to Λ_j^+, but from $-\infty$ to Λ_j^- and from Λ_j^+ to ∞. These linear integral equations are completed by the equations, determining $\Lambda_i^\pm$:

$$\frac{1}{2}\left(\int_{\Lambda_i^+}^{\pi,(\infty)} dx\sigma_{i,L}(x) - \int_{-\pi,(-\infty)}^{\Lambda_i^-} dx\sigma_{i,L}(x)\right) \\ + \frac{\delta_{i,1}}{\pi}\int_{-\Lambda_2^-}^{\Lambda_2^+} dx\sigma_{2,L}(x)\tan^{-1}(4x/U) = -\frac{D_i}{L} , \tag{9.51}$$

$$\int_{\Lambda_i^-}^{\Lambda_i^+} dx\sigma_{i,L}(x) = \frac{M_i}{L} ,$$

for a Hubbard chain (with π in limits for unbound electrons), and similar relations for a t-J chain with necessary changes of the ranges of integrations, see above.

The equations for $\sigma_{i,L}(x)$ can be written in the form

$$\sigma_{i,L}(x) = \sigma_i(x|\Lambda^+_{1,2}, \Lambda^-_{1,2}) + \frac{1}{24L^2}\sum_{j=1}^{2}\Bigg(\frac{\rho_j(x|\Lambda^+_{1,2}, \Lambda^-_{1,2})}{\sigma_{j,L}(\Lambda^+_j)} - \frac{\rho_j(-x|-\Lambda^-_{1,2}, -\Lambda^+_{1,2})}{\sigma_{j,L}(\Lambda^-_j)}\Bigg) , \tag{9.52}$$

where $\sigma(x|\Lambda^+_{1,2}, \Lambda^-_{1,2})$ and $\rho_i(x|\Lambda^+_{1,2}, \Lambda^-_{1,2})$ are the solutions of the following linear integral equations

$$\sigma_i(x|\Lambda^+_{1,2}, \Lambda^-_{1,2}) = \frac{1}{2\pi}\left(\frac{dp_i^0(x)}{dx} + \sum_{j=1}^{2}\int_{\Lambda^-_j}^{\Lambda^+_j} dy K_{ij}(x,y)\sigma_j(y|\Lambda^+_{1,2}, \Lambda^-_{1,2})\right) ,$$

$$\rho_i(x|\Lambda^+_{1,2}, \Lambda^-_{1,2}) = \frac{1}{2\pi}\sum_{j=1}^{2}\Bigg(\frac{\partial K_{ij}(x,\Lambda^+_j)}{\partial x} + \int_{\Lambda^-_j}^{\Lambda^+_j} dy K_{ij}(x,y)\rho_j(y|\Lambda^+_{1,2}, \Lambda^-_{1,2})\Bigg) , \tag{9.53}$$

for a Hubbard chain. For a t-J chain one has to change the ranges of integrations as explained above. Another possibility is to convert the integrals from $-\infty$ to Λ^-_j and from Λ^+_j to ∞ to the ones from Λ^-_j to Λ^+_j for a t-J chain. This can be performed by a Fourier transformation, which implies the formal changes $K_{11} \to -K_{22}$, $K_{22} \to 0$, $p_2^0 \to 0$, $dp_1^0(x)/dx \to K_{12}(x)$, $\varepsilon_1^0(x) \to H - K_{21}(x,0)$, and $\varepsilon_2^0(x) \to 2 + \mu - (H/2)$. Notice, that after such a transformation for a supersymmetric t-J chain one has

$$\int_{\Lambda^-_1}^{\Lambda^+_1} dx\sigma_{1,L}(x) = 1 - \frac{N-M}{L} , \qquad \int_{\Lambda^-_2}^{\Lambda^+_2} dx\sigma_{2,L}(x) = 1 - \frac{N}{L} . \tag{9.54}$$

Then the energy of the state can be written as

$$E = E_0 + L\varepsilon_\infty(\Lambda^+_i, \Lambda^-_i) + \frac{1}{24L}\sum_{i=1}^{2}\left(\frac{e_i(\Lambda^+_i, \Lambda^-_i)}{\sigma_{i,L}(\Lambda^+_i)} + \frac{e_i(-\Lambda^-_i, -\Lambda^+_i)}{\sigma_{i,L}(\Lambda^-_i)}\right) , \tag{9.55}$$

where

$$\varepsilon_\infty(\Lambda^+_i, \Lambda^-_i) = \sum_{j=1}^{2}\int_{\Lambda^-_j}^{\Lambda^+_j} dx\varepsilon_j^0(x)\sigma_j(x|\Lambda^+_{1,2}, \Lambda^-_{1,2}) , \tag{9.56}$$

and

$$e_i(\Lambda_i^+,\Lambda_i^-) = \frac{d\varepsilon_i^0(\Lambda_i^+)}{d\Lambda_i^+} - \sum_{j=1}^{2}\int_{\Lambda_j^-}^{\Lambda_j^+} dx\varepsilon_j^0(x)\rho_j(x|\Lambda_{1,2}^+,\Lambda_{1,2}^-) \ . \qquad (9.57)$$

Naturally, these equations can be re-written in terms of dressed energies

$$\varepsilon_\infty(\Lambda_i^+,\Lambda_i^-) = \frac{1}{2\pi}\sum_{i=1}^{2}\int_{\Lambda_i^-}^{\Lambda_i^+} dx\varepsilon_i(x|\Lambda_{1,2}^+,\Lambda_{1,2}^-)\left(\frac{dp_i^0(x)}{dx}\right) , \qquad (9.58)$$

where the dressed energies $\varepsilon_i(x|\Lambda_{1,2}^+,\Lambda_{1,2}^-)$ satisfy the set of equations

$$\varepsilon_i(x|\Lambda_{1,2}^+,\Lambda_{1,2}^-) = \varepsilon_i^0(x) - \sum_{j=1}^{2}\int_{\Lambda_j^-}^{\Lambda_j^+} dyK_{ij}^t(x,y)\varepsilon_j(y|\Lambda_{1,2}^+,\Lambda_{1,2}^-) , \qquad (9.59)$$

which implies

$$e_i(\Lambda_i^+,\Lambda_i^-) = \frac{\partial\varepsilon_i^0(x|\Lambda_{1,2}^+,\Lambda_{1,2}^-)}{\partial x}\Big|_{x=\Lambda_i^+} \ . \qquad (9.60)$$

In the infinite chain $\varepsilon_\infty(\Lambda_i^+,\Lambda_i^-)$ is minimal with respect to $\Lambda_i^\pm$ at given μ and H. This condition leads to

$$\varepsilon_i(\Lambda_i^\pm|\Lambda_{1,2}^+,\Lambda_{1,2}^-) = 0 , \qquad (9.61)$$

which is the determination of the ground state Fermi points for dressed energies. From the symmetry one can suppose that in the ground state $\Lambda_i^+ = -\Lambda_i^- = \Lambda_i$. The next step is to expand $\varepsilon_\infty(\Lambda_i^+,\Lambda_i^-)$ up to the second order in $(\Lambda_i^\pm \mp \Lambda_i)$. We find

$$\begin{aligned}\varepsilon_\infty(\Lambda_i^+,\Lambda_i^-) &= \varepsilon_\infty(\Lambda_i,-\Lambda_i) + \sum_{j=1}^{2}\frac{\frac{\partial}{\partial x}\varepsilon_j(x|\Lambda_{1,2},-\Lambda_{1,2})|_{x=\Lambda_j}}{\sigma_j(\Lambda_j|\Lambda_{1,2},-\Lambda_{1,2})} \\ &\times\frac{1}{2}\big([\sigma_j(\Lambda_j|\Lambda_{1,2},-\Lambda_{1,2})(\Lambda_j^+-\Lambda_j)]^2 \\ &+[\sigma_j(\Lambda_j|\Lambda_{1,2},-\Lambda_{1,2})(\Lambda_j^-+\Lambda_j)]^2\big) \ . \end{aligned} \qquad (9.62)$$

(We write this equation with the accuracy of L^{-2}.) The reader already knows from Chapter 4 that

$$\frac{1}{\sigma_j(\lambda_j|\Lambda_{1,2},-\Lambda_{1,2})}\frac{\partial}{\partial x}\varepsilon_j(x|\Lambda_{1,2},-\Lambda_{1,2})|_{x=\Lambda_j} = 2\pi v_j^F , \qquad (9.63)$$

where v_j^F are Fermi velocities of low-lying excitations. It is easy to check that for $\Lambda_i^\pm = \pm\Lambda_i$ the equations for $\sigma_i(x|\Lambda_{1,2},-\Lambda_{1,2})$ and $\varepsilon_i(x|\Lambda_{1,2},-\Lambda_{1,2})$

coincide with the standard definitions of densities and dressed energies for the ground state in the thermodynamic limit from previous chapters. Let us denote $\nu_i = M_i/L$, $\delta_i = D_i/L$ and calculate

$$\frac{\partial \nu_i}{\partial \Lambda_j^+} = -\frac{\partial \nu_i}{\partial \Lambda_j^-} = \sigma_j(\Lambda_j|\lambda_j, -\Lambda_j) Z_{ji} \ ,$$
$$\frac{\partial \delta_i}{\partial \Lambda_j^+} = \frac{\partial \delta_i}{\partial \Lambda_j^-} = \sigma_j(\Lambda_j|\lambda_j, -\Lambda_j) X_{ji} \ , \qquad (9.64)$$

where we introduced dressed charge matrices Z_{ij} and X_{ij}. It is not difficult to see from the above equations that $X_{ij} = \frac{1}{2}(Z_{ij}^t)^{-1}$. Dressed charge matrices can be expressed as $Z_{ij} = \xi_{ij}(\Lambda_i)$, where $\xi_{ij}(x)$ satisfy the equations

$$\xi_{ij}(x) = \delta_{ij} + \sum_{l=1}^{2} \int_{-\Lambda_l}^{\Lambda_l} dy K_{il}^t(x, y) \xi_{lj}(y) \ . \qquad (9.65)$$

Again, the coefficients of a dressed charge matrix satisfy the relation $\xi_{ij}(x) = (\partial \varepsilon_i(x)/\partial \mu_j)$, where μ_i are generalized chemical potentials for low-lying excitations. For example, for a repulsive Hubbard chain they are $\mu_1 = -\mu - \frac{H}{2}$ and $\mu_2 = H$, and for an attractive Hubbard chain we have $\mu_1 = -\mu - \frac{H}{2}$ and $\mu_2 = -2\mu$. A dressed charge matrix measures how strong the interaction is in a system. The reader can see that, *e.g.*, for a Hubbard chain for $U = 0$ the dressed charge matrix is the unity matrix. Nondiagonal components of a dressed charge matrix show that despite the fact that quantum numbers $J_{1,2}$ define different states, those states are not independent (the reader knows it, because Bethe ansatz equations for charged excitations and excitations, which carry spin, for densities and dressed energies are connected to each other). If there was a real spin-charge separation in Bethe ansatz solvable models for correlated electrons, then equations for densities and dressed energies for charge-carrying and spin-carrying excitations were independent of each other, and dressed charge matrices were diagonal. This is why the reader has to remember that when one speaks about a spin-charge separation in Bethe ansatz solvable models of correlated electrons, it only means that their low-lying excitations are spread with different velocities, and those excitations, as a rule, carry different charges and spins. It is important to notice that often a transpose definition of dressed charge matrices is used.

It is instructive to present some results for dressed charge matrices for integrable correlated electron models. For a repulsive Hubbard chain for $H = 0$ we have $Z_{11} = 2Z_{12} = \xi(Q)$, $Z_{21} = 0$, and $Z_{22} = 1/\sqrt{2}$. Here for large $U \gg \sin Q$, $\xi(Q) = 1 + 4\ln 2 \sin Q/\pi U$ (for small number of electrons it implies $\sin Q \to \pi N/L$, and for the half-filling $\xi(\pi) = 1$). For $U \gg 1$ one obtains $\xi(Q) = 1 + 4\ln 2\sin(\pi N/L)/\pi U$. For small $U \ll \sin Q$, one gets $\xi(Q) = \sqrt{2}(1 - U/8\pi \sin Q)$. Summarizing, for $H = 0$ one has the component of a dressed charge varying in the interval $1 \le \xi(Q) \le \sqrt{2}$.

For $H = H_s$ diagonal components of the dressed charge matrix of a repulsive Hubbard chain become equal to 1, while $Z_{21} = 0$ and $Z_{12} = (2/\pi)\tan^{-1}[4\sin(\pi N/L)/U]$, *i.e.*, it varies between 0 and 1 as one changes U and/or N. At half-filling the behaviour of a repulsive Hubbard chain is equivalent to the one of a Heisenberg antiferromagnetic spin-$\frac{1}{2}$ model, because charged excitations become gapful.

For an attractive Hubbard chain for $H < H_c$ one has only charged low-lying excitations. The dressed charge varies between 1 and $1/\sqrt{2}$, depending on N and U. On the other hand, at $H = H_s$ the components of a dressed charge matrix are $Z_{11} = 1$, $Z_{21} = 0$, $Z_{12} = -\frac{1}{2}$ and $Z_{22} = 1/\sqrt{2}$.

For a supersymmetric t-J chain the behaviour of a dressed charge matrix at $H = 0$ is similar to the one of a repulsive Hubbard chain. At half filling one has $Z_{11} = 1$, $Z_{21} = 0$, $Z_{12} = \frac{1}{2}$ and Z_{22} is equivalent to the dressed charge of a Heisenberg antiferromagnetic spin-$\frac{1}{2}$ model.

It is also instructive to connect components of a dressed charge matrix with physical values. Denoting χ and χ_c as spin and charge susceptibilities, and $\chi_{mix} = (\partial m^z/\partial\mu)$, we get, *e.g.* for a repulsive Hubbard chain

$$\pi\hat{Z}^{-1}\hat{V}(\hat{Z}^{-1})^t = \begin{pmatrix} \frac{L^2}{N^2\chi_c} + \frac{1}{4\chi} + \frac{1}{\chi_{mix}} & -\frac{1}{2\chi} - \frac{1}{\chi_{mix}} \\ -\frac{1}{2\chi} - \frac{1}{\chi_{mix}} & \frac{1}{\chi} \end{pmatrix}, \tag{9.66}$$

where $\hat{Z}$ is the dressed charge matrix, and

$$\hat{v} = \begin{pmatrix} v_1^F & 0 \\ 0 & v_2^F \end{pmatrix}. \tag{9.67}$$

It is possible to write similar formulas for an attractive Hubbard chain and for a supersymmetric t-J chain.

By using dressed charge coefficients and velocities of low-lying excitations it is easy to write

$$E = E_0 + L\varepsilon_\infty(\Lambda_i, -\Lambda_i) - \frac{\pi}{6L}\sum_{i=1}^{2} v_i^F + \frac{2\pi}{L}\sum_{i=1}^{2} v_i^F \Delta_i \,, \tag{9.68}$$

where

$$\Delta_1 = \left(\sum_{j=1}^{2} Z_{1j}(D_j - \delta_j L)\right)^2 + \frac{1}{4(\det Z)^2}[Z_{22}(M_1 - \nu_1(\mu, H)L)$$
$$- Z_{21}(M_2 - \nu_2(\mu, H)L)]^2 ,$$
$$\Delta_2 = \left(\sum_{j=1}^{2} Z_{2j}(D_j - \delta_j L)\right)^2 + \frac{1}{4(\det Z)^2}[Z_{12}(M_1 - \nu_1(\mu, H)L)$$
$$- Z_{11}(M_2 - \nu_2(\mu, H)L)]^2 . \tag{9.69}$$

We can again introduce particle-hole excitations by removing $J_{i,j}$ from a Dirac sea for low-lying excitations and introducing $J_{i,j}$ outside the sea. In order not to change M_i and D_i, *i.e.*, the total number of quasiparticles and the number of quasiparticles moved from the left Fermi point to the right one in each Dirac sea for low-lying excitations, the number of particles and holes for particle-hole excitations should be equal both in the vicinity of the left and right Fermi points. We characterize holes and particles in the vicinity of $J_i^{\pm}$ as

$$J_{i,p}^{\pm} = J_i^{\pm} \pm n_{i,p}^{\pm} , \qquad J_{i,h}^{\pm} = J_i^{\pm} \mp n_{i,h}^{\pm} , \qquad i = 1, 2 , \tag{9.70}$$

where the numbers $n_{i,p,h}^{\pm} > 0$ are half integers. We then introduce total numbers as

$$n_i^{\pm} = \sum (n_{i,p}^{\pm} + n_{i,h}^{\pm}) , \tag{9.71}$$

where $n_i^{\pm}$ are integers since $J_{i,p}^{\pm} = J_{i,h}^{\pm}$. The expression for the total momentum is then

$$P(M_i, D_i, n_i^{\pm}) = \frac{2\pi}{L}\sum_{i=1}^{2}(M_i D_i + n_i^{+} - n_i^{-}) + 2\sum_{\sigma} P_{\sigma}^{F} D_{\sigma} , \tag{9.72}$$

where the Fermi momenta are

$$P_{\sigma}^{F} = \frac{\pi}{2}\left(\frac{N}{L} \pm 2m^{z}\right) , \tag{9.73}$$

for a repulsive Hubbard chain $D_{\uparrow} = D_1$, $D_{\downarrow} = D_1 + D_2$, for an attractive Hubbard chain $D_{\uparrow} = D_1 + D_2$ and $D_{\downarrow} = D_2$, while for a supersymmetric t-J chain one has $D_{\uparrow} = -D_1 - D_2$ and $D_{\downarrow} = -D_2$, and it is necessary to add to the total momentum the term $2\pi(D_1 + D_2)$. The expression for the total energy is

$$E(M_i, D_i, n_i^{\pm}) = E_0 + L\varepsilon_{\infty}(\Lambda_i, -\Lambda_i) - \frac{\pi}{6L}\sum_{i=1}^{2} v_i^F + \frac{2\pi}{L}\sum_{i=1}^{2} v_i^F[\Delta_i + n_i^+ + n_i^-] . \tag{9.74}$$

Hence, the low-energy state of a correlated electron chain with periodic boundary conditions is characterized by two sets of quantum numbers, M_i, D_i and $n_i^{\pm}$, which define particle excitations, excitations, which manifest transfers from one Fermi point to the other one, and particle-hole excitations for Dirac seas of all low-lying excitations with possible negative energies above the ground state.

This description is modified when one considers finite size corrections for a correlated electron chain with a single impurity. Calculations, similar to the above, yield $\Delta_i \to \Delta_{i,imp}$, where

$$\Delta_{1,imp} = \left(\sum_{j=1}^{2} Z_{1j}^t(D_j - d_{j,imp} - \delta_j L)\right)^2 + \frac{1}{4(\det Z)^2}[Z_{22}(M_1 - m_{1,imp} - \nu_1(\mu, H)L) - Z_{21}(M_2 - m_{2,imp} - \nu_2(\mu, H)L)]^2 ,$$

$$\Delta_{2,imp} = \left(\sum_{j=1}^{2} Z_{1j}^t(D_j - d_{j,imp} - \delta_j L)\right)^2 + \frac{1}{4(\det Z)^2}[Z_{12}(M_1 - m_{1,imp} - \nu_1(\mu, H)L) - Z_{11}(M_2 - m_{2,imp} - \nu_2(\mu, H)L)]^2 , \tag{9.75}$$

where $m_{1,2,imp}$ are related to the valence, n_{imp}, and the magnetization, m_{imp}^z of an impurity. For a repulsive Hubbard chain they are $n_{imp} = m_{1,imp}$, $m_{imp}^z = \frac{1}{2}m_{1,imp} - m_{2,imp}$. For an attractive Hubbard chain and for a supersymmetric t-J chain we have $m_{imp}^z = \frac{1}{2}m_{1,imp}$ and $n_{imp} = m_{1,imp} + 2m_{2,imp}$. As for $d_{j,imp}$, they define shifts of the total momentum of a correlated electron chain caused by an integrable impurity as

$$d_{i,imp} = -\frac{1}{2}\left(\int_{\Lambda_i}^{\infty} dx\sigma_i^{(1)}(x) - \int_{-\infty}^{-\Lambda_i} dx\sigma_i^{(1)}(x)\right) + \frac{1}{4\pi}[x_i(\infty) + x_i(-\infty)] , \tag{9.76}$$

where, for example for an integrable impurity in a supersymmetric t-J chain, considered in Chapter 7, we have

$$x_2(y) = 2\int_{-B}^{B} dx \tan^{-1}[2(y-x)]\sigma_{1,h}^{(1)}(x) + 2\tan^{-1}[2(y-\theta)/(2S+1)] \\ + 2\int_{-Q}^{Q} dz \tfrac{1}{1+(y-z)^2}\sigma_{2h}^{(1)}(z) , \tag{9.77}$$

$$x_1(x) = 2\tan^{-1}[(x-\theta)/S] + 2\int_{-Q}^{Q} dy \tan^{-1}[2(x-y)]\sigma_{2h}^{(1)}(y)$$

for periodic boundary conditions, where $\sigma_i^{(1)}$ satisfies the equation for density of an impurity of order of L^{-1}, *cf.* Chapter 7. Observe that, for a supersymmetric t-J chain, one has to replace $\sigma_i^{(1)} \to \sigma_{ih}^{(1)}$ (*i.e.*, densities of holes). Naturally, the values $m_{j,imp}$ and $d_{j,imp}$ are defined mod 1. They determine shifts of the values $\Delta M_i = M_i - L\nu_i(\mu, H)$ and $\Delta D_i = D_i - L\delta_i$ due to a single impurity. It is important to emphasize that a dressed charge matrix of a correlated electron chain with a single impurity also does not depend on the parameters of the impurity.

For a finite concentration of impurities in correlated electron chains these expressions are changed due to the addition of Dirac seas for low-lying excitations (new Dirac seas) caused by a finite concentration of impurities.

Calculations of finite size corrections for correlated electron chains with open boundary conditions proceed along the same lines as above. The answer is

$$E = E_0 + L\varepsilon_\infty + e_b - \frac{\pi}{24L}\sum_{i=1}^{2} v_i^F + \frac{\pi}{L}\sum_{i=1}^{2} v_i^F \Delta_i^b , \tag{9.78}$$

where e_b is the energy of open boundaries (both, the contributions from open edges and boundary potentials/fields), *cf.* Chapter 6,

$$\Delta_1^b = \frac{1}{2(\det Z)^2}[Z_{22}(M_1 - \Theta_1(p_{1,L}) - \nu_1(\mu,H)L) \\ - Z_{21}(M_2 - \Theta_2(p_{1,L}) - \nu_2(\mu,H)L)]^2 , \\ \Delta_2^b = \frac{1}{2(\det Z)^2}[Z_{12}(M_1 - \Theta_1(p_{1,L}) - \nu_1(\mu,H)L) \\ - Z_{11}(M_2 - \Theta_2(p_{1,L}) - \nu_2(\mu,H)L)]^2) , \tag{9.79}$$

and $\Theta_i(p_{1,L})$ are the contributions from open edges themselves and boundary potentials/fields $p_{1,L}$ of the finite size. For example, for a repulsive

Hubbard chain they are

$$\Theta_i(p_{1,L}) = \frac{1}{2}\left(\int_{-\Lambda_i}^{\Lambda^i} dx \rho_i^{(1)}(x) - 1\right) + \Theta_{i,bs} \ . \tag{9.80}$$

Here $\Theta_{i,bs}$ define the contribution from boundary bound states, determined mod 1. For the boundary potentials/fields $p_{1,L,\uparrow} = \pm p_{1,L,\downarrow}$ at zero homogeneous magnetic field $H = 0$ one obtains

$$\Theta_2(p_{1,L}) = -\frac{\Theta_1(p_{1,L})}{2} - \frac{s(b_{1,L})}{2} \ , \tag{9.81}$$

where $s(x) = 0$ for $p_{1,L,\uparrow} = p_{1,L,\downarrow}$ and for $p_{1,L,\uparrow} = -p_{1,L,\downarrow}$ we have $s(x) = 1$ for $x > 0$ and $s(x) = -1$ for $x < 0$ with

$$b_{1,L} = \frac{U}{4} + \frac{p_{1,L,\uparrow}^2 - 1}{2p_{1,L,\uparrow}^2} \ . \tag{9.82}$$

The role of boundary bound states can be investigated, *e.g.*, for the boundary potential $p_{1,\uparrow} = p_{1,\downarrow} = p$ applied only to the left boundary, *cf.* Chapter 6. One has $\Theta_{1,bs} = \theta(p-1) + \theta(p-p_2)$ and $\Theta_{2,bs} = \theta(p-p_2)$, where $\theta(x)$ are Heaviside step functions, and p_2 is defined in Chapter 6.

Studies of phase shifts Θ_i for an attractive Hubbard chain and a supersymmetric t-J chain can be performed in an analogous way.

9.3 Elements of Conformal Field Theory

The reader knows that any classical system close to the point of the second order (continuous) phase transition, reveals strong precursor fluctuations of the ordered phase. The typical scale, related to those fluctuations, is the *correlation length*, which, for for $H = 0$ is proportional

$$\xi \sim |T - T_c|^{-\nu} \tag{9.83}$$

where T_c is the temperature of the phase transition and $\nu > 0$. Naturally, the correlation length diverges at the phase transition point. Thermodynamic characteristics, like the specific heat, magnetic or charge susceptibilities, *etc.* can exhibit similar divergencies. For example, the specific heat for $H = 0$ behaves when $T \to T_c$ as

$$c_{H=0} \sim \frac{1}{\alpha}\left[\left(\frac{|T - T_c|}{T_c}\right)^{-\alpha} - 1\right] \ . \tag{9.84}$$

The magnetization (per site) for $H = 0$, when T approaches T_c from below, can be expressed as

$$m^z \sim (T_c - T)^\beta , \tag{9.85}$$

and at the critical point $T = T_c$ the magnetization per site is proportional to

$$m^z \sim H^{1/\delta} , \tag{9.86}$$

while the magnetic susceptibility at $H = 0$ behaves as

$$\chi \sim |T - T_c|^{-\gamma} , \tag{9.87}$$

when $T \to T_c$. It is also possible to consider the behaviour of a two-point correlation function at the temperature of the second order phase transition, $T = T_c$ for $H = 0$, which behaves as

$$G^{(2)}(r) \sim \frac{1}{r^{d-2+\eta}} , \tag{9.88}$$

where d is the space dimension. Observe that the magnetic susceptibility is $\chi \propto \int G^{(2)}(r)d^dr$, so that $\chi \sim \xi^{2-\eta}$. For $r \gg \xi$ we have

$$G^{(2)}(r) \sim \frac{\exp(-r/\xi)}{r^{(d-1)/2}} , \tag{9.89}$$

which implies $\xi^2 = \sum r^2 G^{(2)}(r) / \sum G^{(2)}(r)$. Finally, there is a critical exponent, which is related to the time dependence close to the critical point. The typical relaxation time τ_r diverges as the critical point is approached as

$$\tau_r \sim \xi^z . \tag{9.90}$$

(These notations are commonly accepted in the theory of phase transitions; please do not confuse with the previous notations, where the same letters were used.)

Critical exponents are connected by the scaling relations between each other. B. Widom suggested that near T_c the Helmholtz free energy per site can be approximated by some function Ψ of one variable

$$f(T, H) = t^{d/y_t} \Psi(h/t^{y_h/y_t}) , \tag{9.91}$$

where $t = |T - T_c|/T_c$, do not confuse with time variable, and $h = H/H_*$, H_* is some constant (this equation is known as the *Widom scaling hypothesis*). Using the scaling hypothesis, one can derive the relations between the critical exponents α, β, γ and δ. For example, we can calculate the zero-field magnetization

$$m^z(H=0) = -t^{(d-y_h)/y_t}\Psi'(0) \ , \tag{9.92}$$

which yields

$$\beta = \frac{d-y_h}{y_t} \ . \tag{9.93}$$

Similarly one obtains

$$\chi|_{H=H_*} = t^{(d-2y_h)/y_t}\Psi''(0) \ , \tag{9.94}$$

from which it immediately follows that

$$\gamma = \frac{2y_h - d}{y_t} \ . \tag{9.95}$$

Let us now calculate the zero-field specific heat

$$c_{H=0} \approx -\frac{1}{T_c}\frac{\partial^2}{\partial t^2}[t^{d/y_t}\Psi(h/t^{y_h/y_t})]|_{H=0} = -\frac{d-y_t}{y_t^2}t^{(d/y_t)-2}\frac{\Psi(0)}{T_c} \ , \tag{9.96}$$

which implies

$$\alpha = 2 - \frac{d}{y_t} \ . \tag{9.97}$$

One can find, by re-expressing the Widom scaling hypothesis as $f(T,H) = h^{d/y_h}\tilde{\Psi}(h/t^{y_h/y_t})$, where $\tilde{\Psi}(u) = u^{-d/y_h}\Psi(u)$,

$$m^z_{T=T_c} = -H^{(d/y_h)-1}\frac{d\tilde{\Psi}(\infty)}{y_h} \ , \tag{9.98}$$

from which one can derive

$$\delta = \frac{y_h}{d-y_h} \ . \tag{9.99}$$

We can eliminate y_h and y_t to obtain two relations (due to G. S. Rushbrooke and R. B. Griffiths, respectively)

$$\begin{aligned} \alpha + 2\beta + \gamma &= 2 \ , \\ \alpha + \beta(\delta+1) &= 2 \ . \end{aligned} \tag{9.100}$$

L. P. Kadanoff introduced the hypothesis about the behaviour of the two-point correlation function (at $H = 0$, for clarity)

$$G^{(2)}(r,t) \sim \frac{t^{2(d-y_h)/y_t}\Psi(rt^{1/y_t})}{r^{d-2+\eta}} . \tag{9.101}$$

The reader can see that $\Psi(u)$ must be a constant for small u and it has to behave as $\Psi(u) \sim u^{d-2+\eta}$ when u is large, which yields

$$\eta = d + 2 - 2y_h . \tag{9.102}$$

Obviously it follows that $\xi \sim t^{-1/y_t}$, so that

$$\nu = \frac{1}{y_t} . \tag{9.103}$$

It is easy to show (the following relations are due to M. E. Fisher and B. D. Josephson, respectively) that

$$\begin{aligned} \gamma &= (2-\eta)\nu , \\ \nu d &= 2 - \alpha . \end{aligned} \tag{9.104}$$

The last relation is often referred to as a *hyperscaling relation*, since it connects the singularity in the specific heat with the behaviour of a correlation length. Notice that it is invalid when d is sufficiently large: when d exceeds the "upper critical dimension" d_c, all the critical exponents are independent on d, in which regime the exponent of t in the Kadanoff's hypothesis is ν. Scaling relations are universal and depend on the symmetry of the system. Correlation functions diverge as power laws of distance and time at the critical point. One can calculate the exponents of those divergencies from the concrete model under consideration. According to the above, many one-dimensional systems (for which $T_c = 0$) are critical at $T = 0$ (in fact, those which have gapless low-lying excitations).

Let us investigate symmetries of many-electron systems at a critical point. This conformal field theory analysis was pioneered by A. A. Belavin, A. M. Polyakov and A. B. Zamolodchikov. These systems at a critical point are *translationally and rotationally invariant.* (They are also Lorentz-invariant in higher dimensions. However, in one space dimension the Lorentz invariance is, in fact, rotations in the (x,t)-plane.) At a critical point a system is characterized by the *scale invariance* $x \to \lambda x$. The rotational and scale invariances together mean the invariance under the *global conformal symmetry group.* (To remind: simple *conformal transformations* keep invariant angles between two vectors.) The global conformal group is finite dimensional for higher dimensional systems (also taking time into account), as well as the associated Lie algebra of generators of the group. This is why, the finite number of constraints permits an evaluation of only two- or three-point correlation functions, but not higher correlators. As the reader saw above, if one knows how correlation function behave, then it implies, *e.g.*, that one understands the behaviour of thermodynamic characteristics of a system close to a critical point. Therefore, it is important to obtain expressions for correlation functions. It turns out that for one-dimensional dynamical systems the global conformal group is finite dimensional, and, hence, all correlation functions can be found in principle.

Consider the transformation $x \to x + \xi(x)$. It is conformal, if $\xi(x)$ satisfies some constraints. Formally all analytic functions are permitted for a conformal transformation. In higher space dimensions $\xi(x)$ is only the polynomial of the second degree in x. Such a group is called the *local conformal group*. It is wider than the global conformal group. Let us use complex variables, which describe right- and left-moving one-dimensional electrons:

$$z = v\tau + ix \ , \qquad \bar{z} = v\tau - ix \ , \tag{9.105}$$

where v is the characteristic velocity of the field theory and $\tau = it + \mathrm{sign}t$ is the Euclidean time. Conformal transformation implies

$$z \to z + \xi^z(z) = f(z) \ , \qquad \bar{z} \to \bar{z} + \xi^{\bar{z}}(\bar{z}) = \bar{f}(\bar{z}) \ . \tag{9.106}$$

Since $\xi^z(z)$, $f(z)$, $\xi^{\bar{z}}$ and $\bar{f}(\bar{z})$ are analytic, they can be expanded in Laurent series, *e.g.*

$$\xi^z(z) = \sum_{n=-\infty}^{\infty} \xi^n z^{n+1} \ , \tag{9.107}$$

etc. The generators of the local conformal group in this case are

$$l_n(z) = -z^{n+1}\frac{\partial}{\partial z} \ , \qquad \bar{l}_n(\bar{z}) = -\bar{z}^{n+1}\frac{\partial}{\partial \bar{z}} \ . \tag{9.108}$$

These generators satisfy the local conformal algebra known as the *classical Virasoro algebra*:

$$\begin{aligned} &[l_n, \bar{l}_m] = 0 \ , \qquad [l_n, l_m] = (m-n)l_{m+n} \ , \\ &[\bar{l}_n, \bar{l}_m] = (m-n)\bar{l}_{m+n} \ . \end{aligned} \tag{9.109}$$

It is interesting to notice that the global conformal algebra is generated by two sets of operators $l_{0,\pm 1}$ and $\bar{l}_{0,\pm 1}$. Since these two algebras are independent, the global conformal algebra is related to the natural variables for left- and right-moving particles. To remind, in a physical theory one deals with $\bar{z} = z^*$.

In quantum mechanics an infinitesimal symmetry variation of some field ϕ is generated by

$$\delta_\xi \phi = \xi[Q, \phi] , \tag{9.110}$$

where Q is some general conserved charge, associated with the symmetry. Local co-ordinate transformations are generated by general charges. One can construct such a charges from an energy-momentum (stress-energy) tensor T_{ij}. This tensor is symmetric, $T_{ij} = T_{ji}$, because of the rotational invariance. Due to the scale invariance one has $\mathrm{tr}T_{ij} = 0$. There are no additional constraints due to the conformal invariance. These conditions mean that only diagonal components of an energy-momentum tensor do not vanish

$$T(z) = T_{zz}(z) , \qquad \bar{T}(\bar{z}) = \bar{T}_{\bar{z}\bar{z}}(\bar{z}) . \tag{9.111}$$

Consider two operators $A(z)$ and $B(z)$, both being analytic functions of z (similar expressions can be obtained for functions of $\bar{z}$). In what follows we suppose to consider a field theory, containing such operators, in which one can compute all multi-point correlation functions of the operators $A(z)$, $B(z)$ and all others, which occur in the theory. The *operator product expansion* is meant if the operators $A(z)$ and $B(z)$ are assumed to have the following property as $z_1 \to z_2$

$$A(z_1)B(z_2) = \frac{C(z_2)}{(z_1 - z_2)^2} + \frac{D(z_2)}{(z_1 - z_2)} + O(1) , \tag{9.112}$$

where $C(z)$ and $D(z)$ are some operators. Operator product expansion becomes a true equation when inserted in any correlation function with other operators of the considered field theory at positions w_j, when the distance between z_1 and z_2 becomes much smaller than the distances between z_1 and z_2 and the positions w_j of other operators. Notice that an operator product expansion is the local property of the field theory. Then Eq. (9.112) implies

$$[A_m, B_n] = nC_{n+m} + D_{n+m} . \tag{9.113}$$

This expression means that the transforms of the operators A and B are related to the transforms of C and D which multiply the poles of the operator product expansion of A with B. Consider the operator product expansion for the energy-momentum tensor

$$T(z)T(w) = \frac{c/2}{(z-w)^4} + \frac{2}{(z-w)^2}T(w) + \frac{1}{z-w}\frac{\partial}{\partial z}T(z) , \qquad (9.114)$$

where the coefficient c is called a *central charge*. One can derive the algebra of generators from the Laurent series of the energy-momentum tensor

$$T(z) = \sum_{n=-\infty}^{\infty} L_n z^{-n-2} , \qquad (9.115)$$

which is the (quantum) Virasoro algebra

$$[L_n, \bar{L}_m] = 0 , \qquad [L_n, L_m] = (m-n)L_{m+n} + \frac{c}{12}(n-1)n(n+1)\delta_{n,-m} ,$$
$$[\bar{L}_n, \bar{L}_m] = (m-n)\bar{L}_{m+n} + \frac{\bar{c}}{12}(n-1)n(n+1)\delta_{n,-m} , \qquad (9.116)$$

where $\bar{c}$ is also a central charge. The classical Virasoro algebra pertains to the case $c = \bar{c} = 0$. From these formulas it is clear that central charges determine universality classes of the considered class of systems. A classical symmetry cannot be carried out in quantum mechanics because of renormalization effects: for classical Virasoro algebras one has zero central charges.

One can write the generalized charge as

$$Q = \frac{1}{2\pi i}\oint [dzT(z)\xi(z) + d\bar{z}\bar{T}(\bar{z})\bar{\xi}(\bar{z})] . \qquad (9.117)$$

Then a field variation can be written as

$$\delta_{\xi\bar{\xi}}\phi(w,\bar{w}) = \frac{1}{2\pi i}\oint dz[T(z)\xi(z), \phi(w,\bar{w})] + d\bar{z}[\bar{T}(\bar{z})\bar{\xi}(\bar{z}), \phi(w,\bar{w})] . \qquad (9.118)$$

For the special class of fields, known as *primary fields*, one has

$$\delta_{\xi\bar{\xi}}\phi(w,\bar{w})$$
$$= \left(h\frac{\partial}{\partial z}\xi^z(z) + \xi^z(z)\frac{\partial}{\partial z} + \bar{h}\frac{\partial}{\partial \bar{z}}\xi^{\bar{z}}(\bar{z}) + \xi^{\bar{z}}(\bar{z})\frac{\partial}{\partial \bar{z}}\right)\phi(w,\bar{w}) , \qquad (9.119)$$

where h and $\bar{h}$ are real numbers. They are called the *conformal weights* of a primary field $\phi(w,\bar{w})$. In fact, the last formula is the infinitesimal version of the transformation

$$\phi(w,\bar{w}) \to \left(\frac{\partial f}{\partial z}\right)^h \left(\frac{\partial \bar{f}}{\partial \bar{z}}\right)^{\bar{h}} \phi(f(w), \bar{f}(\bar{w})) . \qquad (9.120)$$

One can introduce

$$\Delta = h + \bar{h} \ , \qquad s = h - \bar{h} \ , \tag{9.121}$$

with Δ being the scaling dimension and s being the scaling spin of the primary field $\phi(w, \bar{w})$. In a basis of eigenfunctions of L_0 and $\bar{L}_0$ the operators $(L_0 + \bar{L}_0)$ and $i(L_0 - \bar{L}_0)$ are generators of dilations and rotations, respectively. All other fields are usually called *secondary fields* (*descendants*).

Equation (9.120) denotes how some complex tensor of the rank $(h, \bar{h})$ transforms. Usually, it transforms with integer powers of $(\partial f / \partial z)$ and $(\partial \bar{f} / \partial \bar{z})$ which are numbers of z and $\bar{z}$ indices. For primary fields one could also imagine non-integer exponents. Such non-integer exponents are called *anomalous dimensions*. From this viewpoint scaling dimensions of primary fields are anomalous ones.

By using Eq. (9.119) one can calculate some correlation functions. For instance, consider the variation of a two-point correlation function. It must be invariant under a conformal transformation

$$\begin{aligned} \delta_{\xi\bar{\xi}} G^{(2)}(z_{1,2}, \bar{z}_{1,2}) &\equiv \delta_{\xi\bar{\xi}} \langle \phi_1(z_1, \bar{z}_1) \phi_2(z_2, \bar{z}_2) \rangle \\ &= \langle (\delta_{\xi\bar{\xi}} \phi_1) \phi_2 \rangle + \langle \phi_1 (\delta_{\xi\bar{\xi}} \phi_2) \rangle = 0 \ , \end{aligned} \tag{9.122}$$

which, together with Eq. (9.119), implies

$$G^{(2)}(z_{1,2}, \bar{z}_{1,2}) = \frac{C_{12}}{(z_1 - z_2)^{2h} (\bar{z}_1 - \bar{z}_2)^{2\bar{h}}} \ , \tag{9.123}$$

where $C_{12} \propto \delta_{\Delta_1, \Delta_2}$ is a constant. A three-point correlation function can be derived in a similar way.

One can introduce the operator product expansion of the energy-momentum tensor with a primary field ϕ (known also as *Ward identities*) as

$$T(z)\phi(w, \bar{w}) = \frac{h}{(z - w)^2} \phi(w, \bar{w}) + \frac{1}{z - w} \frac{\partial}{\partial w} \phi(w, \bar{w}) + \cdots \ . \tag{9.124}$$

Operator product expansion is valid at short distances, where the radial ordering takes place. (For right-movers this equation is called holomorphic part of the stress-energy tensor. There is an equivalent equation for left-movers, called antiholomorphic part; in what follows we shall mostly not present those antiholomorphic counterparts to save the space.) Any secondary field has a higher than double-pole singularity in its operator product expansion.

The reader can see from the operator product expansion of the energy-momentum tensor that L_n are generators of transformations of fields associated with the monomial of degree $n+1$ in z. Let us consider the transformation $\xi^z(z) = -\xi_n z^{n+1}$. One has $\delta\phi(z,\bar{z}) = -\xi_n[L_n, \phi(z,\bar{z})]$. Due to the unitarity of the energy-momentum tensor, its generators satisfy the following relation:

$$L_m^\dagger = L_{-m} . \tag{9.125}$$

For the vacuum state $|0\rangle$ the regularity of the energy-momentum tensor means

$$L_{m\geq -1}|0\rangle = 0 , \qquad L^\dagger_{m\leq -1}|0\rangle = 0 . \tag{9.126}$$

The energy-momentum tensor transforms under a local conformal transformation $z' = f(z)$ as

$$T(z) \to T(z') = \left(\frac{dz'}{dz}\right)^2 T(z') + \frac{c}{12}\left(\frac{\frac{\partial^3 z'}{\partial z^3}}{\frac{\partial z'}{\partial z}} - \frac{3}{2}\frac{\left(\frac{\partial^2 z'}{\partial z^2}\right)^2}{\left(\frac{\partial z'}{\partial z}\right)^2}\right) . \tag{9.127}$$

From the first term of this formula the reader can see that the energy-momentum tensor is a field of conformal weight $(2,0)$. The last term of that formula is called the *Schwarz derivative.*

Consider the representations of the Virasoro algebra, which can be constructed from the *highest weight states* $|h\rangle$, (created by the action of a holomorphic primary field ϕ on the vacuum state, *i.e.*, $|h\rangle = \phi(0)|0\rangle$) as

$$L_0|h\rangle = h|h\rangle , \qquad L_{n>0}|h\rangle = 0 , \tag{9.128}$$

i.e., a highest weight state is an eigenfunction of L_0. On the other hand, L_n ($n>0$) are lowering operators, which destroy the highest weight state. L_{-n} ($n>0$) play the role of "creation operators". Acting on the highest weight states they generate *descendant states* $L_{-n_1}\cdots L_{-n_k}|0\rangle$ (with $1 \leq n_1 \leq \cdots \leq n_k$ and the eigenvalue of of L_0 being $h + n_1 + \cdots + n_k$). These states form the basis for the representation vector space. The highest weight state has the lowest eigenvalue. Hence, in a given sector of the considered field theory, it is the ground state. Descendant states are excited states. n is the level related to an operator L_{-n}. The conformal weight of all descendant states on some level N is $h+N$. One usually calls the vector space generated from the highest weight state *Verma module.* We can group all states in a conformal field theory into *conformal towers* (families). Conformal towers consist of a highest weight state and all descendant states. Action of different primary fields on the vacuum state produces different highest weight states. It turns out that conformal towers are the very convenient way of classification of low-energy excitations in a system. All correlation functions of secondary fields can be found from those of only primary fields.

In fact, unitary representations of the Virasoro algebra only exist for certain values of central charges and conformal weights. For example, it is true for $c \geq 1$ and $h \geq 0$, or for

$$
\begin{aligned}
c &= 1 - \frac{6}{m(m+1)} , \qquad m = 3, 4, \dots , \\
h_{p,q}(m) &= \frac{[(m+1)p - mq]^2 - 1}{4m(m+1)} ,
\end{aligned} \tag{9.129}
$$

(known as the *Kac formula*) where $1 \leq p \leq m-1$ and $1 \leq q \leq p$. The most useful from these series is $c = \frac{1}{2}$, which describes the two-dimensional Ising model. Another example, $c = 1$, known as the Gaussian model, is considered in the next chapter.

Above one supposed that fields are defined in the infinite space plane. That means

$$
\langle T(z) \rangle = \sum_{m=-\infty}^{\infty} \langle 0 | \frac{L_m}{z^{m+2}} | 0 \rangle = 0 . \tag{9.130}
$$

If one considers finite systems of the size L with periodic boundary conditions, we can use the exponential transformation

$$
z = \exp(2\pi i u / L) , \qquad u = -i \frac{L}{2\pi} \ln z \tag{9.131}
$$

to map the infinite z-plane onto a strip u of width L with periodic boundary conditions. Under this transformation the Laurent expansion of the energy-momentum tensor becomes a Fourier transformation,

$$
T(u) = \left[T(z) - \frac{c}{12} \left(\frac{\frac{\partial^3 u}{\partial z^3}}{\frac{\partial u}{\partial z}} - \frac{3}{2} \frac{\left(\frac{\partial^2 u}{\partial z^2} \right)^2}{\left(\frac{\partial u}{\partial z} \right)^2} \right) \right] \left(\frac{dz}{du} \right)^2 , \tag{9.132}
$$

which means that

$$
\langle T(u) \rangle = \frac{c}{24} \left(\frac{2\pi}{L} \right)^2 . \tag{9.133}
$$

To remind, the energy-momentum tensor measures the cost of energy because of the change in a metric (the change in the action $\delta\mathcal{A} = -(2\pi)^{-1}\int T_{ij}(\partial\xi_j/\partial x_i)dxd\tau$). Then the change in energy related to a non-conformal transformation $u_1' \to (1+\epsilon)u_1$, $u_2' \to u_2$ (*i.e.*, a horizontal dilation of the u-strip), which changes the length of the system, yields

$$E = E_0 - \frac{c\pi v}{6L} . \tag{9.134}$$

Hence, we can determine the central charge from finite size calculations. Local conformal transformations (not global ones, for which $c = 0$) are not defined in all points of a complex plane. They are also not one-to-one mappings of that complex plane on itself. This is why, one has such a shift in the energy due to the finite size.

The change to open boundary conditions corresponds to the introduction of a cut in the plane from z_0 to z_1. Choosing $0 < iv\tau_0 = z_0 < iv\tau_1 = z_1$ real and mapping the plane to a cylinder *via* the conformal transformation Eq. (9.131) this cut gets mapped onto a seam in the direction of the cylinder. The correction to the ground state energy in the case of open boundary conditions is then

$$E = E_0 - \frac{c\pi v}{24L} . \tag{9.135}$$

Let us study now, following J. L. Cardy, the two-point ground state correlation function of a primary operator $\phi(z,\bar{z})$ with the conformal weights $(h,\bar{h})$. Under the conformal transformation, which maps the infinite z-plane onto a strip u, this correlator transforms as

$$\langle\phi(u,\bar{u})\phi(u',\bar{u}')\rangle = \frac{C(\Delta)(\pi/L)^{2\Delta}}{(\sinh[\pi(u-u')/L])^{2h}(\sinh[\pi(\bar{u}-\bar{u}')/L])^{2\bar{h}}} , \tag{9.136}$$

with $C(\Delta)$ being a constant.

At finite temperatures correlation functions have to satisfy periodic boundary conditions in the Matsubara Euclidean time with the period $L = 2v/T$. This defines the temperature behaviour of the two-point correlation function

$$\langle\phi(u,\bar{u})\phi(u',\bar{u}')\rangle = \frac{C(\Delta)(\pi T/2v)^{2\Delta}}{(\sinh[\pi T(u-u')/2v])^{2h}(\sinh[\pi T(\bar{u}-\bar{u}')/2v])^{2\bar{h}}} . \tag{9.137}$$

On the physical surface $\bar{u} = u^*$ the finite size expression for the two-point correlation function can be expanded as ($u = v\tau + ix$)

$$\langle\phi(u,\bar{u})\phi(u',\bar{u}')\rangle$$
$$= C(\Delta)\left(\frac{2\pi}{L}\right)^{2\Delta}\sum_{N,\bar{N}=0}^{\infty} a_N a_{\bar{N}} \exp[-2\pi v(\Delta + N + \bar{N})(\tau - \tau')/L]$$
$$\times \exp[-2\pi i(s + N - \bar{N})(x - x')/L] \; , \tag{9.138}$$

where N and $\bar{N}$ are characteristics of descendant operators. On the other hand, one can present the correlation function as an action of some operator $\hat{\phi}(u_2)$ on states of a Hilbert space

$$\langle\phi(u,\bar{u})\phi(u',\bar{u}')\rangle = \sum_n \langle 0|\hat{\phi}(u_2)|n,k\rangle\langle n,k|\hat{\phi}(u_2)|0\rangle e^{-(E_n - E_0)(\tau-\tau')} \; , \tag{9.139}$$

where k is the momentum (*i.e.*, the space dependence is proportional to $\exp(ikx_2)$). It follows that

$$E_n = E_0 + \frac{2\pi v}{L}(\Delta + N + \bar{N}) \; ,$$
$$k = k_0 + \frac{2\pi}{L}(s + N - \bar{N}) \; . \tag{9.140}$$

This expression actually relates finite size corrections to the energy and momentum with the conformal dimension and conformal spin of primary fields. k_0 and E_0 are the momentum and the energy of the highest weight state in the conformal tower built by a primary field ϕ, extrapolated to the infinite system.

At $L \to \infty$ the correlation functions of primary fields are then (we write them in terms of x and τ, taking $x' = \tau' = 0$)

$$\langle\phi(x,t)\phi(0,0)\rangle = \frac{C\exp[i(k_0 + \bar{k}_0)x]}{(v\tau + ix)^{2h}(v\tau - ix)^{2\bar{h}}} \; , \tag{9.141}$$

where C is a constant. Correlation functions of secondary (descendant) fields can be obtained by the replacement $(h,\bar{h}) \to (h + N, \bar{h} + \bar{N})$.

For open boundary conditions two-point correlation functions for $L \ll (u - u')$ are given by

$$\langle\phi(u,\bar{u})\phi(u',\bar{u}')\rangle$$
$$= C(\Delta)\sum_n \langle 0|\phi(0)|B;n,k\rangle\langle B;n,k|\phi(0)|0\rangle e^{(E_n^B - E_0)(\tau-\tau')} \; , \tag{9.142}$$

where B changes periodic boundary conditions to open ones. The procedure, similar to the periodic case, yields

$$E_n^B = E_0 + \frac{\pi v}{L}(\Delta + N) \ . \tag{9.143}$$

Summarizing, the conformal field theory gives the possibility to connect characteristics of finite size corrections to the energy and momentum (the latter for periodic boundary conditions) with correlation functions of primary fields and conformal towers of descendant fields, generated by primary fields.

9.4 Asymptotics of Correlation Functions

As we have shown in the previous section, the knowledge of finite size corrections to the energy and the momentum of many one-dimensional quantum systems gives the possibility to find the asymptotic behaviour of correlation functions for primary and secondary fields. Thanks to conformal invariance, the universality class of the considered model can be uniquely described by single dimensionless number, the central charge of the underlying Virasoro algebra. The value of the central charge can be extracted from the finite size correction to the ground state energy. Then, each primary field with the scaling dimension $\Delta = h + \bar{h}$ and scaling spin $s = h - \bar{h}$ gives rise to a tower of excited states. Correlation functions of primary fields for periodic boundary conditions are known to be (see the previous section), *e.g.*, for an infinite chain in the ground state

$$\langle\phi(x,t)\phi(0,0) = \frac{C(\Delta)e^{-2iDP^F x}}{(v^F t + x)^{2h}(v^F t - x)^{2\bar{h}}} + \cdots \ , \tag{9.144}$$

(note that the momentum is not determined uniquely). One has similar relations for a finite size chain in the ground state, and temperature behaviour of correlation functions, namely

$$\langle\phi(x,t)\phi(0,0)\rangle = \frac{C(\Delta)(\pi/L)^{2\Delta}}{(\sin[\pi(v^F t + x)/L])^{2h}(\sin[\pi(v^F t - x)/L])^{2\bar{h}}} + \cdots \ , \tag{9.145}$$

and

$$\langle\phi(x,t)\phi(0,0)\rangle = \frac{C(\Delta)(\pi T/2v^F)^{2\Delta}}{(\sinh[\pi T(x - iv^F\tau)/2v^F])^{2h}(\sinh[\pi T(x + iv^F\tau)/2v^F])^{2\bar{h}}} + \cdots \ . \tag{9.146}$$

Here the coefficients $C(\Delta)$ are expressed in terms of corresponding form-factors $C(\Delta) = (\pi/L)^{-\Delta}|\langle 0|\phi|n\rangle|^2$, where $|n\rangle$ are eigenfunctions related to excitations. It is important to emphasize that these asymptotics are valid for small enough temperatures (where the approximation of dispersion laws of low-lying excitation by a linearized function is correct; also, considered temperatures have to be lower than gaps of other excitations, like bound states, *etc.*) and for large enough sizes of the system L, where finite size corrections in series in L^{-1} are small. The presence of marginal operators in the theory can lead to logarithmic corrections to conformal asymptotics of correlation functions. Such operators appear, *e.g.* for systems with the SU(2) spin symmetry.

Notice, that for excitations with gaps (which need activation energies) their correlation functions decay exponentially, proportional to $\exp(-x/\xi)$, where $\xi \sim v^F/G$ is the correlation length (G is the gap).

Let us denote the following functions

$$\begin{aligned} &G^z(x,t) = \langle S^z(x,t)S^z(0,0)\rangle \ , \qquad G^{\perp}(x,t) = \langle S^-(x,t)S^+(0,0)\rangle \ , \\ &G_{aa}(x,t) = \langle a_{\downarrow}(x,t)a^{\dagger}_{\downarrow}(0,0)\rangle \ , \qquad G_{nn}(x,t) = \langle n(x,t)n(0,0)\rangle \ , \quad (9.147) \\ &G_{sp}(x,t) = \langle a^{\dagger}_{\uparrow}(x,t)a^{\dagger}_{\downarrow}(x,t)a_{\downarrow}(0,0)a_{\uparrow}(0,0)\rangle \ , \end{aligned}$$

where we, in fact, introduced the continuum version of operators related to considered models as $a_{n\sigma} \to a_{\sigma}(x)$, *etc.* The last three correlation functions determine the electron-electron, density-density, and pair-pair (for singlet pairs) correlators, respectively.

The simplest case is the situation for quantum spin chains, a repulsive Hubbard chain and a supersymmetric t-J model at half-filling, all for $H < H_s$, and an attractive Hubbard chain for $H < H_c$. Here one has only one Dirac sea (and, hence, only one kind of low-lying excitations, which produce finite size corrections), while other excitations are related to higher energies, because of their gaps (activation energies). From the calculation of finite size corrections the reader knows that this case pertains to $c = 1$ (*i.e.*, to the Gaussian model). Scaling dimensions of primary fields are Δ, given in the first section of this chapter. Then, conformal weights of primary fields, which determine exponents of correlation functions, are

$$h = \frac{1}{2}\left(\frac{\Delta M}{2Z} + Z\Delta D\right)^2 , \qquad \bar{h} = \frac{1}{2}\left(\frac{\Delta M}{2Z} - Z\Delta D\right)^2 , \qquad (9.148)$$

while for secondary fields one has $h + n^+$ and $\bar{h} + n^-$. Here ΔM and ΔD are connected with excitations above the ground state, see the first section of this chapter. It is necessary to choose the minimal values of the numbers ΔM, ΔD, and $n^{\pm}$, which describe a concrete excitation under consideration. It is important to point out that one requires that all the conformal weights be non-negative, otherwise there would be unphysical divergencies in correlation functions. Observe, also, that the choice of ΔM defines the choice of ΔD, *i.e.*, the latter is not really independent, because of the restrictions on quantum numbers J_j and $J^{\pm}$ (which, in turn, follows from the demand for Bethe ansatz wave functions to be non-degenerate).

Then it is relatively easy to obtain asymptotics of correlation functions. For a Heisenberg-Ising spin-$\frac{1}{2}$ chain ($H < H_s$) we have

$$G^{\perp}(x,t) \approx \frac{C_1}{[x^2-(v^Ft)^2]^{\theta_{\perp}}} + \frac{C_2}{|v^Ft+x|^{\nu}}\mathrm{Re}\left(e^{2iP^Fx}\frac{v^Ft-x}{v^Ft+x}\right) + \cdots ,$$
$$G^{z}(x,t) \approx (m^z)^2 + \frac{2A[(v^Ft)^2+x^2]}{[(v^Ft)^2-x^2]^2} + \frac{B\cos(2P^Fx)}{[x^2-(v^Ft)^2]^{\theta_z}} + \cdots , \tag{9.149}$$

where A, B, $C_{1,2}$ are constants, $\theta_z = Z^2$, $\theta_{\perp} = 1/4\theta_z$, and $\nu = 2\theta_z + (2\theta_z)^{-1}$. These expressions are related to the choice $\Delta M = 1$, $\Delta D = 0$, and $\Delta M = 0$, $\Delta D = 1$, respectively. For the half-filled repulsive Hubbard chain and for a supersymmetric t-J chain for $H < H_s$ one obtains expressions for spin-spin correlation functions, similar to the above case with $v^F \to v_2^F$, and $Z \to Z_{22}$, while the expression for a density-density correlation function coincides with the one for S^z-S^z correlator, but with the change $(m^z)^2 \to 1$. For an attractive Hubbard chain for $H < H_c$ the asymptotic behaviour of spin-singlet pair correlator and density-density correlator are similar to the behaviour of $G^{\perp}(x,t)$ and $G^z(x,t)$, respectively, with the change $(m^z)^2 \to 1$, $v^F \to v_2^F$, and $Z \to Z_{22}$. Here the choice of ΔM_2 and ΔD_2 is similar to the above choice for a quantum spin chain.

In the cases of spin chains with the finite concentration of similar impurities (with the same spins as in the host chain), a repulsive Hubbard chain away from half-filling, a supersymmetric t-J chain, all for $H < H_s$, and an attractive Hubbard chain for $H_c < H < H_s$, the reader knows that the ground states correspond to fillings of two Dirac seas. Hence, there are two kinds of low-lying excitations, which give contributions to finite size corrections to the energy and momentum. These contributions imply that in the conformal limit one has the *semidirect product* of two independent Virasoro algebras, both having central charge equal to 1 (*i.e.*, the semidirect product of Gaussian models). We write about the semidirect product, because Fermi velocities of low-lying excitations, are, generally speaking, different from each other. (In the degenerate case, at which $v_1^F = v_2^F$, models are also conformally invariant with the central charge $c = 2$.)

To find the analog of an equation for asymptotics of correlation functions of primary and secondary fields we can write

$$E = E_0(0) + \frac{2\pi}{L}\sum_{i=1}^{2} v_i^F (h_i + \bar{h}_i + n_i^+ + n_i^-) + \cdots ,$$
$$P = P_0(0) + \tfrac{2\pi}{L}\sum_{i=1}^{2}(h_i - \bar{h}_i + n_i^+ - n_i^-) + 2D_i P_i^F , \tag{9.150}$$

where $E_0(0)$ and $P_0(0)$ are the energy and momentum of the ground state. Comparing these equations with the results of the second section of this chapter we can obtain conformal weights as functions of the parameters of the concrete models (note that the conformal spin cannot be determined uniquely, because of the possible gap of the momentum). Generalizing the expression for correlation function for the present case we get

$$\langle\phi(x,t)\phi(0,0)\rangle = \frac{C(\Delta_{1,2})\exp[-2i(D_1 P_1^F + D_2 P_F^2)x]}{(v_1^F t + x)^{2h_1}(v_1^F t - x)^{2\bar{h}_1}(v_2^F t + x)^{2h_2}(v_2^F t - x)^{2\bar{h}_2}} + \cdots , \tag{9.151}$$

where

$$h_1 = \frac{1}{2}\left(\frac{Z_{22}\Delta M_1 - Z_{21}\Delta M_2}{2\det Z} + Z_{11}\Delta D_1 + Z_{12}\Delta D_2\right)^2 + n_1^+ ,$$
$$\bar{h}_1 = \frac{1}{2}\left(\frac{Z_{22}\Delta M_1 - Z_{21}\Delta M_2}{2\det Z} - Z_{11}\Delta D_1 - Z_{12}\Delta D_2\right)^2 + n_1^- ,$$
$$h_2 = \frac{1}{2}\left(\frac{Z_{12}\Delta M_1 - Z_{11}\Delta M_2}{2\det Z} + Z_{21}\Delta D_1 + Z_{22}\Delta D_2\right)^2 + n_1^+ ,$$
$$\bar{h}_2 = \frac{1}{2}\left(\frac{Z_{12}\Delta M_1 - Z_{11}\Delta M_2}{2\det Z} - Z_{21}\Delta D_1 - Z_{22}\Delta D_2\right)^2 + n_1^-, \tag{9.152}$$

and similar expressions as for the case of only one Dirac sea of low-lying excitation for the ground state correlations for finite chains and low-temperature behaviour of correlation functions. Again, we have to choose the minimal and non-negative exponents. It is important to point out that the choice of ΔM_i defines (but not totally) the choice of ΔD_i, *i.e.*, $\Delta D_1 = (\Delta M_1 + \Delta M_2)/2 \pmod 1$, and $\Delta D_2 = \Delta N_1/2 \pmod 1$.

Let us see what is the concrete choice of ΔM_i and ΔD_i for asymptotics correlation of concrete models. For example, for a repulsive Hubbard chain for $0 \le 0 < H_s$ we choose for $G_{aa}(x,t)$ $\Delta M_1 = 1$, $\Delta M_2 = 1$, $\Delta D_1 = 0$ (mod 1), $\Delta D_2 = \frac{1}{2}$ (mod 1). For the density-density correlator we get $\Delta M_1 = \Delta M_2 = 0$, $\Delta D_1 = \Delta D_2 = 0$ (mod 1). For $G^z(x,t)$ the choice is the same as for density-density correlator. For $G^{\perp}(x,t)$ one takes $\Delta M_1 = 0$, $\Delta M_2 = 1$, $\Delta D_1 = \frac{1}{2}$ (mod 1), $\Delta D_2 = 0$ (mod 1). For pair-pair correlation function the choice is: $\Delta M_1 = 2$, $\Delta M_2 = 1$, $\Delta D_1 = \frac{1}{2}$ (mod 1), $\Delta D_2 = 0$ (mod 1). For example, asymptotics for the density-density correlation function look like

$$G_{nn}(x,t) \approx (N/L)^2 + \frac{C_1 \cos(2P^F_{1\uparrow} x)}{|v_1^F t + x|^{2(Z_{11}-Z_{12})^2} |v_2^F t + x|^{2(Z_{21}-Z_{22})^2}} + \frac{C_2 \cos(2P^F_{1\downarrow} x)}{|v_1^F t + x|^{2Z_{12}^2} |v_2^F t + x|^{2Z_{22}^2}} + \frac{C_3 \cos[2(P^F_\uparrow + P^F_{1\downarrow})x]}{|v_1^F t + x|^{2Z_{11}^2} |v_2^F t + x|^{2Z_{21}^2}} + \frac{C_4[(v_1^F t)^2 + x^2]}{[(v_1^F t)^2 - x^2]^2} + \frac{C_5[(v_2^F t)^2 + x^2]}{[(v_2^F t)^2 - x^2]^2} + \cdots . \tag{9.153}$$

For $H = 0$ it reduces to (notice that $P^F_\uparrow = P^F_\downarrow = \pi N/2L$)

$$G_{nn}(x,t) \approx (N/L)^2 + \frac{C_1' \cos(2P^F x)}{|v_1^F t + x|^{\theta_1/4} |v_2^F t + x|} + \frac{C_3 \cos(4P^F x)}{|v_1^F t + x|^{\theta_1}} + \frac{C_4[(v_1^F t)^2 + x^2]}{[(v_1^F t)^2 - x^2]^2} + \frac{C_5[(v_2^F t)^2 + x^2]}{[(v_2^F t)^2 - x^2]^2} + \cdots , \tag{9.154}$$

$\theta_1 = 2\xi^2(Q)$; see the second section of this chapter. It varies from 2 to 4 as the Hubbard repulsion constant increases from 0 to ∞. In this case $H = 0$ we also have

$$G_{aa}(x,t) \approx \frac{1}{|v_1^F t + x|^{\nu_1} |v_2^F t + x|^{1/2}} \mathrm{Re}\left[A_1 e^{-iP^F x} \left(\frac{x - v_1^F t}{x + v_1^F t} \right)^{1/4} \times \left(\frac{x - v_2^F t}{x + v_2^F t} \right)^{1/4} \right] + \frac{1}{|v_1^F t + x|^{\nu_3} |v_2^F t + x|^{1/2}} \times \mathrm{Re}\left[A_2 e^{-i3P^F x} \left(\frac{x - v_1^F t}{x + v_1^F t} \right)^{3/4} \left(\frac{x - v_2^F t}{x + v_2^F t} \right)^{1/4} \right] + \cdots \tag{9.155}$$

where $\nu_1 = \theta_1^{-1} + \theta_1/16$, and $\nu_3 = \theta_1^{-1} + 9\theta_1/16$. These exponents are monotonic functions of the Hubbard interaction constant U, $\frac{1}{2} \le \nu_1 \le \frac{5}{8}$ for $0 \le U \le \infty$, and $\frac{5}{2} \ge \nu_3 \ge \frac{13}{8}$ for $0 \le U \le \infty$, respectively. Spin-spin correlators are equal to each other for $H = 0$, and their expression coincides with the one for density-density correlator with the change $(N/L) \to 0$. For the correlator of spin-singlet pairs one obtains

$$G_{sp}(x,t) \approx \frac{B_1}{|v_1^F t + x|^{4/\theta_1}|v_2^F t + x|} + \frac{1}{|v_1^F t + x|^{\nu_s}} \mathrm{Re}\left[B_2 e^{-i2P^F x}\frac{x - v_1^F t}{x + v_1^F t}\right] + \cdots \tag{9.156}$$

where $\nu_s = \frac{4}{\theta_1} + \frac{\theta_1}{4}$, which increases from 2 to $\frac{5}{2}$ when U is changed between 0 to ∞.

For an attractive Hubbard chain for $H_c < H < H_s$ and for a supersymmetric t-J chain (for the latter one has to take negative values of ΔM_i because we used holes in our description) one gets for $G_{aa}(x,t)$ $\Delta M_1 = 1$, $\Delta M_2 = 0$, $\Delta D_1 = \frac{1}{2}$ (mod 1), $\Delta D_2 = 0$ (mod 1). For the density-density correlator we choose $\Delta M_1 = \Delta M_2 = 0$, $\Delta D_1 = \Delta D_2 = 0$ (mod 1). For $G^z(x,t)$ the choice is the same as for density-density correlator. For $G^\perp(x,t)$ we take $\Delta M_1 = 1$, $\Delta M_2 = 0$, $\Delta D_1 = \frac{1}{2}$ (mod 1), $\Delta D_2 = 0$ (mod 1). For pair-pair correlation function the choice is: $\Delta M_1 = 0$, $\Delta M_2 = 1$, $\Delta D_1 = \frac{1}{2}$ (mod 1), $\Delta D_2 = 0$ (mod 1).

It is easy to compute the Fourier transforms of correlation functions

$$g(k,\omega) = \int dx \int dt G(x,t) \exp[i(\omega t - kx)] \ . \tag{9.157}$$

In general this integral is not absolutely convergent. The calculation near the singularities $\omega = \pm v_i^F(k - k_0)$, where k_0 determines space oscillations (by the term in the numerator), yields

$$g(k,\omega) = \mathrm{const}[\omega \mp v_1^F(k - k_0)]^{2(h_2+\bar{h}_2+A)-1} \ , \tag{9.158}$$

where $A = h_1$ for the upper sign, and $A = \bar{h}_1$ for the lower sign, for $\omega \approx \pm v_1^F(k - k_0)$, and

$$g(k,\omega) = \mathrm{const}[\omega \mp v_2^F(k - k_0)]^{2(h_1+\bar{h}_1+B)-1} \ , \tag{9.159}$$

where $B = h_2$ for the upper sign, and $B = \bar{h}_2$ for the lower sign, for $\omega \approx \pm v_2^F(k - k_0)$. It turns out that to obtain these expressions one needs to consider the case with the sum of all conformal weights positive and the sum of three of them less than $\frac{1}{2}$, and then continue analytically. For the Fourier transform of equal-time correlators $g(k)$, one has to consider the cases $k > k_0$ and $k < k_0$ separately. Contour integration produces (for $p > 0$)

$$\frac{g(k_0+p)}{g(k_0-p)} = (-1)^{2s} , \tag{9.160}$$

where $s = h - \bar{h}$ is the conformal spin for the considered field (operator), which implies

$$g(k \sim k_0) \propto [\mathrm{sign}(k-k_0)]^{2s}|k-k_0|^{2\Delta_1+2\Delta_2-1} . \tag{9.161}$$

The extra sign will appear for $g_{aa}(k)$.

Very similar consideration is applied in the case of additional Dirac seas present in the ground state due to a finite concentration of similar impurities. Here each additional Dirac sea implies additional factors like $(x \pm vt)$ with their conformal weights and Fermi velocities. The only, but very important difference appears because the filling of Dirac seas, caused by a finite concentration of impurities, is not independent from the filling of Dirac seas of the host, and, hence, related shifts ΔM and ΔD are not independent. It is based on the fact that the same magnetic field and/or the same chemical potential govern the filling of Dirac seas for the host and the ones, caused by a finite concentration of impurities.

In the case of a single impurity, as the reader see from the first two sections of this chapter, an impurity does not renormalize the structure of conformal asymptotics of correlation functions and Fermi velocities, but does renormalize exponents (conformal weights). A phase shift due to an impurity is known to be important for many physical systems *via* the Friedel sum rule.

In the case of open boundary conditions the reader sees that the structure of asymptotics of correlation functions and Fermi velocities are the same as in the case of periodic boundary conditions. However, conformal weights are strongly renormalized, because of the absence of transfers from one Fermi point to the other (backscattering). These changes of exponents of boundary correlation functions are related, *e.g.*, to the orthogonality catastrophe, or X-ray edge singularities due to open edges of a chain. It is interesting to observe that one can measure the strength of bulk interactions in a chain by applying a local (boundary) potential: the shift of a boundary exponent due to the boundary potential depends on how strong the coupling between electrons (or spins) in the chain is.

9.5 Persistent Currents in Correlated Electron Rings

Another issue in which one can apply the knowledge of finite size corrections is the description of *persistent currents* in correlated electron rings.

Persistent currents are thermodynamic characteristics of a quantum ring. They are connected with the Aharonov–Bohm phase shift, which appears when charges move along a loop, pierced by a magnetic flux. Recall, the *Aharonov–Bohm effect* essentially entails the force-free (in the absence of any forces) nonlocal topological influence of electro-magnetic interaction (and in the more general setting, any gauge interaction) on quantum dynamics of charged particles in non-simply connected geometry. According to general principles of quantum mechanics this effect is equal in effect to a variation of the phase of a particle wave function, which causes the interference pattern to shift periodically (with respect to the magnetic field flux) in an experiment. For any closed path enclosing a solenoid this phase variation is equal to

$$\phi_{AB} = \frac{2\pi\Phi}{\Phi_0} , \tag{9.162}$$

where Φ is the magnetic flux in the solenoid, and $\Phi_0 = 2\pi\hbar c/e$, where c is the speed of light in vacuum, and e is the charge of a particle. According to this simple relation, any flux that is a multiple of Φ_0 induces a phase shift that does not have any influence on particle quantum dynamics. In the classical description the effect does not occur for any value of the flux, because the motion of particles takes place in the region outside the solenoid, where electric and magnetic fields are identically zero. In condensed media the Aharonov–Bohm effect normally shows up as magnetic oscillations of kinetic and thermodynamic characteristics of samples in extremely weak magnetic fields, when field-induced forces can be disregarded. Since the electron wave function is not macroscopically coherent for metals in a normal state, Aharonov–Bohm oscillations of thermodynamic variables (persistent currents) are observable in samples of small (mesoscopic) dimensions, smaller than the mean free path between inelastic collisions, and at sufficiently small temperatures. As we showed, electron-electron correlation effects are most conspicuous in the low-dimensional case. Hence, Aharonov–Bohm oscillations can serve as a testing ground for investigations of those electron-electron correlations in conducting loops *via* the period, the initial phase shift and the magnitude of oscillations.

An external magnetic flux yields the nonzero momentum of charged particles. A persistent current is then related to the total orbital moment of all charges in the one-dimensional ring. It is the derivative of the energy of a system in equilibrium with respect to the applied magnetic flux:

$$J(\Phi) = -c\frac{\partial F(\Phi)}{\partial \Phi} , \qquad (9.163)$$

where F is the Helmholtz free energy of the total ring. In the ground state it reduces to the derivative of the ground state energy.

One has to distinguish between persistent currents and *transport currents.* Recall, transport currents are kinetic characteristics of any system, characterized by the resistivity and related to it transition amplitude. In a linear response theory the resistivity is the coefficient connecting the value of a transport current with the value of an applied electric field. Hence, a transport current is the consequence of the difference in potentials applied to the source and drain *cf.* Fig. 9.1 (a). Contrary, a charge persistent current can exist without any applied external electric field: it does not need any source and drain, *cf.* Fig. 9.1 (b).

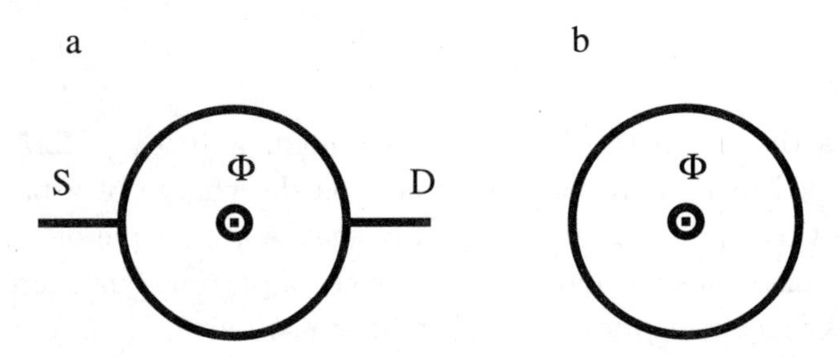

Fig. 9.1 Different geometries for the manifestation of the Aharonov–Bohm effect of an external magnetic flux Φ in a conducting ring: (a) the transport current geometry with the source (S) and drain (D); (b) the persistent current geometry.

When the ring between the source and drain is pierced by an external magnetic flux in the geometry of Fig. 9.1 (a), a transport current is also affected by that flux. Hence, the resistivity of a transport current also becomes flux-dependent. However, such a transport current is not exactly equal to a charge persistent current. This difference in the basic nature of transport and persistent currents produces, *e.g.*, the main difference in the answers, when one considers the effect of a magnetic impurity in a metallic ring, pierced by a flux.

The *Aharonov–Casher effect* is dual to the Aharonov–Bohm one. It is related to the movement of a particle with a magnetic moment (spin) around a two-dimensional electric flux, *e.g.*, the electric flux $F = 4\pi\tau$ generated by a string passing through the center of a ring with linear charge density τ. It produces the phase shift of a particle with spin

$$\phi_{AC} = \frac{2\pi F}{F_0} , \tag{9.164}$$

where $F_0 = 2\pi\hbar c/\mu_B$ is the unit electric flux, μ_B is the Bohr's magneton. Then the spin persistent current is

$$J_s(F) = -c\frac{\partial F(F)}{\partial F} . \tag{9.165}$$

Electro-magnetic fluxes Φ and F can be included into the Hamiltonian of a one-dimensional electron system *via* the standard Peierls factors $\Phi_\uparrow = \pi[(\Phi/\Phi_0)+(F/F_0)]$ and $\Phi_\downarrow = \pi[(\Phi/\Phi_0)-(F/F_0)]$ for electrons with spins up and down, respectively. Then a simple gauge transformation can transfer these phase shifts into *twisted* instead of periodic, boundary conditions.

One can obtain Bethe ansatz equations for correlated electron rings with electro-magnetic fluxes either by using the direct co-ordinate scheme, *cf.* Chapter 4, or introducing the operator

$$T = e^{i\Phi_\uparrow}\frac{1}{2}(I_0 + \sigma_0^z) + e^{i\Phi_\downarrow}\frac{1}{2}(I_0 - \sigma_0^z) \tag{9.166}$$

into the monodromy operator of the associated spin problem (recall, subscript 0 denotes the auxiliary subspace). Then Bethe ansatz equations, *e.g.*, for a Hubbard model for the sets of rapidities k_j ($j = 1, \ldots, N$) and λ_γ ($\gamma = 1, \ldots, M$) become

$$\begin{aligned} e^{i\Phi_\downarrow - i\Phi_\uparrow}\prod_{j=1}^{N}\frac{\lambda_\gamma - \sin k_j + i(U/4)}{\lambda_\gamma - \sin k_j - i(U/4)} &= \prod_{\substack{\beta=1\\ \beta\neq\gamma}}^{M}\frac{\lambda_\gamma - \lambda_\beta + i(U/2)}{\lambda_\gamma - \lambda_\beta - i(U/2)} , \\ e^{ik_jL - i\Phi_\uparrow} &= \prod_{\beta=1}^{M}\frac{\sin k_j - \lambda_\beta + i(U/4)}{\sin k_j - \lambda_\beta - i(U/4)} , \end{aligned} \tag{9.167}$$

and for a supersymmetric t-J chain (for $V = -J/4$, $J = 2$)

$$\begin{aligned} \prod_{j=1}^{N}\frac{\lambda_\alpha - p_j + i/2}{\lambda_\alpha - p_j - i/2}e^{i\Phi_\downarrow - i\Phi_\uparrow} &= \prod_{\beta=1}^{M}\frac{\lambda_\alpha - \lambda_\beta + i}{\lambda_\alpha - \lambda_\beta - i} , \qquad \alpha = 1, \ldots, M , \\ \left[\frac{p_j + i/2}{p_j - i/2}\right]^L e^{-i\Phi_\uparrow} &= \prod_{\alpha=1}^{M}\frac{p_j - \lambda_\alpha + i/2}{p_j - \lambda_\alpha - i/2} , \qquad j = 1, \ldots, N , \end{aligned} \tag{9.168}$$

with the same definitions of the energy as before.

It is easy to see that in the thermodynamic limit the effect of twisted boundary conditions caused by topological Aharonov–Bohm–Casher phase factors is zero. The main contribution from those factors appears to be related with finite size corrections to the energy and the momentum, studied in the second section of this chapter. Taking into account Aharonov–Bohm–Casher topological phases, counting functions of the second section become

$$z_{i,L}(x) = \frac{1}{2\pi}\left(p_i^0(x) - \frac{1}{L}\Phi_i - \frac{1}{L}\sum_{j=1}^{2}\sum_{l=1}^{M_j}\phi_{ij}^0(x, u_{j,l})\right) , \qquad i = 1, 2 , \tag{9.169}$$

where $\Phi_1 = \pi[(\Phi/\Phi_0) + (F/F_0)]$ and for a repulsive Hubbard ring $\Phi_2 = -2\pi F/F_0$, while for an attractive Hubbard ring and for a supersymmetric t-J ring we have $\Phi_2 = 2\pi\Phi/\Phi_0$. The counting functions are periodic in F with the period F_0 and $F_0/2$, and in Φ with the periods Φ_0 and $\Phi_0/2$ (depending on the model, for a repulsive Hubbard case one has $2F/F_0$ periodicity and no $2\Phi/\Phi_0$ one, while for an attractive Hubbard chain and a supersymmetric t-J chain the situation is the opposite). Hence, they remain invariant under replacements $(F/F_0) \to \{\{F/F_0\}\}$, $(2F/F_0) \to \{\{2F/F_0\}\}$, $(\Phi/\Phi_0) \to \{\{\Phi/\Phi_0\}\}$ and $(2\Phi/\Phi_0) \to \{\{2\Phi/\Phi_0\}\}$, where $\{\{x\}\}$ denotes the fractional part of x to the nearest (half)integer. Spin and charge rapidities parametrize each eigenvalue and eigenfunction of the stationary Schrödinger equation, and, therefore, all characteristics of integrable models have to reveal those periodicities also.

Then, proceeding as in the second section we obtain

$$\begin{aligned}\Delta_1 &= \left(\sum_{j=1}^{2} Z_{1j}[(D_j + \tilde{\Phi}_i) - \delta_j L]\right)^2 \\ &\quad + \frac{1}{4(\det Z)^2}[Z_{22}(M_1 - \nu_1(\mu, H)L) - Z_{21}(M_2 - \nu_2(\mu, H)L)]^2 , \\ \Delta_2 &= \left(\sum_{j=1}^{2} Z_{2j}[(D_j + \tilde{\Phi}_i) - \delta_j L]\right)^2 \\ &\quad + \frac{1}{2(\det Z)^2}[Z_{12}(M_1 - \nu_1(\mu, H)L) - Z_{11}(M_2 - \nu_2(\mu, H)L)]^2 .\end{aligned} \tag{9.170}$$

In these equations $\tilde{\Phi}_i$ denote the fractional part of Φ_i to the nearest (half)integer. Finite size correction to the total momentum is

$$P = \frac{2\pi}{L}\sum_{i=1}^{2}[M_i(D_i + \tilde{\Phi}_i) + n_i^+ - n_i^-] + 2\sum_{\sigma} P_\sigma^F D_\sigma \ , \tag{9.171}$$

where the Fermi momenta are for a repulsive Hubbard chain $D_\uparrow = D_1 + \tilde{\Phi}_1$, $D_\downarrow = D_1 + D_2 + \tilde{\Phi}_1 + \tilde{\Phi}_2$, for an attractive Hubbard chain $D_\uparrow = D_1 + D_2 + \tilde{\Phi}_1 + \tilde{\Phi}_2$ and $D_\downarrow = D_2 + \tilde{\Phi}_2$, while for a supersymmetric t-J chain one has $D_\uparrow = -D_1 - D_2 - \tilde{\Phi}_1 - \tilde{\Phi}_2$ and $D_\downarrow = -D_2 - \tilde{\Phi}_2$, and it is necessary to add to the total momentum the term $2\pi(D_1 + D_2 + \tilde{\Phi}_1 + \tilde{\Phi}_2)$. The finite size correction to the energy is

$$E(M_i, D_i, n_i^\pm) = E_0 + L\varepsilon_\infty(\Lambda_i, -\Lambda_i) - \frac{\pi}{6L}\sum_{i=1}^{2} v_i^F + \frac{2\pi}{L}\sum_{i=1}^{2} v_i^F[\Delta_i + n_i^+ + n_i^-] \ . \tag{9.172}$$

It is a trivial exercise for the reader to check that there are no corrections to the energy for open chains, which is the manifestation of the fact that one can totally remove topological phases, related to external electro-magnetic fluxes, from the answers in that case.

From the formulas for the energy and momentum we can see that

- Charge persistent currents (the Aharonov–Bohm effect) in correlated electron rings are determined by a virtual movement of low-lying charge-carrying excitations (for the case of a repulsive Hubbard ring they are connected with unbound electron excitations, while for an attractive Hubbard ring and for a supersymmetric t-J ring with $V = -J/4$ they are connected with unbound electron excitations and spin-singlet pairs, with the interference of those two kinds of oscillations);
- Spin persistent currents (the Aharonov–Casher effect) in correlated electron rings are determined by a virtual movement of low-lying spin-carrying excitations (for the case of a repulsive Hubbard ring they are connected with unbound electron excitations and spinons, with the interference of two kinds of oscillations, while for an attractive Hubbard ring and for a supersymmetric t-J ring with $V = -J/4$ they are connected with only unbound electron excitations);
- Magnitudes of spin and charge persistent currents are determined by Fermi velocities of low-lying excitations and coefficients of a dressed charge matrix;

- The form of oscillations in the ground state is "saw-tooth"-like. It is due to the presence of all harmonics;
- The periodicity of charge and spin persistent currents are determined by charges and spins of those virtual excitations;
- The initial shifts of those persistent currents (*parity effect*) are determined by the values of D_i (*i.e.*, those initial shifts are different for different values of the total number of electrons and the total magnetic moment of a system).

Magnitudes of some oscillations become equal to each other at some values of the external magnetic field, *e.g.*, for $H = 0$ for a repulsive Hubbard ring and supersymmetric t-J chain, or for $H = H_c$ for an attractive Hubbard ring (because some non-diagonal matrix elements of a dressed charge matrix are zero at those critical points). Then the interference of periods of oscillations is not manifested, naturally.

For $H < H_s$ charge persistent currents of a repulsive Hubbard ring reveal the period of oscillations Φ_0, while spin persistent currents manifest the interference pattern of two kinds of periodicities: with F_0 and $F_0/2$. On the other hand, spin persistent currents of an attractive Hubbard ring for $H_c < H < H_s$ and of a supersymmetric t-J ring (for $V = -J/4$) manifest the period of oscillations F_0, while charge persistent currents of those models reveal the interference of oscillations with two periods: Φ_0 and $\Phi_0/2$. For $H < H_c$ in an attractive Hubbard ring there are no mesoscopic (with the magnitude of order of L^{-1}) spin persistent currents, and there is only one period of oscillations of charge persistent currents, equal to $\Phi_0/2$. For $H > H_s$ there are no mesoscopic persistent currents in the ground state. It is important to point out that at crossover points the part of the energy, related to persistent current can display jumps as a function of Φ or F. These "discontinuities" of the energy are, however, microscopic, of order of the uncertainty of the energy according to Heisenberg's principle. These singularities can be interpreted as "supercurrents" necessary to generate the discontinuities of the energy. Any nonzero temperature, naturally, removes those "supercurrents".

Any nonzero temperature strongly reduces magnitudes of persistent currents: They become not mesoscopic, but rather exponentially small. The form of these oscillations becomes harmonic, instead of the "saw-tooth"-like in the ground state, since the temperature suppresses higher harmonic content.

To study the parity effect, let us concentrate on charge persistent currents of a repulsive Hubbard ring at $H = 0$. One has to distinguish three main cases. For odd N charge persistent currents are paramagnetic $J(\Phi) = (2v_1^F\xi^2(Q)/\Phi_0 L)[(\pi/2) - |2\pi\Phi/\Phi_0|]\text{sign}\Phi$ for $\Phi < \Phi_0/2$. There is a crossover of the low-lying levels at $\Phi = 0$ and $\Phi = \pm\Phi_0/2$. For even N charge persistent current are diamagnetic $J(\Phi) = (2v_1^F\xi^2(Q)/\Phi_0 L)(\Phi/\Phi_0)$. There exists a crossover of low-lying levels at $\Phi_c = \Phi_0[(1/4) + (v_2^F/2v_1^F\xi^2(Q))]$ for $N = 4n + 2$ (n is an integer) and at $\Phi = (\Phi_0/2) - \Phi_c$ for $N = 4n$. Summarizing, the parity effect for charge persistent currents in a repulsive Hubbard chain is similar to rings of noninteracting electrons, except of the case $N = 4n$, where the persistent current is changed from a paramagnetic one into a diamagnetic one. Parity effects for spin persistent currents and for persistent currents in other models of correlated electrons can be studied analogously.

Let us consider how a single impurity can affect persistent currents. The analysis, totally equivalent to the above shows, that a single impurity changes neither magnitudes of oscillations (*i.e.*, velocities of low-lying excitations and dressed charge matrices do not depend a single impurity), nor periods of oscillations. The only parameter, which gets renormalized due to an impurity is the initial phase shift of oscillations of persistent currents. This effect is related to the way how an integrable impurity can be introduced into a correlated electron ring without destroying the exact solvability, because the impurity introduces a chirality into the problem, *i.e.*, it renormalizes the momentum of the ring, *cf.* Chapter 7.

Now, let us examine how a finite concentration of impurities can affect persistent currents. The reader already knows that a finite concentration of impurities can yield additional Dirac seas in the ground state. Hence, the effect of such impurities is related to the onset of new, additional oscillation patterns, related to virtual movements of low-lying excitations due to those impurities-induced Dirac seas, and with subsequent interferences of those new oscillations with the previous ones.

It is also interesting to mention so called *microscopic oscillations* of persistent currents. These oscillations appear only for small rings with small number of electrons and very strong interactions. Let us consider, *e.g.*, Bethe ansatz equations for a repulsive Hubbard ring for $U \gg 1$ for $F = 0$ and $\Phi \neq 0$. In this limit one can neglect $\sin k_j$ in comparison with $U/4$ and λ_γ. Then it follows that

$$Lk_j = 2\pi\left(J_j + \frac{\Phi}{\Phi_0} + \frac{1}{N}\sum_{\gamma=1}^{M} J_\gamma\right) , \tag{9.173}$$

with the energy

$$E = E_0 - 2\frac{\sin(\pi N/L)}{\sin(\pi/L)}$$
$$\times \cos\left[\frac{2\pi}{L}\left(\frac{\Phi}{\Phi_0} + \frac{1}{N}\sum_{\gamma=1}^{M} J_\gamma + \frac{J^+ + J^-}{2}\right)\right] , \qquad (9.174)$$

where J^+ and J^- are the maximal and minimal values of J_j (*i.e.*, $(J^+ + J^-)/2 = D_1$). The energy for $\Phi \neq 0$ can be minimized by choosing the set of J_γ such that $\sum_{\gamma=1}^{M} J_\gamma = -p$ for

$$\frac{2p-1}{2N} < \frac{\Phi}{\Phi_0} + \frac{I^+ + I^-}{2} < \frac{2p+1}{2N} . \qquad (9.175)$$

Hence, the energy as a function of Φ becomes a quasiperiodic (with a "period" N^{-1}) sequence of quasiparabolic segments (they become strictly parabolic for large N). These are microscopic oscillations of a charge persistent current. Obviously, they disappear when U becomes smaller or N becomes larger. The nature of these oscillations is obvious: one excites additional spinon oscillations (which carry nonzero momentum p) to minimize the energy lost caused by the external flux. It means that microscopic oscillations of persistent currents appear only in correlated electron chains, where at least two low-lying collective excitations with different velocities can exist. Similar microscopic oscillations of charge and spin persistent currents appear in other one-dimensional models of correlated electrons.

The magnitude of oscillations of persistent currents is related to the stiffness constant D ($E(\Phi) = E(0) + D\Phi^2/L^{2-d} + O(\Phi^4)$, where d is the dimension of the space) of a transport current in the ground state. It is easy to show that

$$D = \frac{1}{L^d}\left[-\frac{\langle 0|T|0\rangle}{2d} - \sum_{n\neq 0}\frac{|\langle 0|j_x|n\rangle|^2}{E_n - E_0}\right] , \qquad (9.176)$$

where T is the kinetic energy of electrons. Let us switch on the vector potential $A_x \exp(-i\omega t)$, *i.e.*, we get the electric field $E_x = (i\omega/c)A_x \exp(-i\omega t)$. Then the imaginary part of the ac conductivity of a chain is

$$\mathrm{Im}\sigma_{xx}(\omega) = \frac{2e^2}{\hbar^2\omega L^d}\left[-\frac{\langle 0|T|0\rangle}{2d} - \mathcal{P}\sum_{n\neq 0}\frac{|\langle 0|j_x|n\rangle|^2(E_n - E_0)}{(E_n - E_0)^2 - (\hbar\omega)^2}\right] , \qquad (9.177)$$

where $\mathcal{P}$ denotes the principal part. It is easy to show that

$$\begin{aligned}\lim_{\omega\to 0}\omega\mathrm{Im}\sigma_{xx}(\omega) &= \frac{2e^2}{\hbar^2}D\ ,\\ \lim_{\omega\to\infty}\omega\mathrm{Im}\sigma_{xx}(\omega) &= -\frac{2e^2}{\hbar^2 dL^d}\langle 0|T|0\rangle\ .\end{aligned} \tag{9.178}$$

It is well-known that the high-frequency behaviour of the real part of the conductivity is related to the imaginary part *via* the f-sum rule

$$\int_{-\infty}^{\infty} d\omega\mathrm{Re}\sigma_{xx}(\omega) = -\frac{\pi e^2}{\hbar^2 dL^d}\langle 0|T|0\rangle\ . \tag{9.179}$$

On the other hand, the low-frequency behaviour implies that

$$\mathrm{Re}\sigma_{xx}(\omega) = \frac{2\pi e^2}{\hbar}\left(D\delta(\hbar\omega) + \frac{1}{L^d}\sum_{n\neq 0}|\langle 0|j_x|n\rangle|^2\delta[(E_n-E_0)^2-(\hbar\omega)^2]\right)\ . \tag{9.180}$$

This expression means that nonzero D (related to the Drude weight), is connected with the magnitude of the charge persistent current in the ground state. Hence, a nonzero D implies infinite dc conductivity in the ground state.

Let us consider how a magnetic impurity (for simplicity we shall study the case $S=\frac{1}{2}$) in a correlated electron chain can affect the ground state magnetoresistivity. We can suppose that due to the contact interaction the impurity-host S-matrix of the real system is momentum independent, and only a function of energy. This assumption is, naturally, correct in the long-wave limit. The impurity-host scattering matrix can be characterized by scattering phase shifts, and the magnetoresistivity due to an impurity can be expressed in terms of phase shifts for electrons at the Fermi levels of low-lying excitations. The T-matrix ($S = 1 - iT = \exp(2i\delta)$, where the phase shift is given mod 2π) can be defined by the one-electron zero temperature Green's function from

$$G_{k,k',\sigma}(\omega) = G^0_{k,\sigma}(\omega) + G^0_{k,\sigma}(\omega)T(\omega)G^0_{k',\sigma}(\omega)\ . \tag{9.181}$$

Hence, the propagator of this equation at $\omega = 0$ yields the phase shift at the Fermi level. Time evolution is given by an additional electron (hole) state at the Fermi level, because the propagator annihilates one electron (hole) at $t = 0$ and creates it at time t. When an electron (hole) propagates through a chain it causes the change in a phase $L[E(N)-E(N-1)]$. Let us assume that a chain has the width much larger than the atomic scale, but much smaller than the length of the ring L. Then the scattering gives the formula for the ground state magnetoresistivity due to a magnetic impurity

$$R = \frac{R_0}{\sum_\sigma \frac{1}{\sin^2 \delta_\sigma^F}} , \tag{9.182}$$

where $\delta^F \sigma = (\pi/2)(n_{imp} \mp 2m^z_{imp})$, and R_0 characterizes the scattering of a correlated electron system in the unitary limit. For a fixed band filling the magnetoresistivity decreases when increasing the magnetization of an impurity. This is why, the magnetoresistivity, *i.e.*, the characteristic of the magnitude of a transport current, strongly depends on the presence of a magnetic impurity in a chain, while such an impurity affects only the initial phase of a charge persistent current in a ring.

Summarizing, in this chapter we presented the calculation of finite-size corrections to characteristics of exactly solvable models of quantum spins and correlated electrons with periodic and open boundary conditions, with and without inhomogeneities. We reminded the reader the main features of the description of critical phenomena and conformal field theory. The latter was used to calculate the asymptotic behaviour of low-energy correlation functions of considered models. Also, another finite-size effect, the behaviour of persistent currents for periodic correlated electron chains is studied.

The calculation of finite size corrections for Bethe ansatz-solvable models was pioneered in [de Vega and Woynarovich (1985)]. The reader can find calculations of finite size corrections for quantum correlated chains, *e.g.*, in [Woynarovich, Eckle and Truong (1989); Woynarovich (1989); Klümper, Batchelor and Pearce (1991); Pearce and Klümper (1991); Kawakami and Yang (1991); Bariev, Klümper, Schadschneider and Zittartz (1993); Bariev (1994)], homogeneous chains with periodic (twisted) boundary conditions, in [Alcaraz, Barber, Batchelor, Baxter and Quispel (1987); Asakawa and Suzuki (1995); Frahm and Zvyagin (1997a); Bedürftig and Frahm (1997)] for chains with open boundary conditions, see also [Frahm and Zvyagin (1997b); Zvyagin (2002); Zvyagin (2003)] for quantum chains with impurities. The dressed charge technique was introduced in [Korepin (1979)]. I can suggest the use of, *e.g.*, the well-known books [Ma (1976); Cardy (1996); Kadanoff (2000); Sachdev (1999)] for the description of critical phenomena in the vicinity of a phase transition. The scaling hypothesis was introduced in [Widom (1965a); Widom (1965b)], see also [Rushbrooke (1963)], and developed in [Griffiths (1965); Kadanoff (1966);

Josephson (1967a); Josephson (1967b); Fisher (1969)]. The conformal field theory was pioneered in [Belavin, Polyakov and Zamolodchikov (1984)]. The reader can find more about the conformal field theory from, *e.g.*, the monographs [Di Franchesco, Mathieu and Sénéshal (1997); Gogolin, Nersesyan and Tsvelik (1998); Korepin, Bogoliubov and Izergin (1993)]. Conformal field theory for finite systems was developed in [Cardy (1986a); Cardy (1986b)]. Correlation functions for quantum spin and correlated electron models were calculated, *e.g.*, in [Bogoliubov, Izergin and Reshetikhin (1987); Bogolyubov and Korepin (1989); Bogolyubov and Korepin (1990); Frahm and Korepin (1990); Frahm and Korepin (1991); Kawakami and Yang (1991)] for periodic systems and [Asakawa and Suzuki (1995); Bedürftig and Frahm (1997); Frahm and Zvyagin (1997b)] for chains with open boundary conditions. For more accurate calculations of correlation functions for integrable models, than by using the conformal field theory, consult [Korepin, Bogoliubov and Izergin (1993)]. The Aharonov–Bohm and Aharonov–Casher effects were introduced in [Aharonov and Bohm (1959)] and [Aharonov and Casher (1984)], respectively. First calculations of persistent currents (in superconductors) can be found in [Byers and Yang (1962)]. For the behaviour of a charge persistent current in a non-interacting metallic ring see, *e.g.*, [Cheung, Gefen, Riedel and Shin (1988)]. Persistent current in a correlated electron chain was first calculated in [Zvyagin (1990b)]. Calculations of spin persistent currents (the Aharonov–Casher effect) in a correlated electron chain are presented in [Zvyagin and Krive (1992)]. The description of a parity effect and microscopic oscillations of charge persistent currents can be found, *e.g.*, in [Yu and Fowler (1992)]. The reader can find a review of properties of charge persistent currents in correlated electron rings in [Zvyagin and Krive (1995)]. The connection between charge persistent currents and ac optical conductivity (Drude weight) was proposed in [Shastry and Sutherland (1990)], and we closely follow their description here. For the behaviour of persistent currents in correlated quantum rings with a single impurity see [Zvyagin (2003)] (there the reader can also find studies of the ground state magnetoresistivity in a correlated electron ring with a magnetic impurity), and for correlated rings with a finite concentration of magnetic impurities see, *e.g.*, [Zvyagin and Schlottmann (1995)].

Chapter 10

Beyond the Integrability: Approximate Methods

In this very short chapter we shall briefly review several approximate methods which are used in the theory of one-dimensional quantum chains. Many recent excellent review articles and books considered these approximate methods, and the purpose of the introduction of this short chapter is only for the "completeness" of the impression of the reader.

10.1 Scaling Analysis

The first (and the simplest) class of methods used in the approximate description of quantum correlated chains is connected with the renormalization group approach. Such a study shows that exponents for characteristics of low-dimensional systems are non-integer in general, in contrast to simple perturbation theories, which of course are not legitimate. An application of scaling relations provides a simple tool to understand some essential aspects of the behaviour of quantum correlated chains under a relevant perturbation. To remind, the response of a classical Helmholtz free energy f_{cl} and the correlation function ξ_{cl} of a classical critical d-dimensional system perturbed by a relevant operator $\delta\mathcal{H}'$ with the renormalization group eigenvalue $y^{-1} > 0$ near a critical point is

$$\Delta f_{cl} \propto \delta^{dy} , \qquad \xi_{cl} \propto \delta^{-y} , \tag{10.1}$$

where d is the space dimension. A quantum critical d-dimensional system formally behaves in a scaling regime equivalently to a $(d+z)$-dimensional classical system, where z is the *dynamical critical exponent*. To remind, the divergence of correlation functions for quantum critical points implies divergencies not only in space, but also in time, because the real space in which one has to consider quantum critical systems is $(d+1)$, where 1 is due to time. It should take a longer time to propagate across the distance of correlation length. One can introduce the "correlation length" (actually a relaxation time) in time direction τ_r which diverges as $\tau_r \sim \xi^z$, *cf.* the previous chapter. As a consequence of divergences in both, ξ and τ_r, various physical quantities in the critical region close to a quantum critical point have a dynamic scaling form,

$$\Psi(k,\omega) = \xi^{\Delta}\Psi(k\xi, \omega\tau_r) , \tag{10.2}$$

where the observable Ψ is measured at the wave vector k and frequency ω (Δ is the scaling dimension): *i.e.*, close to the critical point there is no other characteristic length scale than ξ and no other characteristic time scale than τ_r. At the scale-invariant critical point correlation lengths are divergent, and, hence, the only characteristic length is $2\pi/k$, and the only characteristic frequency is $\omega(k) \sim (v^F k)^z$. This implies the simpler scaling form

$$\Psi(k,\omega)_c = k^{-\Delta}\tilde{\phi}([v^F k]^z/\omega) . \tag{10.3}$$

The behaviour of a two-point correlation function $G^{(2)}(r) \sim r^{-d+2-\eta}$ implies for the Fourier transform $G^{(2)}(k) \sim k^{-2+\eta}$. The Fourier transform of a correlation function for $(d+1)$-dimensional problem is then

$$G^{(2)}(k,\omega) \sim \left[\sqrt{(v^F k)^{2z} - \omega^2}\right]^{-2+\eta} , \tag{10.4}$$

i.e., there is no other characteristic frequency than $(v^F k)^z$ itself, thus, collective modes have become overdamped and the system is in an incoherent diffusive regime.

That is why, the ground state energy and the gap of low-lying excitations of a d-dimensional quantum critical system are formally proportional to the free energy and the inverse correlation function of a $(d+z)$-dimensional classical critical system, respectively. The renormalization group eigenvalue y^{-1} is related to the scaling dimension Δ of the particular operator by

$$\Delta + y^{-1} = d + z . \tag{10.5}$$

For a conformally invariant quantum critical chain we have $d = z = 1$, *i.e.*,

$$y = (2 - \Delta)^{-1} . \tag{10.6}$$

Hence the renormalization of the ground state (internal) energy per site of a quantum critical chain and the gap for low-lying excitations G (which is equal to zero at the unperturbed point) to a relevant perturbation are

$$\Delta E_q \propto -\delta^{2/(2-\Delta_e)} \ , \ G_q \propto \delta^{1/(2-\Delta_e)} \ , \tag{10.7}$$

respectively, where the subscript q implies the quantum situation, and Δ_e is the minimal scaling exponent for energy-energy correlation functions (*i.e.*, if the Hamiltonian of the quantum critical chain with the nearest-neighbour interactions is $\mathcal{H} = \sum_j \mathcal{H}_{j,j+1}$, the energy-energy correlation function is $\langle \mathcal{H}_{r,r+1}\mathcal{H}_{0,1}\rangle$). Here we ignored logarithmic corrections. They can be present due to marginal operators in the renormalization group sense. For example, they appear when one studies systems with the SU(2) spin symmetry. To find the scaling dimension for a quantum critical chain (as a function of, *e.g.*, coupling constants, band fillings, hopping integrals, external uniform magnetic field, *etc.*) we can use the conformal field theory, *cf.* the previous chapter. According to the conformal field theory approach asymptotics of correlation functions of primary fields in the ground state are known to be

$$\langle \phi_{\Delta^\pm}(x,t)\phi_{\Delta^\pm}(0,0)\rangle \sim \frac{\exp(2iDP^F x)}{(v^F\tau + ix)^{2h}(v^F\tau - ix)^{2\bar{h}}} \ , \tag{10.8}$$

where v^F and P^F are the Fermi velocity and the Fermi momentum, respectively. Scaling dimensions and spins for each primary field are determined by $\Delta_\phi = h + \bar{h}$ and $s_\phi = h - \bar{h}$. Conformal weights $(h, \bar{h})$ can be calculated according to the finite size analysis of the low energy physics of a critical quantum model. For critical correlated electron models the low energy physics in general case corresponds to a semidirect product of two conformal field theories. At total and half-filling of the band for a repulsive Hubbard chain the gap can be opened in the spectrum of charged low-lying excitations, or, for an attractive Hubbard chain the gap for unbound electron excitations persists for $H < H_c$. Also, in the spin-saturated phase, for $H > H_s$, spin-carrying excitations are gapped. Thus, in such cases the semidirect product reduces to one conformal field theory.

It means that for a quantum critical chain the gap in spectrum of low-lying excitations due to a relevant perturbation is

$$G \approx v^F \left(\frac{\delta}{v^F}\right)^{1/(2-\Delta_e)} \ , \tag{10.9}$$

and the ground state energy correction reads

$$\Delta E \approx -v^F \left(\frac{\delta}{v^F}\right)^{2/(2-\Delta_e)} \qquad (10.10)$$

For example, for spin-$\frac{1}{2}$ chains the dimerization (a relevant perturbation) produces exponents for a spin gap (for $H = 0$) from $\frac{2}{3}$ for the isotropic antiferromagnetic Heisenberg chain ($\Delta_e = \frac{1}{2}$) to $\frac{1}{2}$ for the isotropic XY chain (which is related to free fermions, $\Delta_e = 1$), *cf.* Chapters 2 and 9. In such a way exponents describe the strength of interactions in a quantum critical system.

Now, few words about logarithmic corrections. As the reader knows from the previous chapter, for some models (*e.g.*, those, which respect the SU(2) spin symmetry), finite size corrections due to low-lying excitations can be written as

$$E = E_0 + \frac{2\pi v^F}{L}\left[\Delta + \frac{b}{\ln L} + \cdots\right] . \qquad (10.11)$$

According to the conformal field theory the amplitude b determines multiplicative logarithmic corrections to the two-point correlation function,

$$\langle\phi(x)\phi(0)\rangle \approx \frac{1}{[x(\ln x)^b]^{2\Delta}} \cdot \qquad (10.12)$$

Hence, the correction to the gap of low-lying excitations of a quantum critical system (which has logarithmic finite size corrections) due to a relevant perturbation δ appears to be

$$G \sim \left(\frac{\delta}{(\ln\delta)^{b\Delta_e}}\right)^{\frac{1}{(2-\Delta_e)}} . \qquad (10.13)$$

At nonzero temperatures one has to use the scaling form

$$\Psi(T,\omega) = f(\hbar\omega/T, \delta/T^{1/z\nu}) , \qquad (10.14)$$

where δ measures the distance to the quantum critical point. In the regime $\hbar\omega \ll T$ one expects the behaviour of the scaling function of a single scaling variable $\delta/T^{1/z\nu}$. The effect of deviation from the critical value is rescaled by the factor of $T^{1/z\nu}$. The transition appears sharper as the temperature is lowered. On the other hand, for $\hbar\omega \gg T$ the scaling is dominated by ω, and the scaling function is independent of T, and, hence, the scaling variable reduce to $(\hbar\omega/T)^{-1/z\nu}\delta/T^{1/z\nu} \sim \delta/\omega^{1/z\nu}$.

10.2 Bosonization

There are many ways of the introduction of bosonization of Fermi fields. D. C. Mattis and E. H. Lieb were pioneers of this way of consideration of one-dimensional condensed matter models. Usually theorists distinguish between the *Abelian bosonization* and *non-Abelian bosonization* for electron systems. The goal of the diagonalization of the Hamiltonian of interacting electron system with the help of bosonization is to find relevant (and simple) quantum numbers, which (approximately) parametrize low-energy eigenvalues and eigenfunctions of the stationary Schrödinger equation.

The bosonization is related to the presentation of the Hamiltonian of an interacting system as a quadratic form of current operators (*e.g.*, in the Sugawara construction). It is connected to the important example of the application of the conformal field theory, the Gaussian model, whose action is the action of the bosonic field $\Phi(z,\bar{z})$:

$$\mathcal{A} = \frac{g}{2\pi}\int dzd\bar{z}\left(\frac{\partial\Phi}{\partial z}\right)\left(\frac{\partial\Phi}{\partial\bar{z}}\right) , \tag{10.15}$$

where g is the coupling constant. This model is critical and conformal invariant. The solution of equations of motion can be given in terms of right- and left-moving fields (holomorphic and antiholomorphic parts) as $\Phi(z,\bar{z}) = [\phi(z)+\bar{\phi}(\bar{z})]/2\sqrt{g}$. The correlation functions are

$$\langle\phi(z)\phi(w)\rangle = -\ln(z-w) , \tag{10.16}$$

and similar for the antiholomorphic part. One can see that the fields $\phi(z)$ are not conformal fields, but their derivatives are. To show how this comes about we can construct the energy-momentum tensor

$$\begin{aligned} T(z) &= -\frac{1}{2} : \ [\partial\phi(z)/\partial z)]^2 \ : \\ &\equiv -\frac{1}{2}\lim_{a\to 0}\left(\frac{\partial\phi[z+(a/2)]}{\partial z}\frac{\partial\phi[z-(a/2)]}{\partial z} - 1/d^2\right) . \end{aligned} \tag{10.17}$$

From the operator product expansion of ϕ with T one gets

$$T(z)\frac{\partial\phi(w)}{\partial w} = \frac{1}{(z-w)^2}\frac{\partial\phi(w)}{\partial w} + \frac{1}{z-w}\frac{\partial^2\phi(w)}{\partial w^2} + \cdots , \tag{10.18}$$

from which the reader can see that $(\partial\phi(w)/\partial w)$ is a primary field with the conformal weight $(1,0)$. It is possible to identify these derivatives with some currents $J(w) = i(\partial\phi(w)/\partial w)$. The current-current correlation function is

$$\langle J(z)J(w)\rangle = \frac{1}{(z-w)^2} . \tag{10.19}$$

Then one can write the Laurent series for currents

$$J(z) = \sum_{n=-\infty}^{\infty} \frac{J_n}{z^{n+1}} , \ J_n = \frac{1}{2\pi i}\oint dz z^n J(z) . \tag{10.20}$$

For periodic boundary conditions J_n are the Fourier components of the current. They satisfy the algebra

$$[J_n, J_m] = n\delta_{n,-m} , \tag{10.21}$$

which is known as U(1) Kac–Moody algebra (the reader can see that it is bosonic up to a factor n, which can be absorbed into the re-definition of J_n). This algebra in mathematics is often called an affine Lie algebra. After transforming back one obtains

$$[J(x_1), J(x_2)] = -i\frac{\partial}{\partial x_1}\delta^{(1)}(x_1 - x_2) , \tag{10.22}$$

where the right hand side is known as a Schwinger term, or a chiral U(1) anomaly. The modes with $n < 0$ can be considered as "creation operators" and those with $n > 0$ are "annihilation operators". U(1) Kac–Moody algebra is the special case of some more general Lie algebra of the generators J_n^a

$$[J_n^a, J_m^b] = if^{abc}J_{n+m}^c + kn\delta_{a,b}\delta_{n,-m} , \tag{10.23}$$

where f^{abc} are structure constants of the algebra and integer k is the level of Kac–Moody algebra. The central charge of the associated Virasoro algebra is related to k as

$$c = \frac{3k}{k+2} . \tag{10.24}$$

The reader can see that for U(1) Kac–Moody algebra the central charge is $c = 1$.

The Kac–Moody generators are useful in classifying excitations of the Gaussian model. The Hamiltonian (sometimes called as *Sugawara Hamiltonian*) is related to the energy-momentum tensor

$$T(x) = \frac{1}{2} : J(x)J(x) : + \frac{\pi^2}{6L^2} \tag{10.25}$$

via

$$\mathcal{H}_S = \frac{1}{2\pi}\int dx[T(x) + \bar{T}(x)] \ . \tag{10.26}$$

It is obvious from the first formula that $c = 1$. The generators of the Virasoro algebra are given by

$$\begin{aligned}\frac{2\pi}{L}L_m &= \frac{1}{2\pi}\int_0^\infty dx e^{i2\pi mx/L}T(x) \\ &= \frac{\pi}{L}\sum_{n=-\infty}^{\infty} : J_n J_{m-n} : -\delta_{m,0}\frac{c\pi}{12L} \ ,\end{aligned} \tag{10.27}$$

where $c = 1$. Then it is possible to derive the commutation relation between L_m and J_n as

$$[L_m, J_n] = -nJ_{n+m} \ . \tag{10.28}$$

It is important to notice that

$$[\mathcal{H}_S, J_m] = -\frac{2\pi}{L}mJ_m \ , \tag{10.29}$$

which reveals that the transform of the current operator, J_m, acts as a "creation operator" for $m < 0$ and as a "annihilation" one for $m > 0$. The spectrum is harmonic because of the linearized dispersion law. This is why the operators J_m can be used to generate the conformal tower of descendant states from the highest weight state.

For complicated correlated electron chains the goal of the bosonization procedure is to present, *e.g.*, in the framework of the Sugawara construction, the Hamiltonian of electrons as a quadratic form of current operators, which satisfy Kac–Moody algebra. Then conformal properties of the latter permit to know the structure of low-energy excitations.

To characterize electron states we can start with the introduction of the number operators

$$\hat{N}_\sigma = \sum_k : a^\dagger_{k,\sigma}a_{k,\sigma} :\equiv \sum_k [a^\dagger_{k,\sigma}a_{k,\sigma} - {}_0\langle 0|a^\dagger_{k,\sigma}a_{k,\sigma}|0\rangle_0] \ . \tag{10.30}$$

These operators count the number of electrons with spin σ with respect to the free electron reference ground state $|0\rangle_0$ of the Hamiltonian

$$\mathcal{H}_0 = \sum_{k,\sigma} v^F k : a^\dagger_{k,\sigma}a_{k,\sigma} : \ , \tag{10.31}$$

where $k \equiv P - P^F$ measures the deviation from the Fermi momentum P^F. The dispersion has been linearized about the Fermi energy E^F as $\varepsilon(P) \approx v^F k + E^F$. Columns denote the normal ordering with respect to the free Dirac sea of the vacuum state $|0\rangle_0$, which is defined as

$$a_{k>0,\sigma}|0\rangle_0 = a^{\dagger}_{k\leq 0,\sigma}|0\rangle_0 = 0 \ . \tag{10.32}$$

The operators $a^{\dagger}_{k,\sigma}$ and $a_{k,\sigma}$ satisfy standard anticommutation relations. The values of k are quantized as $k = (2\pi/L)[n_k - P_0/2]$, where n_k are integers. $P_0 = 0, 1$, since in the ground state the chemical potential (and, hence, P^F) has to either coincide with a degenerate level for $P_0 = 0$, or lie between two of them, $P_0 = 1$. In both cases the energy level spacing is $2\pi v^F/L$, *cf.* the previous chapter. All sums over k have to be unbounded, since we want to use unbounded fermion momentum spectrum. For this purpose one can take the effective bandwidth D (do not confuse with the number of particles transfered from one Fermi point to the other one) to be infinite, but then introduce an (ultraviolet) cut-off. It is possible to denote the (non-unique) N-electron ground state as

$$|N\rangle = |N_{\uparrow}\rangle \otimes |N_{\downarrow}\rangle \tag{10.33}$$

in such a way that $\hat{N}_{\sigma}|N\rangle = N_{\sigma}|N\rangle$.

Now we can introduce one-dimensional chiral Fermi fields as

$$\psi_{\sigma}(x) = \sqrt{\frac{2\pi}{L}} \sum_{n_k} e^{ikx} a_{k,\sigma} \ , \tag{10.34}$$

where x belongs to the interval $-L/2$ and $L/2$. The reader can check that these fields satisfy standard anticommutation relations

$$\{\psi_{\sigma}(x), \psi^{\dagger}_{\sigma'}(x')\} = 2\pi\delta(x - x')\delta_{\sigma,\sigma'} \ . \tag{10.35}$$

$P_0 = 0$ pertains to periodic boundary conditions for $\psi_{\sigma}(x)$, while $P_0 = 1$ corresponds to antiperiodic ones. We can also define bosonic electron-hole creation operators as

$$b^{\dagger}_{q,\sigma} = \frac{i}{\sqrt{n_q}} \sum_{n_k} a^{\dagger}_{k+q,\sigma} a_{k,\sigma} \tag{10.36}$$

where n_q are positive integers ($q = 2\pi n_q/L > 0$). These operators create "density excitations" (or currents, see below) with the momentum q

for electrons with spin σ. The reader can check that they satisfy Bose commutation relations

$$[b_{q,\sigma}, b^\dagger_{q',\sigma'}] = \delta_{q,q'}\delta_{\sigma,\sigma'} \ . \tag{10.37}$$

The important property of these Bose operators is that they commute with $\hat{N}_\sigma$:

$$[b_{q,\sigma}, \hat{N}_{\sigma'}] = 0 \ . \tag{10.38}$$

We can choose from all states $|N\rangle$ a unique state $|N\rangle_0$, called the N-particle ground state because all other states of $|N\rangle$ have higher energies. This state contains no holes, *i.e.*,

$$b_{q,\sigma}|N\rangle_0 = 0 \ , \tag{10.39}$$

valid for any positive q and σ. Observe that $|N\rangle_0$ has lower energy than the vacuum state $|0\rangle_0$. For $P_0 = 0$ the states $a_{0,\sigma}|0\rangle_0$ are degenerate with $|0\rangle_0$. Any N-electron state can be written as an action of some function of $b^\dagger_{q,\sigma}$ onto $|N\rangle_0$.

Now we can introduce bosonic fields

$$\phi_\sigma(x) = -\sum_{q>0} \frac{1}{\sqrt{n_q}} \left(e^{-iqx} b_{q,\sigma} + e^{iqx} b^\dagger_{q,\sigma}\right) e^{-aq/2} \ , \tag{10.40}$$

where $a \sim (P^F)^{-1}$ is a short-distance (ultraviolet) cut-off, introduced to avoid possible divergencies, if the bandwidth D goes to infinity. The fields $(\partial\phi_\sigma(x)/\partial x)$ are canonically conjugated to $\phi_\sigma(x)$, because of the following commutation relations:

$$\left[\phi_\sigma(x), \frac{\partial\phi_{\sigma'}(x')}{\partial x'}\right] = 2\pi i \delta_{\sigma,\sigma'} \left(\frac{a}{\pi[a^2 + (x - x')^2]} - \frac{1}{L}\right) \ . \tag{10.41}$$

To complete the classification of states one needs to introduce the *Klein factors* (sometimes called *ladder operators*), defined as

$$\begin{aligned} U^\dagger_\sigma f(b^\dagger)|N\rangle_0 &= f(b^\dagger) a^\dagger_{N_\sigma+1,\sigma}|N\rangle_0 \ , \\ U_\sigma f(b^\dagger)|N\rangle_0 &= f(b^\dagger) a_{N_\sigma,\sigma}|N\rangle_0 \ , \end{aligned} \tag{10.42}$$

so that U_σ $(U^\dagger_\sigma)$ commutes with any function of $b^\dagger$ and removes (adds) an electron with spin σ from (to) the top-most filled level of $|N\rangle_0$ in a way that it decreases (increases) the electron number with spin σ by one. They satisfy the following Clifford algebra:

$$\begin{aligned}
&U_\sigma U_\sigma^\dagger = U_\sigma^\dagger U_\sigma = 1 \ , \\
&\{U_\sigma, U_{\sigma'}\} = 0 \ , \qquad \{U_\sigma^\dagger, U_{\sigma'}^\dagger\} = 0 \ , \ \sigma \neq \sigma' \ , \\
&\{U_\sigma, U_{\sigma'}^\dagger\} = 2\delta_{\sigma,\sigma'} \ , \qquad [U_\sigma, \hat{N}_{\sigma'}] = \delta_{\sigma,\sigma'} U_\sigma \ , \\
&[U_\sigma, b_{q,\sigma'}^\dagger] = [U_\sigma, b_{q,\sigma'}] = 0 \ .
\end{aligned} \tag{10.43}$$

One can explicitly express Klein factors as a function of fermionic and bosonic fields, introduced above, as

$$U_\sigma = \sqrt{a}\psi_\sigma(0)e^{i\phi_\sigma(0)} \ . \tag{10.44}$$

Using the above operators one can *bosonize* fermion fields as

$$\begin{aligned}
\psi_\sigma(x) &= \frac{1}{\sqrt{a}} U_\sigma e^{-i(2\hat{N}_\sigma - P_0)\pi x/L} e^{-i\phi_\sigma(x)} \ , \\
\rho_\sigma(x) =: \psi_\sigma^\dagger(x)\psi_\sigma(x) &:= \frac{\partial\phi_\sigma(x)}{\partial x} + \frac{2\pi\hat{N}_\sigma}{L} \ ,
\end{aligned} \tag{10.45}$$

so that the Hamiltonian of free electrons with the linearized about Fermi points dispersion law can be written as

$$\begin{aligned}
\mathcal{H}_0 &= \sum_\sigma \frac{\pi v^F}{L}\hat{N}_\sigma(\hat{N}_\sigma + 1 - P_0) + \sum_{q>0,\sigma} v^F q b_{q,\sigma}^\dagger b_{q,\sigma} \\
&= \sum_\sigma \frac{\pi v^F}{L}\hat{N}_\sigma(\hat{N}_\sigma + 1 - P_0) + \int_{-L/2}^{L/2} \frac{dx}{2\pi}\frac{v^F}{2} : \left(\frac{\partial\phi_\sigma(x)}{\partial x}\right)^2 : \ .
\end{aligned} \tag{10.46}$$

Using the bosonization rules it is easy to express correlation function of Fermi fields as a function of those for Bose fields

$$\langle \mathcal{T}\psi_\sigma(z)\psi_{\sigma'}^\dagger(0)\rangle = \frac{\delta_{\sigma,\sigma'}\mathrm{sign}(\tau)}{a} e^{\langle \mathcal{T}\phi_\sigma(z)\phi_\sigma(0) - \phi_\sigma(0)\phi_\sigma(0)\rangle} \ , \tag{10.47}$$

where $\mathcal{T}$ defines the time-ordering operator for the Euclidean time τ and we used the property of any function of Bose operators $\hat{B} = \sum_{q>0}(\lambda_d f_q^\dagger + \tilde{\lambda}_q b_q)$, governed by the free bosonic Hamiltonian with the linear dispersion law

$$\langle \exp(\lambda\hat{B})\rangle = \exp(\langle \hat{B}^2\rangle \lambda^2/2) \ . \tag{10.48}$$

It is possible to define the *physical Fermi field* as

$$\Psi_{ph,\sigma}(x) = \sqrt{\frac{2\pi}{L}} \sum_{P=-\infty}^{\infty} e^{iPx} a_{P,\sigma}$$
$$= \sqrt{\frac{2\pi}{L}} \sum_{k=-P^F}^{\infty} \left(e^{-i(P^F+k)x} a_{-k-P^F,\sigma} + e^{i(P^F+k)x} a_{P^F+k,\sigma} \right) , \tag{10.49}$$

where $P > 0$ pertain to right-moving electrons and $P < 0$ correspond to left-movers. They can be approximately viewed as two separate "species" as $a_{k,L/R,\sigma} = a_{\mp(k+P^F),\sigma}$, with the dispersion law $\varepsilon_{k,L/R} = \varepsilon[\mp(k+P^F)]$. Then one can factor out rapidly fluctuating phase factors $\exp(\mp iP^F x)$ (which is, naturally, valid only for large enough P^F) and express $\Psi_{ph,\sigma}(x)$ in terms of slowly varying (on the scale of $(P^F)^{-1}$) fields $\tilde{\psi}_{L/R,\sigma}$ as

$$\Psi_{ph,\sigma}(x) \approx e^{-iP^F x} \tilde{\psi}_{L,\sigma}(x) + e^{iP^F x} \tilde{\psi}_{R,\sigma}(x) ,$$
$$\tilde{\psi}_{L/R,\sigma}(x) = \sqrt{\frac{2\pi}{L}} \sum_{k=-\infty}^{\infty} e^{\mp ikx} a_{k,L/R,\sigma} . \tag{10.50}$$

It means that we extended the single-particle Hilbert space by introducing additional states at the bottom of the Fermi sea, below $\varepsilon(P = 0)$, *e.g.*, $\varepsilon_{k,L/R} = \varepsilon(0) + v^F(k + P^F)$ for $k < -P^F$. Usually $E^F \gg 1$ these additional states need very high energies for their excitation, and do not change the low-energy physics close to Fermi points. Then one can define left- and right-moving boson fields, and operators $\tilde{\rho}_{L/R,\sigma}(x)$ as above.

The simple model Hamiltonian of interacting spinless fermions can then be presented as

$$\mathcal{H} = \mathcal{H}_0 + \mathcal{H}_{int} = iv^F \int_{-L/2}^{L/2} \frac{dx}{2\pi} : \left[\psi_L^\dagger(x) \frac{\partial}{\partial x} \psi_L(x) \right.$$
$$\left. - \psi_R^\dagger(x) \frac{\partial}{\partial x} \psi_R(x) \right] : + \int_{-L/2}^{L/2} \frac{dx}{2\pi} : \left[g_2 \tilde{\rho}_L(x) \tilde{\rho}_R(x) \right.$$
$$\left. + \frac{g_4}{2} (\tilde{\rho}_L^2(x) + \tilde{\rho}_R^2(x)) \right] := \frac{v}{4} \int_{-L/2}^{L/2} \frac{dx}{2\pi} : \left[\frac{1}{g} (\tilde{\rho}_L(x) + \tilde{\rho}_R(x))^2 \right.$$
$$\left. + g(\tilde{\rho}_L(x) - \tilde{\rho}_R(x))^2 \right] : , \tag{10.51}$$

where

$$v = \sqrt{(v^F + g_4)^2 - g_2^2}\ ,\ g = \sqrt{\frac{v^F + g_4 - g_2}{v^F + g_4 + g_2}}\ . \qquad (10.52)$$

Interactions between electrons are described by such a Hamiltonian that are often called *forward scattering interactions.*

This Hamiltonian can be easily diagonalized by using the bosonization and Bogolyubov transformation

$$\begin{aligned}
\mathcal{H} &= \frac{\pi v}{L}\left(\frac{1+g^2}{g}\sum_{L,R}\left[\frac{\hat{N}_{L/R}^2}{2} + \sum_q n_q b_{q,L/R}^\dagger b_{q,L/R}\right]\right. \\
&\quad \left. + \frac{1-g^2}{g}\left[\hat{N}_L\hat{N}_R - \sum_q n_q(b_{q,L}b_{q,R} + b_{q,R}^\dagger b_{q,L}^\dagger)\right]\right) \\
&= \frac{2\pi v}{L}\sum_{\pm}\left(g^{\pm}\hat{\mathcal{N}}_{\pm}^2 + \sum_q n_q B_{q,\pm}^\dagger B_{q,\pm}\right) \\
&= v\sum_{\pm}\left[\frac{2\pi}{L}g^{\pm}\hat{N}_{\pm}^2 + \frac{1}{2}\int_{-L/2}^{L/2}\frac{dx}{2\pi} :\ \left(\frac{\partial\Phi_{\pm}(x)}{\partial x}\right)^2\ :\right]\ , \qquad (10.53)
\end{aligned}$$

where $\hat{\mathcal{N}}_{\pm}$ is number operator in the new basis (after the Bogolyubov transformation) and

$$\begin{aligned}
\hat{N}_{\pm} &= \frac{\hat{N}_L \mp \hat{N}_R}{2}\ , \\
\Phi_{\pm}(x) &= -\sum_{q>0}\frac{e^{-aq/2}}{\sqrt{n_q}}\left[e^{-iqx}B_{q,\pm} + e^{iqx}B_{q,\pm}^\dagger\right]\ , \qquad (10.54) \\
B_{q,\pm,} &= \frac{1}{\sqrt{8g}}[(1+g)(b_{q,L} \mp b_{q,R}) \pm (1-g)(b_{q,L}^\dagger \mp b_{q,R}^\dagger)]\ .
\end{aligned}$$

Equivalently, one can define currents

$$J_{L/R,k} = \int_{-L/2}^{L/2}\frac{dx}{2\pi} :\ \tilde{\psi}_{L/R}(x)^\dagger\tilde{\psi}_{L/R}(x)\ :\ . \qquad (10.55)$$

Then the Hamiltonian of, *e.g.*, free spinless fermions can be written as the quadratic form of the normal ordered operators for chiral (Noether) currents

$$\mathcal{H}_0 = \mathcal{H}_{0,L} + \mathcal{H}_{0,R} = \frac{v^F}{2}\int_{-L/2}^{L/2}\frac{dx}{2\pi}$$
$$\times\ (:\ J_L(x)J_L(x)\ :+:\ J_R(-x)J_R(-x)\ :) = \frac{2\pi v^F}{L}$$
$$\times\left(\frac{1}{2}(J_{R,0}J_{R,0} + J_{L,0}J_{L,0}) - \frac{1}{12} + \sum_{k=1}^{\infty}(J_{R,-k}J_{R,k} + J_{L,-k}J_{L,k})\right)\ . \tag{10.56}$$

The last line implies that one has two (non-interacting) theories with central charges being equal to 1. The reader can compare this result with the one for the Sugawara construction of the Gaussian model. It turns out that the expression for the Hamiltonian of free fermions seems quartic in fermion operators, since $J_{R/L}$ are quadratic in fermion operators. However, there is no contradiction with the definition of $\mathcal{H}_0$, because we used the normal ordering operation. Using similar calculations as for the Gaussian model we can show that

$$[\mathcal{H}_{0,L/R}, J_{L/R,m}] = -\frac{2\pi v^F}{L}mJ_{L/R,m}\ . \tag{10.57}$$

Since we know the commutation relations for $J_{L/R}$, it is an easy task to calculate the spectrum of the Sugawara construction of the Hamiltonian, in which zero modes define the spectra of primary states, and the ones with $k \geq 1$ pertain to descendants. These descendant states are particle-hole excitations, while different zero modes determine different Kac–Moody conformal towers. The eigenvalues of $J_{L/R,0}$ are often called the charge of conformal tower. On the other hand, they measure the total number of fermions. So, the low-energy excitations of the Hamiltonian of free spinless fermions are representations of the U(1) Kac–Moody algebra, related to the Gaussian (free bosonic) field theory. The energy spectrum of the free spinless fermion Hamiltonian in the Sugawara form after bosonization can be written as

$$E = \frac{2\pi v^F}{L}\left(-\frac{1}{12} + \frac{Q_R^2 + Q_L^2}{2} + m_R + m_L\right)\ , \tag{10.58}$$

where $Q_{L/R}$ are non-negative integers, related zero modes, and $m_{L/R}$ are non-negative integers, related to particle-hole excitations (descendants).

On the other hand, the reader saw above that the inclusion of forward scattering interactions renormalizes velocities of low-lying excitations ($v^F \to v$) and introduces the parameter g. The latter one is related, as the reader remembers from the previous chapter, to the conformal weights (or scaling dimensions and spins) of primary operators. From this viewpoint, the Sugawara construction of the bosonized form of the low-energy Hamiltonian of spinless fermions with only forward scattering interactions gives the easy way to construct asymptotics of correlation function exponents *via* the results of the conformal field theory.

Let us turn now to the consideration of electrons (fermions with spin σ). The Abelian bosonization uses the fact that the Hamiltonian of electrons with spins is often invariant under $\mathrm{U}(1)_{\sigma=\uparrow}\times\mathrm{U}(1)_{\sigma=\downarrow}$ symmetry, *i.e.*, the conservation of the total charge and the z-projection of the total spin moment. This symmetry implies

$$\tilde{\psi}_{L/R,\sigma} \to \exp(i\alpha_\sigma)\tilde{\psi}_{L/R,\sigma} \ . \tag{10.59}$$

Using this fact we can write $b_{q,c/s} = (b_{q,\uparrow} \pm b_{q,\downarrow})/\sqrt{2}$ (and similar for $b_{q,c/s}$, which imply

$$\begin{aligned} \phi_{q,c/s} &= (\phi_{q,\uparrow} \pm \phi_{q,\downarrow})/\sqrt{2} \ , \\ \hat{N}_{q,c/s} &= (\hat{N}_{q,\uparrow} \pm \hat{N}_{q,\downarrow})/\sqrt{2} \ . \end{aligned} \tag{10.60}$$

For interactions conserving the total charge and the z-projection of the total spin moment due to the linearization of the dispersion law low-lying excitations, which describe dynamics of charge and spin degrees of freedom, the latter are separated from each other (spin-charge separation). Then the Hamiltonian in the bosonized form can be written as a sum of mutually commuting parts, one of which describes the charge low-energy dynamics, and the other one describes the spin dynamics of the correlated electron chain. In fact, for only forward scattering interactions one can use *e.g.*, above formulas for spinless fermions for charge and spin part of the Hamiltonian. Then, for such interactions, the Hamiltonian of interacting electron model is diagonalized using Abelian bosonization.

It turns out, however, that there exist electron-electron interactions like

$$\mathcal{H}_{bac} = g_1 \sum_{\sigma} \sum_{k_1,k_2,q} a^{\dagger}_{k_1,R,\sigma} a_{k_1-q,L,\sigma} a^{\dagger}_{k_2,L,-\sigma} a^{\dagger}_{k_2+q,R,-\sigma} \ , \tag{10.61}$$

which are usually called *backward scattering.* They can be expressed in the bosonization language as

$$\mathcal{H}_{bac} \propto g_1 \sum_{\sigma} \int_{-L/2}^{L/2} \frac{dx}{2\pi} \cos(\sqrt{8}\phi_s) \ , \tag{10.62}$$

where ϕ_s is the linear combination of $\phi_{\pm,\sigma}$. The most important difference of that expression from forward scattering interactions is that creation and annyhilation fermion operators related to right- and left-movers cannot be written in the form of pair products of $\tilde{\rho}_{L/R}$. Such a Hamiltonian cannot be simply diagonalized as for the forward scattering. Backward scattering interactions usually produce gaps for low-lying excitations (*i.e.*, they are relevant perturbations in the renormalization group sense).

The non-Abelian bosonization uses the fact that the Hamiltonian of electrons with spins is, in fact, invariant under larger group U(1)×SU(2), which has non-Abelian component. The aim of the non-Abelian bosonization is to provide bosonization scheme in which SU(2) symmetry is preserved at all steps, unlike the Abelian one. The U(1) symmetry implies

$$\tilde{\psi}_{L/R,\sigma} \to \exp(i\alpha)\tilde{\psi}_{L/R,\sigma} \tag{10.63}$$

($\alpha_\uparrow = \alpha_\downarrow = \alpha$). The U(1) Noether current, which measures the total electron density, is

$$J_{L/R,k} = \sum_{\sigma} \int_{-L/2}^{L/2} \frac{dx}{2\pi} :\ \tilde{\psi}_{L/R,\sigma}(x)^\dagger \tilde{\psi}_{L/R,\sigma}(x)\ : \ . \tag{10.64}$$

On the other hand, the non-Abelian SU(2)-symmetry implies

$$\begin{aligned} \tilde{\psi}_{L/R,\sigma} &\to U_\sigma^{\sigma'} \tilde{\psi}_{L/R,\sigma'} \ , \\ \tilde{\psi}^\dagger_{L/R,\sigma} &\to (U^\dagger)_\sigma^{\sigma'} \tilde{\psi}^\dagger_{L/R,\sigma'} \ , \end{aligned} \tag{10.65}$$

where $U_\sigma^{\sigma'}$ is SU(2)-symmetric matrix. The SU(2) Noether current is

$$\vec{J}_{L/R,k} = \sum_{\sigma,\sigma'} \int_{-L/2}^{L/2} \frac{dx}{2\pi} :\ \tilde{\psi}_{L/R,\sigma}(x)^\dagger \vec{\sigma}_{\sigma,\sigma'} \tilde{\psi}_{L/R,\sigma'}(x)\ : \ , \tag{10.66}$$

where the components of $\vec{\sigma}$ are Pauli matrices.

The task of the non-Abelian bosonization is to find a Sugawara construction for the Hamiltonian, quadratic in the Abelian U(1) currents J (charge densities) and non-Abelian SU(2) currents $\vec{J}$ (spin densities). The reader can check, that one cannot write holomorphic and antiholomorphic parts of the Hamiltonian (as it was easily done for spinless fermions) only as the quadratic form of Abelian U(1) charge currents. However, it remedied by taking into account the non-Abelian SU(2) spin currents. After some simple algebra we can write the Sugawara construction for free spinful electrons

$$\begin{aligned}\mathcal{H}_0 &= \mathcal{H}_{0,L,c} + \mathcal{H}_{0,R,c} + \mathcal{H}_{0,L,s} + \mathcal{H}_{0,R,s} \\ &= \frac{v^F}{4}\int_{-L/2}^{L/2}\frac{dx}{2\pi}\Big(: J_L(x)J_L(x) : + : J_R(-x)J_R(-x) : \\ &\quad + : \vec{J}_L(x)\vec{J}_L(x) : + : \vec{J}_R(-x)\vec{J}_R(-x) : \Big) . \end{aligned} \tag{10.67}$$

All four parts of the Hamiltonian are mutually commuting (the commutation of spin and charge parts of the Hamiltonian follows from the fact that $[J_{L,R}, \vec{J}_{L/R}] = 0$). One can define the holomorphic part for the energy-momentum tensor for charge and spin as

$$\begin{aligned} T_c(x) &= \frac{1}{4} : J_R(x)J_R(x) : + \frac{\pi^2}{6L^2} , \\ T_s(x) &= \frac{1}{3} : \vec{J}_R(x)\vec{J}_R(x) : + \frac{\pi^2}{6L^2} , \end{aligned} \tag{10.68}$$

and similar for the antiholomorphic one. Using their Fourier transforms one can get for right-movers (holomorphic part)

$$\begin{aligned} \mathcal{H}_{0,R,c} &= \frac{2\pi v^F}{L}\left(L_{m=0,c} - \frac{1}{24}\right) \\ &= \frac{2\pi v^F}{L}\left[\left(\frac{1}{4}J_{R,0}J_{R,0} - \frac{1}{24}\right) + \frac{1}{2}\sum_{m=1}^{\infty} J_{R,-m}J_{R,m}\right] , \end{aligned} \tag{10.69}$$

and

$$\begin{aligned} \mathcal{H}_{0,R,s} &= \frac{2\pi v^F}{L}\left(L_{m=0,c} - \frac{1}{24}\right) \\ &= \frac{2\pi v^F}{L}\left[\left(\frac{1}{3}\vec{J}_{R,0}\vec{J}_{R,0} - \frac{1}{24}\right) + \frac{2}{3}\sum_{m=1}^{\infty} \vec{J}_{R,-m}\vec{J}_{R,m}\right] , \end{aligned} \tag{10.70}$$

and similar for antiholomorphic part (left-movers). The commutation relations for Fourier transforms for charge U(1) operators are

$$[J_{L/R,n}, J_{L/R,m}] = 2n\delta_{n,-m} \tag{10.71}$$

(it is the U(1) Kac–Moody algebra), with

$$[L_{c,n}, J_{R,m}] = -mJ_{R,n+m} \tag{10.72}$$

(and similar for antiholomorphic part). The last formula implies again that charge-carrying Fourier transforms of current operators $J_{L/R,m}$ act as raising and lowering ("creation" for $m < 0$ and "annihilation" for $m > 0$) operators of the energy of the charge part of the Hamiltonian. Since we know the commutation relations for charge currents, it is an easy task to write down the spectrum of the Sugawara construction of the charge part of the Hamiltonian, in which zero modes define the spectra of primary charge states, and the ones with $m \geq 1$ pertain to descendants. These descendant states are charge particle-hole excitations, while different charge zero modes determine different charge Kac–Moody conformal towers.

For spin SU(2) operators the commutation relations for Fourier transforms are $(\alpha, \beta, \gamma = x, y, z)$

$$[J_{L/R,n,\alpha}, J_{L/R,m,\beta}] = i\epsilon_{\alpha\beta\gamma} J_{L/R,n+m,\gamma} + \frac{1}{2} n \delta_{\alpha,\beta} \delta_{n,-m} \tag{10.73}$$

(it is the SU(2) Kac–Moody algebra), notice that

$$[J_{L/R,0,\alpha}, J_{L/R,0,\beta}] = i\epsilon_{\alpha\beta\gamma} J_{L/R,0,\gamma} \ , \tag{10.74}$$

i.e., they form the standard SU(2) spin algebra, and

$$[L_{s,n}, J_{R,m,\alpha}] = -mJ_{R,n+m,\alpha} \tag{10.75}$$

(and similar for antiholomorphic part). Spin-carrying Fourier transforms of current operators $J_{L/R,m,\alpha}$ act as "creation" for $m < 0$ and "annihilation" for $m > 0$ operators of the energy of the spin part of the Hamiltonian. Again, since we know the commutation relations for spin currents, it is an easy task to write down the spectrum of the Sugawara construction of the spin part of the Hamiltonian, in which zero modes define the spectra of primary spin states, and the ones with $m \geq 1$ pertain to descendants. These descendant states are spin particle-hole excitations, while different spin zero modes determine different spin Kac–Moody conformal towers.

Summarizing, the energy spectrum of the free electron Hamiltonian in the Sugawara form after non-Abelian bosonization can be written as

$$E = \frac{2\pi v^F}{L}\left(-\frac{1}{6} + \frac{Q_R^2 + Q_L^2}{4} + \frac{j_R(j_R+1) + j_L(j_L+1)}{3} + m_{R,c} + m_{L,c} + m_{R,s} + m_{L,s}\right) \ , \tag{10.76}$$

where $Q_{L/R}$ and $j_{L/R}$ are non-negative integers, related to charge and spin zero modes, respectively, and $m_{L/R,c,s}$ are non-negative integers, related to particle-hole charge and spin excitations. (Notice that for free electrons the only allowed combination of U(1) and SU(2) primary states is $Q_{L/R} = 2j_{L/R}$ (mod 2) with $j_{L/R} = 0, \frac{1}{2}$.) Inclusion of forward scattering interactions renormalizes velocities of low-lying charge and spin excitations, $v^F \to v_{c,s}$, and introduces two parameters of interactions $g_{s,c}$ (the SU(2) symmetry fixes the value of g_s). The latter are related to the conformal weights (or scaling dimensions and spins) of primary charge and spin operators. Again, the Sugawara construction of the bosonized form of the low-energy Hamiltonian of spinful electrons with only forward scattering interactions in the non-Abelian scheme permits to easily construct asymptotics of correlation function exponents *via* the results of the conformal field theory.

In fact, by using the bosonization procedure we (approximately) mapped the Hamiltonian of a one-dimensional correlated electron system to the Hamiltonian of a free Bose gas with the linear dispersion law, which describes particle-hole excitations, plus zero modes, which describe changes of the number of electrons about the right and left Fermi points (or, in the other basis, which is often used, the change of the total number of electrons close to Fermi points and transfers from the right to the left Fermi point). This is very similar to the answer, obtained in the previous chapter, where we used exact results. The difference is because in the bosonization approximation spin and charge zero modes yield independent contributions to the low energy states, *i.e.*, for the bosonized picture there is a spin-charge separation. On the other hand, as the reader remembers, there is no exact spin-charge separation in the exact Bethe ansatz description, because of non-diagonal components of dressed charge matrices. Also, it turns out that the bosonization deals with weak enough interactions, comparing to the Fermi velocities of free electrons.

The class of Hamiltonians, similar to what we studied above, *i.e.*, with only forward scattering, was named by F. D. M. Haldane as a *Luttinger liquid* (or *Tomonaga–Luttinger liquid*). Physicists believe that the low-energy physics of metallic phases of one-dimensional correlated electron systems is often well described by the Luttinger liquid picture. The disadvantage of this (very universal for some class of models) approach is that one ever starts the filling of Dirac seas from the situation of the free electron gas. Hence, this description seems to be good for models, in which low-energy excitations have the same structure as free electrons, *i.e.*, unbound electron excitations. It is impossible to derive the bosonized Hamiltonian for systems, in which low-energy excitations have a different structure, *e.g.*, for pairs, though one can introduce the Luttinger liquid Hamiltonian for them phenomenologically, and, then, apply the powerful machinery of the bosonization. Also, it is necessary to keep in mind that the bosonization picture is limited for weak electron-electron interactions.

Summarizing, in this chapter we briefly presented several approximate theoretical methods, used to describe non-integrable quantum many-body models: the scaling approach and the bosonization. These methods can serve as complementary ones to the exact methods, studied in the previous chapters of this book.

The reader can find the description of the renormalization group approach and scaling in, *e.g.*, [Ma (1976); Cardy (1996); Kadanoff (2000); Sachdev (1999)]. The description of quantum phase transitions can be found in [Sondhi, Girvin, Carini and Shahar (1997); Sachdev (1999)]. Calculations of logarithmic corrections for quantum chains due to marginal operators are given in [Affleck, Gepner, Schulz and Ziman (1989)]. The bosonization approach for correlated electron systems was pioneered in [Mattis and Lieb (1965)]; see also [Tomonaga (1950); Luttinger (1963)]. It was first used for calculations of correlation functions of correlated quantum chains in [Luther and Peschel (1974)]. Approximate description of low-dimensional correlated electron models was reviewed in [Sólyom (1979)]. The conception of the Luttinger (Tomonaga–Luttinger) liquid was introduced by Haldane in [Haldane (1981)]. The reader can find the modern reviews of the bosonization procedure, *e.g.*, in the well-known book [Gogolin, Nersesyan and Tsvelik (1998)] and the review article [van Delft and Schoeller (1998)].

Bibliography

Aharonov, Y. and Bohm, D. (1959). Significance of electromagnetic potentials in the quantum theory, *Phys. Rev.* **115**, pp. 485–491.

Aharonov, Y. and Casher, A. (1984). Topological quantum effects for neutral particles, *Phys. Rev. Lett.* **53**, pp. 319–321.

Alcaraz, F. C., Barber, M. N., Batchelor, M. T., Baxter, R. J. and Quispel, G. R. W. (1987). Surface exponents of the quantum XXZ, Askin-Teller and Potts models, *J. Phys. A: Math. Gen.* **20**, pp. 6397–6409.

Anderson, P. W. (1959). New approach to the theory of superexchange interactions, *Phys. Rev.* **115**, pp. 2–13.

Andrei, N., Furuya, K. and Lowenstein, J. W. (1983). Solution of the Kondo problem, *Rev. Mod. Phys.* **55**, pp. 331–402.

Andrei, N. and Johannesson, H. (1984). Heisenberg chain with impurities (An integrable model), *Phys. Lett. A* **100**, pp. 108–112.

Asakawa, H. and Suzuki, M. (1995). Finite-size corrections in the XXZ model and the Hubbard model with boundary fields, *J. Phys. A: Math. Gen.* **29**, pp. 225–245.

Affleck, I., Gepner, D., Schulz, H. J. and Ziman T. (1989). Critical behaviour of spin-s Heisenberg antiferromagnetic chains: analytic and numerical results, *J. Phys. A: Math. Gen.* **22**, pp. 511–529.

Babujian, H. M. (1983) Exact solution of the isotropic Heisenberg chain with arbitrary spins: Thermodynamics of the model, *Nucl. Phys. B* **215**, pp. 317–336.

Bahder, T. B. and Woynarovich F. (1986). Gap in the spin excitations and magnetization curve of the one-dimensional attractive Hubbard model, *Phys. Rev. B* **33**, pp 2114–2121.

Bariev, R. Z. (1994). Correlation functions of a one-dimensional anisotropic t-J model, *Phys. Rev. B* **49**, pp. 1474–1476.

Bariev, R. Z., Klümper, A., Schadschneider, A. and Zittartz, J. (1993). Excitation spectrum and critical exponents of a one-dimensional integrable model of fermions with correlated hopping, *J. Phys. A: Math. Gen.* **26**, pp. 4863–4873.

Baxter, R. J. (1982). *Exactly Solved Models in Statistical Mechanics*, New-York, Academic Press.

Bedürftig, G., Eßler, F. H. L. and Frahm, H. (1996). Integrable impurity in the supersymmetric t-J model, *Phys. Rev. Lett.* **77**, pp. 5098–5101.

Bedürftig, G. and Frahm, H. (1997). Spectrum of boundary states in the open Hubbard chain, *J. Phys. A: Math. Gen.* **30**, pp. 4139–4149.

Belavin, A. A., Polyakov, A. M. and Zamolodchikov, A. B. (1984). Infinite conformal symmetry in two-dimensional quantum field theory, *Nucl. Phys. B* **241**, pp. 333–380.

Bethe, H. (1931). Theorie der Metalle. Erster Teil. Eigenwerte und Eigenfunktionen der lineären atomischen Kette, *Z. Phys.* **71**, pp. 205–226.

Bogoliubov, N. M., Izergin, A. G. and Reshetikhin N. Y. (1987). Finite-size effects and infrared asymptotics of the correlation functions in two dimensions, *J. Phys. A: Math. Gen.* **20**. pp. 5361–5369.

Bogolyubov, N. M. and Korepin, V. E. (1989). The role of quasi-one-dimensional structures in high-T_c superconductivity, *Int. J. Mod. Phys.* **3**, pp. 427–439.

Bogolyubov, N. M. and Korepin, V. E. (1990). Correlation functions of the one-dimensional Hubbard model, *Teor. Mat. Fiz.*, **82**, pp. 331–348 (in Russian) [*Theor. Math. Phys.*, **82**, pp. 231–243].

Byers, N. and Yang, C. N. (1962). Theoretical considerations concerning quantized magnetic flux in superconducting cylinders, *Phys. Rev. Lett.*, **7**, pp. 46–49.

Cardy, J. L. (1986a). Operator content of two-dimensional conformally invariant theories, *Nucl. Phys. B* **270**, pp. 186–204.

Cardy, J. L. (1986b). Effect of boundary conditions on the operator content of two-dimensional conformally invariant theories, *Nucl. Phys. B* **275**, pp. 200–218.

Cardy, J. L. (1996). *Scaling and Renormalization in Statistical Physics*, Cambridge, Cambridge University Press.

Cherednik, I. V. (1984). Factorizing particles on a half-line and root systems, *Teor. Mat. Fiz.* **61**, pp. 35–44 (in Russian) [*Theor. Math. Phys.* **61**, pp. 977–983].

Cheung, H.-F., Gefen, Y., Riedel, E. K. and Shin, W.-H. (1988). Persistent current in small one-dimensional metal rings , *Phys. Rev. B* **37**, pp. 6050–6062.

de Sa, P. A. and Tsvelik, A. M. (1995). Anisotropic spin-1/2 Heisenberg chain with open boundary conditions, *Phys. Rev. B* **52**, pp. 3067–3070.

de Vega, H. J. and Woynarovich, F. (1985). Method for calculation finite size corrections in Bethe ansatz systems — Heisenberg chain and 6-vertex model, *Nucl. Phys. B* **251**, pp. 439–456.

de Vega, H. J. and Woynarovich, F. (1992). New integrable quantum chains combining different kinds of spins, *J. Phys. A: Math. Gen.* **25**, pp. 4499–4516.

Di Franchesco, P., Mathieu, P. and Sénéshal, D. (1997). *Conformal Field Theory*, New-York, Springer-Verlag.

Eßler, F. H. L. (1996). The supersymmetric t-J model with a boundary, *J. Phys. A: Math. Gen.* **29**, pp. 6183–6203.

Eßler, F. H. L. and Korepin, V. E. (1992). Higher conservation laws and algebraic Bethe ansätze for the supersymmetric t-J model, *Phys. Rev. B* **46**, pp. 9147–9162.

Eßler, F. H. L., Korepin, V. E. and Schoutens, K. (1992). Completeness of the SO(4) extended Bethe ansatz for the one-dimensional Hubbard model, *Nucl. Phys. B* **384**, pp. 431–458.

Foerster, A. and Karowski, M. (1993a). Algebraic properties of the Bethe ansatz for an spl(2,1)-supersymmetric t-J model, *Nucl. Phys. B* **396**, pp. 611–638.

Foerster, A. and Karowski, M. (1993b). The supersymmetric t-J model with quantum group invariance, *Nucl. Phys. B* **408**, pp. 512–534.

Foerster, A., Links, J. and Tonel, A. P. (1999). Algebraic properties of an integrable t-J model with impurities, *Nucl. Phys. B* **552**, pp. 707–726.

Fisher, M. E. (1969). Rigorous inequalities for critical-point correlation exponents, *Phys. Rev.* **180**, pp. 594–600.

Frahm, H. and Korepin, V. E. (1990). Critical exponents for the one-dimensional Hubbard model, *Phys. Rev. B* **42**, pp. 10533–10565.

Frahm, H. and Korepin, V. E. (1991). Correlation functions of the one-dimensional Hubbard model in a magnetic field, *Phys. Rev. B* **43**, pp. 5653–5662.

Frahm, H. and Zvyagin, A. A. (1997a). Nonlinear boundary oscillations in strongly correlated electron quantum wires, *Phys. Rev. B* **55**, pp. 1341–1344.

Frahm, H. and Zvyagin, A. A. (1997b). The open spin chain with impurity: An exact solution, *J. Phys.: Condensed Matter* **9**, pp. 9939–9946.

Gaudin, M. (1967). Un systeme a une dimension de fermions en interaction, *Phys. Lett. A* **24**, pp. 55–56.

Gaudin, M. (1971). Boundary energy of a Bose gas in one dimension, *Phys. Rev. A* **4**, pp. 386–394.

Gaudin, M. (1983). *La Fonction d'Onde de Bethe pour les Modèles Exact de la Méchanique Statistique*, Paris, Masson.

Göhmann, F. (2001). Algebraic Bethe ansatz for gl(1|2) generalized model and Lieb-Wu equations, *Nucl. Phys. B* **620**, pp. 501–518.

Gogolin, A. O., Nersesyan, A. A. and Tsvelik, A. M. (1998). *Bosonization and Strongly Correlated Systems*, Cambridge, Cambridge University Press.

Griffiths, R. B. (1965). Thermodynamic inequality near the critical point for ferromagnets and fluids, *Phys. Rev. Lett.* **14**, pp. 623–624.

Griffiths, R. B. (1969). Nonanalytic behavior above the critical point in a random Ising ferromagnet, *Phys. Rev. Lett.* **23**, pp. 17–19.

Gutzwiller, M. C. (1963). Effect of correlation on thr ferromagnetism of transition metals, *Phys. Rev. Lett.* **10**, pp. 159–162.

Ha, Z. N. C. (1996). *Quantum Many-Body Systems in One-Dimensional Solvable Models*, Singapore, World Scientific.

Haldane, F. D. M. (1981). 'Luttinger-liquid theory' of one-dimensional quantum fluids: I. Properties of the Luttinger model and their extension to the general 1D interacting spinless Fermi gas, *J. Phys. C: Solid. State Phys.* **14**, pp. 2585–2609.

Haldane, F. D. M. (1983). Non-linear field theory of large-spin Heisenberg antiferromagnets — semi-classically quantized solitons of the one-dimensional easy-axis Neel state, *Phys. Rev. Lett.* **50**, pp. 1153–1156.

Haldane, F. D. M. (1991). "Fractional statistics" in arbitrary dimensions: A generalization of the Pauli principle, *Phys. Rev. Lett.* **67**, pp. 937–940.

Hohenberg, P. C. (1967). Existence of long-range order in one and two dimensions, *Phys. Rev.* **158**, pp. 383–386.

Hubbard, J. (1963). Electron correlations in narrow energy band, *Proc. Roy. Soc. A* **276**, pp. 238–257.

Hulthén, L. (1938). Über das Austauchsproblem eines Kristalles (Dissertation), *Arkiv Mat. Astr. Fysik* **26A**, pp. 1–106.

Inoue, M. and Suzuki, M. (1988). The ST-transformation approach to analytic solutions of quantum systems. 2. Transfer matrix and Pfaffian methods, *Progr. Theor. Phys.* **79**, pp. 645–664.

Ising, E. (1925). Beitrag zur Theorie des Ferromagnetismus, *Z. Phys.* **31**, pp. 253–258.

Izyumov, Y. A. and Skryabin Y. N. (1990). *Statistical Mechanics of Magnetically Ordered Systems*, New-York, Plenum.

Jimbo, M. (1990). *Yang–Baxter Equation in Integrable Systems*, Singapore, World Scientific.

Jordan, P. and Wigner, E. (1928). Über das Paulische Äquivalenzverbot, *Z. Phys.* **47**, pp. 631–651.

Josephson, B. D. (1967a). Inequality for the specific heat I. Derivation, *Proc. Phys. Soc.* **92**, pp. 269–275.

Josephson, B. D. (1967b).Inequality for the specific heat II. Application to critical phenomena , *Proc. Phys. Soc.* **92**, pp. 276–284.

Jüttner, G., Klümper, A. and Suzuki, J. (1997). Exact thermodynamics and Luttinger liquid properties of the integrable *t-J* model, *Nucl. Phys. B* **486**, pp. 650–674.

Jüttner, G., Klümper, A. and Suzuki, J. (1998). The Hubbard chain at finite temperatures: ab initio calculations of Tomonaga-Luttinger liquid properties, *Nucl. Phys. B* **522**, pp. 471–502.

Kadanoff, L. P. (1966). Scaling laws for Ising models near T_c, *Physics* **2**, pp. 263–272.

Kadanoff, L. P. (2000). *Statistical Physics: Statics, Dynamics and Renormalization*, Singapore, World Scientific.

Kanamori, J. (1963). Electron correlation and ferromagnetism of transition metals, *Progr. Theor. Phys.* **30**, pp. 275–289.

Katsura, S. (1962). Statistical mechanics of the anisotropic linear Heisenberg model, *Phys. Rev.* **127**, pp. 1508–1518.

Kawakami, N. and Yang. S.-K. (1991). Luttinger liquid properties of highly correlated electron systems in one dimension, *J. Phys.: Condens. Matter* **3**, pp. 5983–6008.

Kleiner, V. Z. and Tsukernik, V. M. (1975). Paramagnetic resonance of an impurity atom in a one-dimensional spin system, *Fiz. Metal. Metalloved.* **39**, pp. 947–951 (in Russian) [*Phys. Met. and Metallogr.* **39**, pp. 40–44].

Kleiner, V. Z. and Tsukernik, V. M. (1980). Static magnetic properties of a spin chain with an impurity, *Fiz. Nizk. Temp.* **6**, pp. 332–337 (in Russian) [*Sov. J. Low Temp. Phys.* **6**, pp. 158–161].

Klümper, A. (1993). Thermodynamics of the anisotropic spin-1/2 Heisenberg chain and related quantum chains, *Z. Phys.* **91**, pp. 507–519.

Klümper, A. (1998). The spin-1/2 Heisenberg chain: Thermodynamics, quantum criticality and spin-Peierls exponents, *Eur. Phys. J. B* **5**, pp. 677–685.

Klümper, A., Batchelor, M. T. and Pearce, P. A. (1991). Central charges of the 6-vertex and 19-vertex models with twisted boundary conditions, *J. Phys. A: Math. Gen.* **24**, p. 3111–3133.

Klümper, A. and Zvyagin, A. A. (1998). Exact thermodynamics of disordered impurities in quantum spin chains, *Phys. Rev. Lett.* **81**, pp. 4975–4979.

Klümper, A. and Zvyagin, A. A. (1998). Disordered magnetic impurities in uniaxial critical quantum spin chains, *J. Phys.: Condensed Matter* **12**, pp. 8705–8726.

Koma, T. (1990). An extension of the thermal Bethe ansatz, *Progr. Theor. Phys.* **83**, pp. 655–659.

Kontorovich, V. M. and Tsukernik V. M. (1967). Magnetic properties of a spin array with two sublattices, *Zh. Eksp. Teor. Fiz.* **53**, pp. 1167-1176 (in Russian) [*Sov. Phys. JETP* **26**, pp. 687–691].

Korepin, V. E. (1979). Direct calculation of the S-matrix in the massive Thirring model, *Teor. Mat. Fiz.* **41**, pp. 169–189 (in Russian) [*Theor. Math. Phys.* **76**, pp. 953–967].

Korepin, V. E., Bogoliubov N. M. and Izergin, A. G. (1993). *Quantum Inverse Scattering Method and Correlation Functions*, Cambridge, Cambridge University Press.

Korepin, V. E. and Eßler F. H. L. (eds.) (1994). *Exactly Solvable Models of Strongly Correlated Electrons*, Singapore, World Scientific.

Lai, C. K. (1974). Lattice gas with nearest-neighbor interaction in one dimension with arbitrary statistics, *J. Math. Phys.* **15**, pp. 1675–1676.

Landau, L. D. and Lifshits, E. M. (1980). *Statistical Physics, Part 1*, (Course of Theoretical Physics, Vol. 5), New-York, Pergamon Press.

Lee, K. J. B. and Schlottmann, P. (1988). Thermodynamic Bethe ansatz equations for the Hubbard chain with an attractive interaction, *Phys. Rev. B* **38**, pp. 11566–11571.

Lee, K. J. B. and Schlottmann, P. (1989). Low-temperature properties of the Hubbard chain with an attractive interaction, *Phys. Rev. B* **40**, pp. 9104–9112.

Lieb, E. H., Schulz, T. and Mattis, D. (1961). Two soluble models of an antiferromagnetic chain, *Ann. Phys.* **16**, pp. 417–466.

Lieb, E. H. and Wu, F. Y. (1968). Absence of Mott transition in an exact solution of the short-range one-band model in one dimension, *Phys. Rev. Lett.*, **20**, pp. 1445–1448.

Lüsher, M. (1978). Quantum nonlocal charges and absence of particle production in 2-dimensional nonlinear sigma model, *Nucl. Phys. B* **135**, pp. 1–19.

Luther, A. and Peschel, I. (1974). Single-particle states, Kohn anomaly and pairing fluctuations in one dimension, *Phys. Rev. B* **9**, pp. 2911–2919.

Luttinger, J. M. (1963) An exactly soluble model of a many-fermion system, *J. Math. Phys.* **4**, pp. 1154–1162.

Ma, S.-K. (1976). *Modern Theory of Critical Phenomena*, Massachusets, Reading, Benjamin.

Mahan, G. D. (1990). *Many-Particle Physics*, New-York, Plenum Press.

Mattis, D. C. (1965). *Theory of Magnetism*, New-York, Harper and Row.

Mattis, D. C. and Lieb, E. H. (1965). Exact solution of a many-fermion system and its associated boson field, *J. Math. Phys.* **6**, pp. 304–312.

McCoy, B. M. and Wu, T. T. (1973). *The Two-Dimensional Ising Model*, Cambridge (Mass.), Harvard University Press.

Mermin, N. D. and Wagner, H. (1966). Absence of ferromagnetism or antiferromagnetism in one- or two-dimensional isotropic Heisenberg models, *Phys. Rev. Lett.* **17**, pp. 1133–1136.

Mintchev, M., Ragoucy, E. and Sorba, P. (2002). Scattering in the presence of a reflecting and transmitting impurity, *Phys. Lett. B* **547**, pp. 313–320.

Nagaosa, N. (1998). *Quantum Field Theory in Strongly Correlated Electronic Systems*, Berlin, Springer.

Orbach, R. (1958). Linear antiferromagnetic chain with anisotropic coupling, *Phys. Rev.* **112**, pp. 309–316.

Pearce, P. A. and Klümper, A. (1991). Finite-size corrections and scaling dimensions of solvable lattice models — An analytic method, *Phys. Rev. Lett.* **66**, pp. 974–977.

Pfeuty, P. (1970). The one-dimensional Ising model with a transverse field, *Ann. Phys.* **57**, pp. 79–90.

Pikin, S. A. and Tsukernik V. M. (1966). The thermodynamics of linear chains in a transverse magnetic field, *Zh. Eksp. Teor. Fiz.* **50**, pp. 1377–1380 (in Russian) [*Sov. Phys. JETP* **23**, pp. 914–916].

Pines, D. and Nozières, P. (1989) *The Theory of Quantum Liquids*, Redwood City, Addison-Wesley.

Pokrovsky, V. L. and Talapov, A. L. (1979). Ground state, spectrum and phase diagram of two-dimensional incommensurate crystals, *Phys. Rev. Lett.* **42**, pp. 65–67.

Ramos, P. B. and Martins, M. J. (1997). Algebraic Bethe ansatz approach for the one-dimensional Hubbard model, *J. Phys. A: Math. Gen.* **30**, pp. L195–L202.

Rushbrooke, G. S. (1963). On the thermodynamics of the critical region for the Ising problem, *J. Chem. Phys.* **39**, pp. 842–843.

Sachdev, S. (1999). *Quantum Phase Transitions*, Cambridge, Cambridge University Press.

Sacramento, P. D. (1994). Spin and charge susceptibilities of the attractive Hubbard chain, *Physica C* **235**, pp. 2159–2160.

Sacramento, P. D. (1995). Thermodynamics of the attractive Hubbard chain, *J. Phys.: Condens. Matter* **7**, pp. 143–150.

Scheunert, M., Nahm, W. and Rittenberg, V. (1977). Irreducible representations of the osp(2,1) and spl(2,1) graded Lie algebras, *J. Math. Phys.* **18**, pp. 155–162.

Schlottmann, P. (1987). Integrable narrow-band model with possible relevance to heavy-fermion systems, *Phys. Rev. B* **36**, pp. 5177-5185.

Schlottmann, P. (1989). Some exact results for dilute mixed-valent and heavy fermion systems, *Phys. Rep.* **181**, pp. 1–119.

Schlottmann, P. (1991). Impurity-induced critical behaviour in antiferromagnetic Heisenberg chains, *J. Phys.: Condens. Matter*, **3**, pp. 6617–6634.

Schlottmann, P. (1994). Closing of the spin gap and ferromagnetism induced by magnetic impurities, *Phys. Rev. B* **49**, pp. 9202–9205.

Schlottmann, P. (1997). Exact results for highly correlated electron systems in one dimension, *Int. J. Mod. Phys.* **11**, pp. 355–667.

Schlottmann, P. (1998a). Closing of the spin gap and ferromagnetism induced by magnetic impurities, *Phys. Rev. B* **57**, pp. 10638–10643.

Schlottmann, P. (1998b). Integrable one-dimensional heavy fermion lattice model, *Nucl. Phys. B* **525**, pp. 697–720.

Schlottmann, P. (1999). Non-Fermi-liquid behavior of impurity spins in the anisotropic Heisenberg chain, *Nucl. Phys. B* **552**, pp. 727–747.

Schlottmann, P. (2000). Properties of magnetic impurities embedded into anisotropic Heisenberg chain with spin gap, *Nucl. Phys. B* **565**, pp. 535–554.

Schlottmann, P. and Sacramento, P. D. (1993). Multichannel Kondo problem and some applications, *Adv. Phys.* **42**, pp. 641–682.

Schlottmann, P. and Zvyagin, A. A. (1997a). Integrable supersymmetric t-J model with magnetic impurity, *Phys. Rev. B* **55**, pp. 5027–5036.

Schlottmann, P. and Zvyagin, A. A. (1997b). Kondo impurity band in a one-dimensional correlated electron lattice, *Phys. Rev. B* **56**, pp. 13989–13998.

Schulz, H. (1985). Hubbard chain with reflecting ends, *J. Phys. C: Solid State Phys.* **18**, pp. 581–601.

Schulz, H. (1987). An exactly solvable Kondo model with quadratic band energy, *J. Phys. C: Solid State Phys.* **20**, pp. 2375–2403.

Shastry, B. S. and Sutherland B. (1990). Twisted boundary conditions and effective mass in Heisenberg-Ising and Hubbard rings, *Phys. Rev. Lett.* **65**, pp. 243–246.

Shiba, H. (1970). Magnetic susceptibility at zero temperature for the one-dimensional Hubbard model, *Phys. Rev. B*, **6**, pp. 930–938.

Sklyanin, E. K. (1988). Boundary conditions for integrable quantum systems, *J. Phys. A: Math. Gen.* **21**, pp. 2375–2389.

Sólyom, J. (1979). Fermi gas model of one-dimensional conductors, *Adv. Phys.* **28**, pp. 201–303.

Sondhi, S. L., Girvin, S. M., Carini, J. P. and Shahar, D. (1997). Continuous quantum phase transitions, *Rev. Mod. Phys.* **69**, pp. 315–333.

Sutherland, B. (1975). Model for a multicomponent quantum system, *Phys. Rev. B* **12**, pp. 3795–3805.

Suzuki, M. (1985). Transfer-matrix method and Monte-Carlo simulations in quantum spin systems, *Phys. Rev. B* **31**, pp. 2957–2965.

Suzuki, M. and Inoue, M. (1987). The ST-transformation approach to analytic solutions of quantum systems. 1. General formulations and basic limit theorems, *Progr. Theor. Phys.* **78**, pp. 787–799.

Takahashi, M. (1999). *Thermodynamics of One Dimensional Solvable Models*, Cambridge, Cambridge University Press.

Takhtadzhan, L. A. and Faddeev, L. D. (1979). The quantum method for the inverse problem and the XYZ Heisenberg model, *Uspekhi Mat. Nauk* **34**, pp. 13–63 (in Russian). [*Russian Math. Surveys* **34**, pp. 11–68].

Takhtajan, L. A. (1982). The picture of low-lying excitations in the isotropic Heisenberg chain of arbitrary spins, *Phys. Lett. A* **87**, pp. 479–482.

Tjon, J. A. (1970). Magnetic relaxation in an exactly soluble model, *Phys. Rev. B* **2**, pp. 2411–2421.

Tsvelick, A. M. and Wiegmann, P. B. (1983). Exact results in the theory of magnetic alloys, *Adv. Phys.* **32**, pp. 453–713.

Tsvelick, A. M. and Wiegmann, P. B. (1984). Solution of the n-channel Kondo problem (scaling and integrability), *Z. Phys.* **54**, pp. 201–206.

Tomonaga, S. (1950). Remarks on Bloch method of sound waves applied to many-fermion problems *Progr. Theor. Phys.* **5**, pp. 544–569.

Uimin, G. V. (1970). One-dimensional problem for $S = 1$ with modified antiferromagnetic Hamiltonian, *Pis'ma v Zh. Eksp. Teor. Fiz.* **12** , pp. 332–335 (in Russian) [*JETP Lett.* **12**, pp. 225–228].

van Delft, J. and Schoeller, H. (1998). Bosonization for beginners — Refermionization for experts, *Ann. der Phys.* **4**, pp. 225-305.

van Hove, L. (1952). The occurence of singularities in the elastic frequency distribution of a crystal, *Phys. Rev.* **89**, pp. 1189–1193.

Widom, B. (1965a). Surface tension and molecular correlations near the critical point, *J. Chem. Phys.* **43**, pp. 3892–3898.

Widom, B. (1965b). Equation of state in the neighborhood of the critical point, *J. Chem. Phys.* **43**, pp. 3899–3905.

Wiegmann, P. B. (1981). Exact solution of the $s - d$ exchange model (Kondo problem), *J. Phys. C: Solid State Phys.* **14**, pp. 1463–1478.

Wiegmann, P. B. and Tsvelick, A. M. (1983). Exact solution of the Anderson model: I, *J. Phys. C: Solid State Phys.* **16**, pp. 2281–2319.

Woynarovich, F. (1989). Finite-size effects in a non-half-filled Hubbard chain, *J. Phys. A: Math. Gen.* **22**, pp. 4243–4256.

Woynarovich, F., Eckle, H.-P. and Truong, T. T. (1989). Non-analytic finite-size corrections in the one-dimensional Bose gas and Heisenberg chain, *J. Phys. A: Math. Gen.* **22**, pp. 4027–4043.

Yamanaka, M., Oshikawa, M. and Affleck, I. (1997). Magnetization plateaux in spin chains: "Haldane gap" for half-integer spins, *Phys. Rev. Lett.* **79**, pp. 1110–1113.

Yang, C. N. (1967). Some exact results for the many-body problem in one dimension with repulsive delta-function interaction, *Phys. Rev. Lett.* **19**, pp. 1312–1315.

Yang, C. N. and Yang, C. P. (1966a). One-dimensional chain of anisotropic spin-spin interactions I. Proof of Bethe's hypothesis for ground state in a finite system, *Phys. Rev.* **150**, pp. 321–327.

Yang, C. N. and Yang, C. P. (1966b). One-dimensional chain of anisotropic spin-spin interactions II. Properties of the ground state energy per site for an infinite system, *Phys. Rev.* **150**, pp. 327–339.

Yang, C. N. and Yang, C. P. (1966c). One-dimensional chain of anisotropic spin-spin interactions III. Applications, *Phys. Rev.* **151**, pp. 258–264.

Yang, C. N. and Yang, C. P. (1969). Thermodynamics of a one-dimensional system of bosons with repulsive delta-function interactions, *J. Math. Phys.* **10**, pp. 1115–1122.

Yu, N. and Fowler, M. (1992). Persistent current of a Hubbard ring threaded with a magnetic flux, *Phys. Rev. B* **45**, pp. 11795–11804.

Yue, R. and Deguchi, T. (1997). Magnetic susceptibility and low-temperature specific heat of integrable 1D Hubbard model under open boundary conditions, *J. Phys. A: Math. Gen.* **30** pp. 8129–8138.

Zvyagin, A. A. (1990a). Magnetic characteristics of a multisublattice spin chain, *Fiz. Tv. Tela* **32**, pp. 314–315 (in Russian) [*Sov. Phys. Solid State* **32**, pp. 181–182].

Zvyagin, A.A. (1990b). To the theory of current states in the Hubbard chain, *Fiz. Tv. Tela* **32**, pp. 1546–1547 (in Russian) [*Sov. Phys. Solid State* **32**, pp. 905–906].

Zvyagin, A. A. (1997). Magnetic impurity in an open correlated electron chain, *Phys. Rev. Lett.* **79**, pp. 4641–4644.

Zvyagin, A. A. (2000). Universal low energy behavior of disordered quantum spin chains: Exact analytic results, *Phys. Rev. B* **62**, pp. R6069–R6072.

Zvyagin, A. A. (2001a). Exact solution for a disordered correlated electron model, *Phys. Rev. B* **63**, 033101.

Zvyagin, A. A. (2001b). Bethe ansatz solvable multichain quantum systems, *J. Phys. A: Math. Gen.* **34**, pp. R21–R53.

Zvyagin, A. A. (2002). Non-Fermi-liquid behavior: Exact results for ensembles of magnetic impurities, *Fiz. Nizk. Temp.* **28**, pp. 1274–1291 [*Low Temp. Phys.* **28**, pp. 907–920].

Zvyagin, A. A. (2003). Persistent currents in a lattice correlated electron ring with a magnetic impurity, *Eur. Phys. J. B* **32**, pp .351–360.

Zvyagin, A. A. and Johannesson, H. (1998). Hidden Kondo effect in a correlated electron chain, *Phys. Rev. Lett.* **81**, pp. 2751–2754.

Zvyagin, A. A. and Krive, I. V. (1992). Aharonov–Casher effect in the repulsive Hubbard model, *Zh. Eksp. Teor. Fiz.* **102**, pp. 1376–1380 (in Russian) [*Sov. Phys JETP* **75**, pp. 745–747].

Zvyagin, A. A. and Krive, I. V. (1995). Persistent currents in one-dimensional systems of strongly correlated electrons, *Fiz. Nizk. Temp.* **21**, pp. 687–716 (in Russian) [*Low Temp. Phys.* **21**, pp. 533–555].

Zvyagin, A. A. and Schlottmann, P. (1995). Aharonov–Casher effect in the Heisenberg spin chain with many impurities, *Phys. Rev. B* **52**, pp. 6569–6574.

Zvyagin, A. A. and Schlottmann, P. (1997). Magnetic impurity in the one-dimensional Hubbard model, *Phys. Rev. B* **56**, pp. 300–306.

Zvyagin, A. A. and Segal, Y. Y. (1995). Low temperature magnetization behavior features of one-dimensional alternating spin chain with impurity, *Fiz. Nizk. Temp.* **21**, pp. 1068–1074 (in Russian) [*Low Temp. Phys.* **21**, pp. 822–826].

Index

L-operator, 118
R-matrix, 116
f-sum rule, 21
t-J model
  supersymmetric t-J model, 97, 98

Abelian bosonization
  non-Abelian bosonization, 339
Aharonov–Bohm effect, 3, 323
Aharonov–Casher effect, 3, 324
algebraic Bethe ansatz, 115
Anderson hybridization impurity model, 234
anisotropic XY model, 35
anisotropic t-J model, 100
anomalous dimensions, 311
auxiliary (sub)space, 118

backscattering, 348
Bethe ansatz, 3, 52
Bethe ansatz equations, 53
bipartite lattice, 44
Bogolyubov inequality, 17
Bogolyubov transformation, 36
bound state, 56
boundary bound states, 150
boundary conditions
  periodic boundary conditions
    open boundary conditions, 12
boundary fields
  potentials, 149

central charge, 310
charge stiffness, 90
chemical potential, 6
chiral anomaly, 340
co-ordinate Bethe ansatz, 49
conformal dimension, 272
conformal field theory, 3, 307
conformal symmetry, 1
conformal towers, 312
conformal transformation, 307
conformal weights, 310
constraints, 8
continuity equation, 21
convolution, 58
correlation length, 304
counting function, 285
critical exponents, 306
critical systems, 307

density of rapidities, 58
density of states, 13
dimerized XY model, 39
dressed charge
  dressed charge matrix, 288, 292, 299
dressed energy, 59
driving term, 61
dynamical critical exponent, 335

energy level spacing, 342
energy-momentum tensor
  stress-energy tensor, 309

equations of state, 9
Euler–Maclaurin formula, 286, 296
exact integrability, 120
exchange interaction, 25
exclusion statistics, 112
extensive variables
  intensive variables, 6

Fermi edge singularities, 178
Fermi energy, 13
Fermi level
  surface
    point, 12
Fermi liquid, 1
Fermi sea
  Dirac sea, 3
Fermi velocity, 3
ferromagnetic phase
  spin-saturated phase, 34
fluctuation-dissipation theorem, 20
forward scattering, 346
free boundary conditions
  free edges, 149
fusion hierarchy
  Y-system, 267

Gaussian distribution, 274
Gaussian model, 313
Gibbs distribution, 8
global conformal symmetry, 307
graded Bethe ansatz, 137
Grassmann parities
  grading, 137
Griffiths singularities, 273

Haldane hypothesis, 46
Hamiltonian, 7
Heisenberg model, 25
Heisenberg–Ising Hamiltonian, 27
Helmholtz free energy, 5
highest weight states, 312
Hohenberg theorem, 20
Hubbard model, 74
Hubbard model with repulsion
  attraction, 88
Hubbard operators, 98
hybridization impurity, 4
hyperscaling, 307

impurity L-operator, 201
impurity bound states, 192
integrals of motion, 120
internal energy, 5
intertwining relations, 118
Ising model, 27
isotropic XY model, 27

Jordan–Wigner transformation
  generalized transformation, 30, 38

Kac–Moody algebra, 340
Klein factors, 343
Kondo impurity model, 217
Kondo temperature, 210

Laurent series, 308
Lieb–Schultz–Mattis theorem, 43
local conformal symmetry, 308
local Kondo temperatures, 269
logarithmically normal distribution, 274
Lorentzian distribution, 274

magnetic anisotropy, 27
magnetic impurity, 4
magnetization plateaux, 43
mathematical vacuum, 82
McCoy–Wu model, 264
Mermin–Wagner theorem, 18
microscopic persistent currents, 329
monodromy matrix
  operator, 81, 119
multi-sublattice spin chain, 38

natural variables, 5
nested Bethe ansatz, 74
non Fermi liquid, 2
non Fermi liquid behaviour, 216

open boundary conditions, 149
operator product expansion, 309

parity effect, 328
partition function, 8
permutation operator, 80, 116
persistent current, 3
persistent currents, 323
Pokrovsky–Talapov phase transition, 113
primary fields, 310

quadrimerized chain, 42
quantum (sub)space, 118
quantum critical line, 249
quantum critical point, 335
quantum determinant, 167
quantum inverse scattering method, 116
quantum phase transitions, 20
quantum transfer matrix, 264

rapidities, 3
rapidity, 53
reflection equations, 165
response function, 20
Ruderman–Kittel–Kasuya–Yosida interaction, 243

scale invariance, 307
scaling dimension, 311
scaling relations, 305
scaling spin, 311
second order phase transition, 34
secondary fields
  descendants, 311, 312
short-distance cut-off, 343
Sommerfeld coefficient, 15
Sommerfeld expansion, 14
spectral parameter, 81
spectral weight function, 20
spin gap, 40, 95, 111
spin ladders, 46
spin-charge separation, 1, 299, 348, 352
spin-singlet pair, 86
spinon, 56
string, 56
string hypothesis, 56, 57
strong disorder, 274
SU(2)-symmetric spin model
  Takhtadjan-Babujian model, 126
SU(3)-symmetric t-J chain
  spin-1 chain, 100
Sugawara construction, 339, 340
supertrace
  graded trace, 138
surface energy, 163, 293

thermodynamic Bethe ansatz, 60
thermodynamic limit, 19
thermodynamic potential, 5
Tomonaga–Luttinger liquid, 4
  Luttinger liquid, 1, 352
topological spin current
  chirality, 205
transfer matrix, 28, 120
transport currents, 324
triangular matrix, 82
twisted boundary conditions, 325
two-particle scattering matrix, 77
two-point correlation function, 305

unbound electron excitation, 86

van Hove features
  van Hove singularities, 17
Virasoro algebra
  classical
    quantum, 308, 310

weak disorder, 275
Widom scaling hypothesis, 305
Wiener–Hopf method, 66
Wilson ratio, 211

Yang–Baxter relations, 81, 116

Zeeman interaction, 10